Ion Exchange in Analytical Chemistry

Authors

Harold F. Walton
Professor Emeritus
Department of Chemistry and Biochemistry
and
Senior Research Associate
Cooperative Institute for Research in Environmental Science
University of Colorado
Boulder, Colorado

and

Roy D. Rocklin
Staff Research Chemist
Dionex Corporation
Sunnyvale, California

CRC Press
Boca Raton Ann Arbor Boston

Library of Congress Cataloging-in-Publication Data

Walton, Harold F. (Harold Frederick), 1912-
Ion exchange in analytical chemistry / by Harold F. Walton and Roy D. Rocklin.
p. cm.
Includes bibliographical references.
ISBN 0-8493-6199-0
1. Ion exchange chromatography. I. Rocklin, Roy David. II. Title.
QD79.C453W35 1990
543'.0893--dc20

89-70839
CIP

This book represents information obtained from authentic and highly regarded sources. Reprinted material is quoted with permission, and sources are indicated. A wide variety of references are listed. Every reasonable effort has been made to give reliable data and information, but the authors and the publisher cannot assume responsibility for the validity of all materials or for the consequences of their use.

Direct all inquiries to CRC Press, Inc., 2000 Corporate Blvd., N.W., Boca Raton, Florida 33431.

International Standard Book Number 0-8493-6199-0

Library of Congress Card Number 89-70839
Printed in the United States

PREFACE

Several books have recently appeared on ion chromatography in response to the phenomenal growth of that field. Most of the books are written as user manuals with specific reference to proprietary columns and equipment. This book is not another manual of ion chromatography. One of the authors, however, is at the forefront of research in the field and has contributed a long chapter on ion chromatography, as well as sections of other chapters.

Twenty years ago the senior author wrote a book with the same title as this one, *Ion Exchange In Analytical Chemistry,* in collaboration with William Rieman, III. The new book is in no sense a second edition or revision of Rieman and Walton's book. The field has changed radically since that book was published, largely because of the high-performance methods of separation and detection that followed the landmark paper by Small, Stevens, and Bauman, ''Novel Ion Exchange Chromatographic Method Using Conductimetric Detection,'' that appeared in 1975.

The time has come for a new evaluation of the role of ion exchange and ion-exchanging materials in chemical analysis. This book tries to provide such an evaluation. It describes up-to-date advances in high-performance ion-exchange chromatography and its applications to analytes that range in complexity from the simplest inorganic ions to biological macromolecules. It also reviews methods of separation and preconcentration of metallic species that were developed in the 1950s and depend on large separation factors. These methods are no longer in the spotlight of current interest, but they have their uses and are intrinsically interesting. One chapter of the book reviews the basic theory of ion exchange; another describes the synthesis and properties of ion-exchanging materials. It is assumed that the reader knows the fundamentals of chromatography.

I would like to express my deepest gratitude to the Cooperative Institute for Research in Environmental Science (CIRES) of the University of Colorado and to its Director, Robert E. Sievers, for facilities and help in manuscript preparation, and to Maria Neary for drafting services. Roy Rocklin is indebted to the Dionex Corporation for facilities, drafting, and permission to refer to recent research.

Harold F. Walton
November 1989

AUTHORS

Harold F. Walton, Ph.D., is a professor emeritus of the Department of Chemistry and Biochemistry of the University of Colorado at Boulder. Since his retirement from full-time duty in 1982, he has been a Senior Research Associate of the Cooperative Institute for Research in Environmental Science.

He received the bachelor's and doctor's degrees from the University of Oxford, England, in 1934 and 1937. After a year at Princeton University and two years in the water-conditioning industry, he taught at Northwestern University, and joined the faculty of the University of Colorado in 1947. He served a term as department chairman.

Dr. Walton has taught in several countries as a visiting professor starting as a Fulbright grantee at the University of Trujillo, Peru, in 1966. He has taught short courses in Costa Rica, Venezuela, France, and Australia, and has taken part in the United Nations International Atomic Energy Agency's program of technical cooperation.

He received the Colorado Section Award of the American Chemical Society in 1977, the Stearns Award for distinguished service at the University of Colorado in 1981, and the Dal Nogare Award presented by the Chromatography Forum of the Delaware Valley in 1988. He is a member of the American Chemical Society, American Association for the Advancement of Science, and the Royal Society of Chemistry (London), and is a corresponding member of the Chemical Society of Peru.

He is the author or co-author of seven books, including *Ligand Exchange Chromatography*, jointly with V. A. Davankov and J. D. Navratil, and *Ion Exchange in Analytical Chemistry*, jointly with W. Rieman, III. He prepared an audio course on ion-exchange chromatography for the American Chemical Society in 1986 and has contributed chapters to several cooperative works.

Dr. Walton has published over 100 research papers on the topics of chromatography and ion exchange, as well as electrochemistry, colloidal electrolytes, uranium ore deposition, and atomic absorption spectrometry.

Roy D. Rocklin, Ph.D., is a Staff Research Chemist in the research and development department of Dionex Corporation. He received a bachelor's degree in chemistry form the University of California, Santa Cruz, in 1975, and a Ph.D. in analytical chemistry from the University of North Carolina in 1980. He is a member of the American Chemical Society.

Since joining Dionex in 1980, he has concentrated on the theory and application of electrochemical detection in liquid and ion chromatography. His major research areas have been the detection of cyanide, sulfide, and other ions detectable using a silver electrode, and the detection of carbohydrates, alcohols, and amines using the new technique of pulsed amperometric detection. He has also contributed to the development and commercial introduction of electrochemical detectors used for conductimetric, DC amperometric, and pulsed amperometric detection.

Before rejoining the research and development department, Dr. Rocklin worked in the Dionex marketing department as Applications Manager from 1985 until 1988. There, he developed applications using all forms of ion chromatography, with emphasis on the theory and application of ion-exchange gradient elution.

Dr. Rocklin has published a number of articles on the topics of electrochemical detection, ion-exchange gradient elution, and ion chromatography.

To Terry

TABLE OF CONTENTS

Chapter 1

INTRODUCTION

Ion exchange is the reversible exchange of ions of similar charge, positive or negative, between an ionic solution and an insoluble solid that is in contact with it. For the solid to exchange ions it must have ions of its own, and there must be a pathway for these ions to move back and forth. In the simplest structures the ions are located on the surface of the solid, as is implied in Figure 1. More commonly the exchangeable ions are in channels, or pores on the molecular scale within the solid.

There are two kinds of ions, positive and negative. Ions of one kind are attached to the solid by covalent bonds. These are the *fixed ions*. Balancing them are ions of opposite charge that are not attached by covalent bonds directed in space, but by electrostatic force. These are the *counter-ions, mobile ions*, or *exchangeable ions*. The process of ion exchange is shown schematically in Figure 1. There the fixed ions are charged negatively. The exchangeable ions are shown as sodium ions. A potassium ion from the surrounding solution enters the exchanger (whose boundary is shown by the dashed line) and displaces a sodium ion, or a calcium ion enters; the calcium ion has a double charge, therefore it must displace two sodium ions. Electrical neutrality is always maintained. The exchanges are always reversible. At the bottom of the figure two sodium ions are seen displacing one calcium ion. What is illustrated here is *cation exchange*; solids having fixed positive ions, instead of the fixed negative ions shown in Figure 1, would undergo *anion exchange*. The solid structure suffers no permanent change, and the exchanges of mobile ions, back and forth, can go on indefinitely.

I. HISTORY

Ion exchange was first observed in soil. In the middle of the 19th century an English gentleman farmer, H. S. Thompson,[1] and the consulting chemist to the Royal Agricultural Society, J. T. Way,[2] wanted to know why water-soluble fertilizers like potash (potassium chloride) and ammonium sulfate stayed in soil and did not wash out when it rained. Way put some soil in a box with a hole in the bottom and washed it thoroughly with rain water. Then he poured over the soil a solution of ammonium sulfate, and washed it down with more rain water, catching the water that drained from the outlet at the bottom. In this water he found no ammonium sulfate, no matter how long he washed, but instead he found calcium sulfate. If he poured potassium chloride solution into the soil, he recovered calcium chloride at the outlet, but no potassium salt. If he continued to wash the soil with rain water, only pure water came out. The ammonium and potassium salts stayed hidden in the box of soil until he poured a solution of calcium or sodium chloride. Then, ammonium or potassium chloride appeared at the outlet. Only the "base" parts of the salts, what we would call the cations, were retained and released; the "acid" components, sulfate, chloride, or nitrate, passed through the soil unchanged. Further study showed that the exchanges took place according to chemical equivalents. One chemical equivalent of potassium chloride, 74 parts by weight, displaced one chemical equivalent of calcium chloride, or 55 parts by weight.

Thus, Way and Thompson described the fundamentals of ion exchange: exchanges of ions of like charge sign, exchange according to chemical equivalents, and reversibility. They did not call it "ion exchange", however, they called it "base exchange". An interesting historical point is that ion exchange was discovered before ions were. At any rate, Arrhenius did not publish his theory of complete ionization until 1887, 36 years after the work of Way and Thompson.

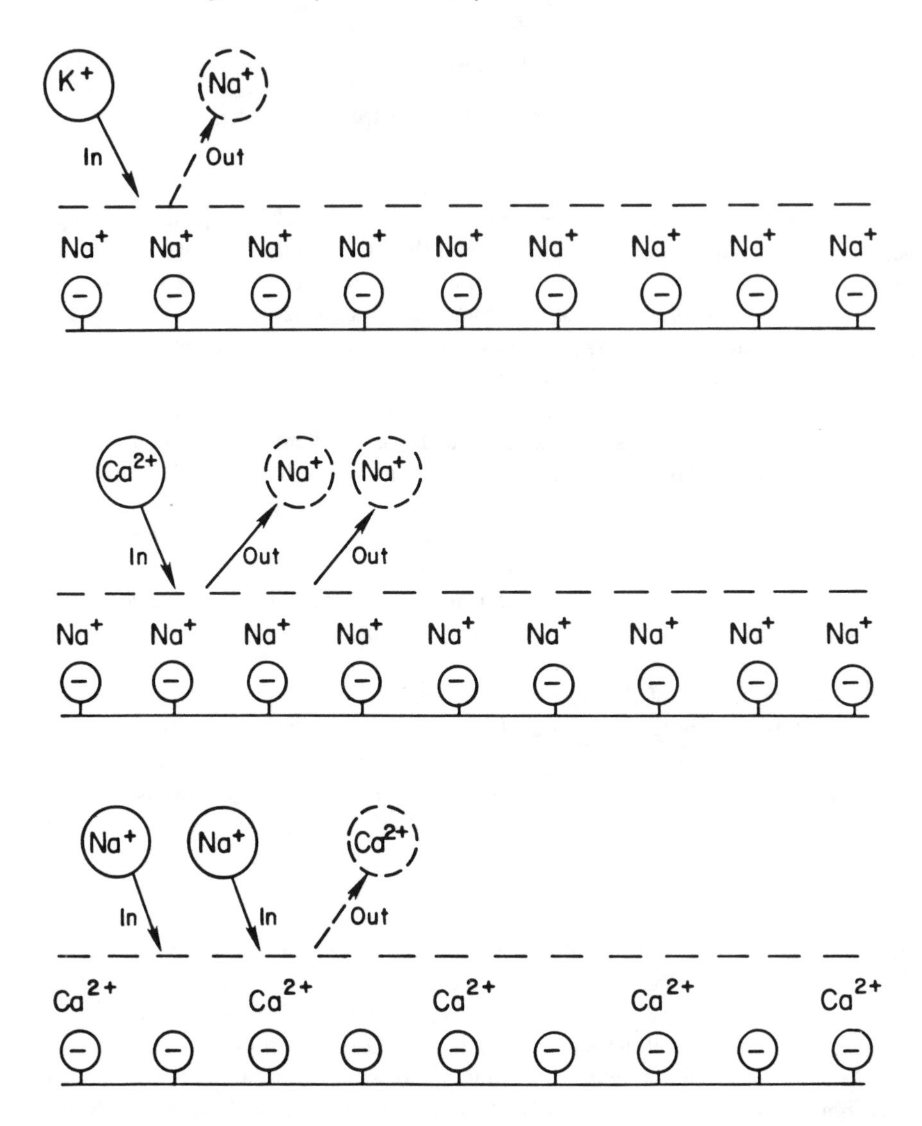

FIGURE 1. Ion exchange: schematic. The minus signs with circles represent the fixed ions; the continuous horizontal lines represent the molecular framework of the exchanger; the horizontal dashed lines represent the phase boundary, which is sharp in a gel-type exchanger (see Chapter 3), but may be diffuse in an electrical double layer formed at a charged surface. (From Walton, H. F., ACS Audio Course C-93, American Chemical Society, Washington, D.C., 1987. With permission.)

Ion exchange was put to industrial use in 1905, when Gans, Harm, and Ruempler in Germany, made artificial ion exchangers by melting aluminates and silicates together. More efficient ion exchangers were later made by precipitating sodium aluminosilicate gels from aqueous solution (see Chapter 2). They were used for softening water, exchanging the harmful ions of calcium and magnesium in hard water for the less objectionable sodium ions. Today, water softening and water conditioning are by far the most important uses of ion exchange. When synthetic organic ion exchangers were produced after 1935, it became possible to remove salts completely from water by combining cation and anion exchange. Unlike the inorganic exchangers, the organic ion exchangers are resistant to acids and alkalies; therefore, hydrogen and hydroxyl ions can be exchanged just like any other ions. By cation exchange the sodium, calcium, and magnesium ions in natural water are replaced by H^+; by anion exchange the chloride, sulfate, and bicarbonate ions are replaced by OH^-; then H^+, and OH^- combine to form water.

Commercial water softening embodies these important features of ion exchange; reversibility and regeneration, and the use of ion exchangers in columns. Ion exchanges are reversible, yet they may be driven to completion in either direction by packing granules of ion exchanger containing one kind of mobile ions, let us say ions A, into a tube or column, and then passing a solution of a second kind of ion, say ions B. Fresh ions B continually enter the column and as they move down (in gravity flow) they keep encountering fresh ions A in the exchanger. The ion-exchange equilibrium is continually driven in one direction, so that soon the upper part of the column contains only ions B, yet the solution flowing from the bottom of the column contains no detectable B at all. (This statement may be tested with radioactive tracers. Pack a column with a cation-exchange resin containing replaceable hydrogen ions, pass a sodium salt solution containing radioactive Na-24, and the column effluent will have no detectable radioactivity until nearly all the hydrogen ions have been displaced.) Eventually, of course, the ions B appear at the bottom of the column, and after a short while ions A are completely displaced and the column contains only B. The column may then be reconverted to ions A by passing an excess of a salt of A. The exchange can be driven to completion, back and forth, any number of times as desired. The breaking through of the displacing ions at the column exit, and the form of the concentration-time curve, are discussed in Chapter 9. Breakthrough is sharper, the smaller the exchanger particles, the greater the affinity of the exchanging ions for the exchanger, and in general, the greater the "plate number" of the column.

In water softening, the column of cation exchanger originally contains only sodium ions, Na^+. Hard water is passed which contains Ca^{2+}. Calcium ions displace sodium ions from the column and eventually appear at the column exit. When this happens the flow of hard water is shut off, and a concentrated solution or "brine" of common salt, NaCl, is passed. The excess of sodium ions drive the calcium ions out to waste, thus "regenerating" the column and making it ready to soften more water. The cycle of softening and regeneration can be repeated indefinitely

The process can be represented by a chemical equation:

$$Ca^{2+} + 2NaRes = 2Na^+ + CaRes_2$$

This is the chemical reaction that is driven back and forth. Note that one calcium ion in the solution is replaced by two sodium ions during softening. This means that the reaction to the right is favored by dilute solutions, while the reverse reaction of regeneration is favored by concentrated solution. This is exactly the effect that is desired, for it makes the softening-regeneration cycle more efficient.

The first person to make a systematic study of ion exchange in chemical analysis was Olof Samuelson in Sweden in 1939.[3,4] He had organic ion exchangers to work with, the phenolsulfonic acid-formaldehyde condensation polymers (see Chapter 2), but he had only cation-exchange resins, not anion-exchange resins. Good anion exchangers were not produced until later. He could, however, work in acid solutions with resins containing H^+ as their replaceable cations. We call these "cation-exchange resins in the hydrogen form". One of the first things he did was to verify that the displacement of hydrogen ions by sodium ions from a resin column was quantitative. He then showed that metal ions could be adsorbed on the resin column and displaced again without loss.

It often happens in chemical analysis that certain anions interfere with the determination of certain cations and vice versa. Thus, phosphate ions in solution interfere with the determination of most metal ions, and even the alkali metals interfere, by coprecipitation, with the exact measurement of phosphate and sulfate. Until the advent of ion chromatography (Chapter 4) sulfate was always determined by precipitation of barium sulfate. The coprecipitation of other ions, both cations and anions, on barium sulfate was notorious.

Samuelson saw that ion exchange could separate anions from cations, in the sense that the anions and cations could be made to change partners, so that they would not interfere with each other's analytical determination. One could take a solution containing a mixture of salts and let it flow through a column of cation exchanger carrying hydrogen ions, i.e., in the hydrogen form. The cations of the mixture would be held back on the column while the anions flowed on, to be collected as a solution of their acids. Later the adsorbed cations in the ion exchanger could be recovered from the column by passing a dilute acid such as hydrochloric or nitric. The two solutions, one containing the cations of interest and the other containing the anions of interest, could be analyzed separately without interference from one another. In this way Samuelson could recover and measure small amounts of alkali metal salts in solutions from vanadium catalysts that contained much vanadate anions and he could accurately measure sulfate in impure aluminum sulfate crystals. Another thing that Samuelson did which is important and used today was to measure the total electrolyte content of salt solutions or hygroscopic, moist, solid salts, by passing measured amounts of the salt solutions through the hydrogen-charged cation-exchange resin column, washing the column well with distilled water, and titrating the released acid with standard alkali. With slight modifications this procedure is useful today to measure the total ionic content of natural waters.

These applications do not require ion-exchange selectivity, but only the ability to exchange one kind of cation for another. Yet it was only a very few years after Samuelson's pioneer work that the lanthanide ions were separated by ion exchange, a task that required very high selectivity and performed separations in hours that had previously required years using fractional crystallizations. Today such separations are performed in minutes. This achievement will be described below. It depended not so much on the selectivity of ion exchange as on differences in the stability of metal-ligand complexes formed in solution.

Samuelson's later work included studies of the partition of solvent mixtures between ion exchangers and liquids, and extensive studies of the chromatography of sugars and carbohydrates on anion- and cation-exchange resins with alcohol-rich mobile phases. Both these topics will be mentioned later (Chapter 3).

II. SCOPE OF THIS BOOK

The main application of ion exchange to analytical chemistry today is, beyond all question, high-resolution, high-performance ion-exchange chromatography. It is simply called, *ion chromatography*. Several books have appeared recently on ion chromatography. Ion exchange is a broader field, however. Ion-exchanging materials have properties and applications that go beyond the actual exchange of ions, such as their use in ligand exchange (Chapter 6) and as reversed-phase adsorbents of nonionic organic compounds. Separations are made by ion exchange that are not strictly chromatographic, yet have their use in chemical analysis. Chapters 8 and 9 describe some of them. The industrial uses of ion exchange and ion exchangers are very important. Conferences on ion-exchange technology attract the process engineers, while conferences on chromatography attract the analytical chemists. It is a pity that the two groups do not communicate more with one another. In this book we have mentioned ion-exchange technology, hydrometallurgy, e.g., but in general, we have limited ourselves to analytical applications. The analytical applications of ion exchange range from simple inorganic anions and cations to transition metals, lanthanides and actinides, to chiral separations of optically active organic compounds, and to amino acids, proteins, and biological macromolecules. We have said a little about all of these topics. A comprehensive treatment would need a multi-volume encyclopedia. Citing references, we have not strived for completeness, but we have tried to list publications that would lead the reader back to related papers and earlier papers by the same author. Applications we have described are illustrative, not complete.

We have assumed that readers are familiar with the elements of chromatography, i.e., retention, band spreading and resolution, the relation of retention to distribution ratio, and the theoretical-plate concept. Theoretical aspects of ion exchange and ion-exchange chromatography are discussed in Chapter 3.

REFERENCES

1. **Thompson, H. S.,** On the absorbent power of soils, *J. R. Agric. Soc. Engl.*, 11, 68, 1850.
2. **Way, J. T.,** On the power of soils to absorb manure, *J. R. Agric. Soc. Engl.*, 11, 313, 1850.
3. **Samuelson, O.,** Ueber die Verwendung von basenaustauschenden Stoffen in der analytishen Chemie. I., *Fresenius Z. Anal. Chem.*, 116, 328, 1939.
4. **Samuelson, O.,** Ueber die Verwendung von basenaustauschenden Stoffen in der analytischen Chemie. II., *Svensk. Kem. Tidscr.*, 51, 195, 1939.

Chapter 2

ION EXCHANGING MATERIALS

I. CONDENSATION POLYMERS

The first synthetic organic ion exchangers were described by Adams and Holmes in 1935.[1] These authors mixed 4 parts of 40% formaldehyde with 1 part of concentrated hydrochloric acid, then added 2 parts of a polyhydric phenol, such as catechol, resorcinol, or hydroquinone, dissolved in 20 parts of water. The mixture set to a gel which was washed, dried, and broken into particles 0.5 mm or less. The particles were packed into columns and were found to absorb metal cations, replacing them in solution by hydrogen ions. Anion-exchanging gels were prepared by using aniline or metaphenylene diamine in place of the phenols.

Hydrochloric acid was the catalyst; the phenols or aromatic amines condensed with formaldehyde, eliminating molecules of water and forming crosslinked structures like this:

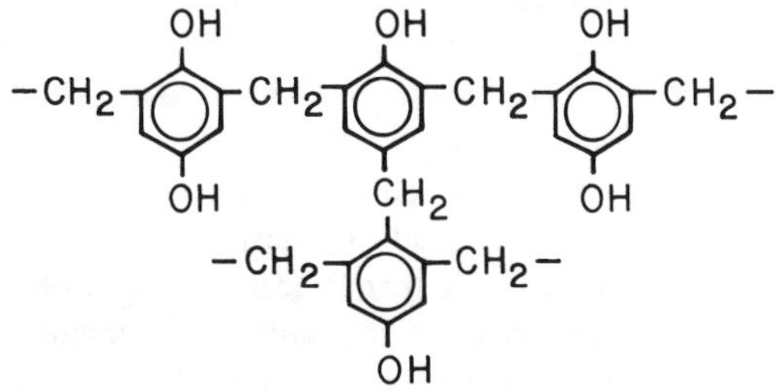

Because of their appearance and because they were like the phenol-formaldehyde plastics that in 1935 were known as ''synthetic resins'', the new products were called ''ion-exchange resins''. The name has stuck, and the ion-exchanging polymers used today are still called ion-exchange resins.

The ion-exchanging power of the gels prepared by Adams and Holmes derives, of course, from the phenolic groups, –OH. These groups are weakly ionized, with pK values around 9 or 10. Strongly acidic gels could be made by condensing phenolsulfonic acid with formaldehyde. These carry two kinds of acidic, ion-exchanging groups, phenolic –OH and sulfonic $-SO_3H$. A resin with functional carboxyl groups, –COOH, was prepared by condensing 3,5-dihydroxybenzoic acid with formaldehyde.[2]

Condensation polymers like these are little used today, because they are chemically unstable, compared with the addition polymers to be described in the next section. Their synthesis is hard to control, and the irregular, nonuniform granules are not suited to high-performance chromatography. Nevertheless, they have certain industrial uses, especially in sugar refining. Condensation polymers with special functional groups have analytical application too. The compound 8-hydroxyquinoline, mixed with cresol or resorcinol to provide crosslinking, may be condensed with formaldehyde to give a chelating resin that collects traces of heavy-metal ions from water.[3,4] Another chelating resin is made by condensing salicylic acid with phenol and formaldehyde.[5]

II. ADDITION POLYMERS

A. STYRENE POLYMERS

By far the most important ion-exchanging polymers today are those based on polystyrene.[6] It will help to understand their structure and properties if we describe how they are made.

The starting point is a mixture of styrene and divinylbenzene (DVB):

$CH=CH_2$ $\qquad$ $CH=CH_2$

$CH=CH_2$

Both these compounds are liquids at room temperature. Usually they are mixed in the proportion 92 mol of styrene to 8 mol of DVB, though this proportion may be varied at will. A small amount of a chain-initiating catalyst like benzoyl peroxide is added and the liquid mixture is poured into a tank of warm water that contains a very small proportion of surfactant (below the critical micelle concentration). The liquid breaks up into drops that are smaller, the more surfactant is added and the faster the stirring. Control of these conditions helps to control the droplet size. Polymerization proceeds, and the liquid droplets become beads of transparent solid, almost perfect spheres. This process is called "bead polymerization" to distinguish it from "emulsion polymerization" in which enough surfactant is used to form micelles and a water-soluble chain initiator, like hydrogen peroxide, is added. The result is a stable, milky dispersion of very fine solid particles called a "latex". Latex particles are used as coatings for some of the packings used in high-performance ion-exchange chromatography, but for most purposes beads are desired.

The styrene molecules join together to form linear chains, while the molecules of DVB provide the all-important crosslinks that join the polystyrene chains together. The chemical structure of the styrene DVB co-polymer can be represented thus:

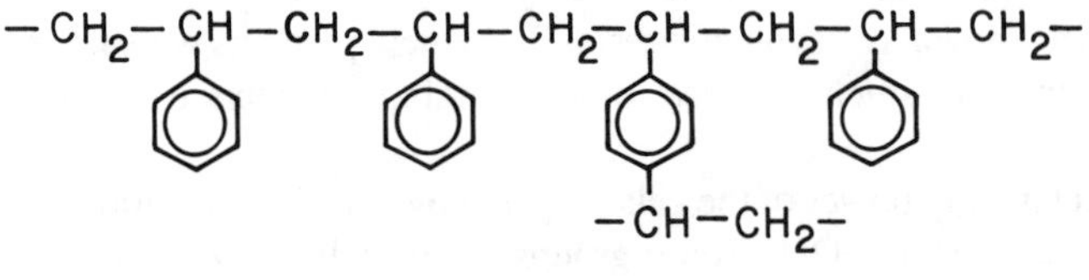

$$-CH_2-CH-CH_2-CH-CH_2-CH-CH_2-CH-CH_2-$$

$$-CH-CH_2-$$

This imaginary structure has 25% DVB or 25% crosslinking. The degree of crosslinking best suited to most applications is 8%, that is 8 mol of DVB to 92 mol of styrene. However, polymers are made commercially with degrees of crosslinking ranging from 2 to 16%. The lower crosslinking allows faster diffusion, especially of large ions and molecules, but lightly crosslinked polymers are very soft, and they swell and shrink excessively when one ion is substituted for another. Resins with higher crosslinking are harder and less easily deformed; they show greater selectivity and carry more ions per unit volume; however, diffusion in and out of the beads is slower.

At this point let us note some complexities that do not appear in the very simplified formula shown above. First, DVB has three isomers, the meta-isomer predominating.[7] Second, technical-grade DVB always contains ethyl vinyl benzene, up to 50%; however, 8% DVB means 8 mol of actual DVB to 92 mol of pure styrene. Third, DVB reacts somewhat faster than styrene, so that the first polymer to be formed contains more DVB and is more highly crosslinked than the polymer that is formed last.[8] Thus, the properties of the polymer beads depend on the conditions of formation and are difficult to control. Much effort has gone into the production of polymer beads suitable for high-performance chromatography.

After polymerization, the next step is to introduce the ionic functional groups. A substance is added that will penetrate the beads and cause them to swell, like nitrobenzene. To introduce sulfonic acid groups one now adds concentrated or fuming sulfuric acid with a catalyst such as a silver salt. Chlorsulfonic acid is also used. To get uniform sulfonation one allows the sulfonating agent to diffuse into the bead at a low temperature, then raises the temperature to make the reaction proceed. The product is a strong acid that will exchange its hydrogen ions for other cations and is therefore called a cation exchanger.

The beads of ion exchanger placed in water swell and become like droplets of concentrated electrolyte solutions. Internally they are fairly homogeneous and are called gel-type or conventional ion exchangers, to be distinguished from macroporous exchangers which will be described later.

To make an anion exchanger the procedure is more complicated. First the polymer is treated with chlormethyl ether, then the product is made to react with a tertiary amine, such as trimethylamine:

$$-CH(C_6H_5)-CH_2- + CH_3OCH_2Cl \longrightarrow -CH(C_6H_4CH_2Cl)-CH_2- \xrightarrow{+(CH_3)_3N} -CH(C_6H_4CH_2N(CH_3)_3^+)-CH_2- \; Cl^-$$

The product is a quaternary ammonium chloride that can exchange its chloride ions for other anions, including the hydroxide ion. The resin hydroxide is a very strong base indeed. If a base is desired that is not quite so strong, another tertiary amine may be used. Dimethylethanolamine gives a cationic functional group that holds hydroxide ions more strongly. The quaternary hydroxide is a weaker base than the trimethylammonium hydroxide, but it is still a fairly strong base.

Two complications should be mentioned. First, chlormethyl ether is a very powerful carcinogen. The hazards of handling it can be reduced by preparing it *in situ* from paraformaldehyde and hydrochloric acid.[9] Second, the chlormethyl group on one polymer chain can attack an aromatic ring on another polymer chain through the Friedel-Crafts reaction, causing additional crosslinking:

$$-CH(C_6H_4CH_2Cl)-CH_2- + C_6H_5-CH(-CH_2-) \longrightarrow -CH(C_6H_4-CH_2-C_6H_4-CH(-CH_2-))-CH_2- + HCl$$

This is more likely to happen with higher DVB crosslinking. It is hard to make a uniform, monofunctional anion exchanger. Titrating a commercial strong-base resin hydroxide with standard acid, and following the pH with a glass electrode, one always finds two inflection points, one for the strong quaternary-base function, the other for weakly basic groups, presumably tertiary amines. These can arise from thermal decomposition of quaternary ammonium hydroxides.

Chlormethylated polystyrene is the starting point for many ion exchangers and reactive polymers. It condenses with ammonia and with primary and secondary amines, with elimination of HCl, giving products that are weak bases. Amino acids are attached in the Merrifield synthesis of peptides; attachment of L-proline gives an asymetric, chiral, chelating ion exchanger that is used for the chromatographic separation of optical isomers of amino acids (see Chapter 6). A similar reaction gives rise to the important chelating ion exchanger, Chelex®-100:

$$-CH(CH_2-)-C_6H_4-CH_2-N(CH_2COOH)_2$$

This is better prepared[10] by a two-step process, first reacting the chlormethylated polymer with ammonia to give P–CH_2NH_2 (where P = the polymer), then reacting the amino polymer with chloracetic acid:

$$-CH(CH_2-)-C_6H_4-CH_2Cl + NH_3 \longrightarrow -CH(CH_2-)-C_6H_4-CH_2NH_2 \xrightarrow{+ClCH_2COOH} -CH(CH_2-)-C_6H_4-CH_2N(CH_2COOH)_2$$

A large family of chelating ion exchangers can be prepared from polystyrene by nitration, followed by reduction to aminopolystyrene, followed by treatment with sodium nitrite and hydrochloric acid to give the diazo resin:[11,12]

$$-CH(CH_2-)-C_6H_5 \xrightarrow[H_2SO_4]{HNO_3} -CH(CH_2-)-C_6H_4-NO_2 \xrightarrow{Na_2S_2O_4} -CH(CH_2-)-C_6H_4-NH_2 \xrightarrow{HNO_2} -CH(CH_2-)-C_6H_4-N{\equiv}N^+$$

The diazo resin can then be coupled with a great variety of compounds, particularly phenolic compounds, to add functional groups having special selectivities for different metals, corresponding to the reactions of organic reagents for metals in aqueous solution. One of the most useful of these products is an azo-coupled 8-hydroxyquinoline:[12]

$$-CH(CH_2-)-C_6H_4-N{=}N-C_9H_5N(OH)$$

Reacting the diazo resin with sodium arsenite solution gives the arsonic acid:[13,14]

$$P \cdot NN^+ + AsO_2^- + H_2O = P{-}N{=}N{-}AsO(OH)_2$$

An attractive feature of the diazo coupling is that it can easily be broken by reduction with sodium dithionite, $Na_2S_2O_4$. An amino group remains P–NH_2. This can be diazotized and coupled with a new ligand if desired.

Another way to attach functional groups to crosslinked polystyrene is to react the polymer with acetic anhydride or acetyl chloride in the presence of aluminum chloride, thus attaching the group $COCH_3$, then oxidize with alkaline permanganate to give –COONa.[15,16] At this stage one has functional carboxyl groups attached to polystyrene. These groups may be converted to –COCl to which other groups can be attached, including iminodiacetate chelating groups. The phosphonic acid group, $PO(OH)_2$, is added to polystyrene by reaction with PCl_3 and $AlCl_3$ catalyst.

It will be evident that by doing appropriate chemistry, many different functional groups can be attached to crosslinked polystyrene. A detailed review of the preparation of chelating polymers is that by Sahni and Reedijk.[10] There is a vast literature on polymers with chelating groups. These materials are selective for particular metals or groups of metals, and their main use is in recovering valuable metals from waste streams or in hydrometallurgy. Especially noteworthy in these applications are the "Srafion" resins that contain isothiouronium and sulfoguanidine groups:[17]

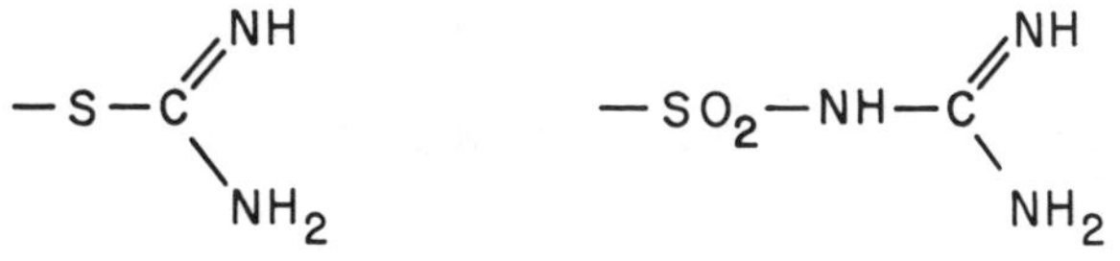

$$-S-C(=NH)NH_2 \qquad\qquad -SO_2-NH-C(=NH)NH_2$$

These resins are selective for gold and the platinum metals, those with d^8 electrons. Platinum(IV) does not have the d^8 configuration, but the resin reduces Pt(IV) to Pt(II), which does have that electron configuration and is strongly adsorbed. The polystyrene base of Srafion NMRR is 2% crosslinked. A macroporous version of this resin is called Monivex; because of its porous structure it is faster acting than Srafion and easier to regenerate.[18] To recover gold and platinum from Srafion it is necessary to burn the resin, for the metals are so tightly held, but they can be stripped from the macroporous resin by a solution of thiourea.

The main analytical use of chelating resins is to collect and preconcentrate metal ions present in low concentrations, particularly in environmental analysis, and to separate minor metallic constituents from concentrated solutions of other substances, e.g., salt brines. This topic will be discussed in Chapter 9. The need for highly selective resins or absorbents in chemical analysis has receded as the techniques of high-performance ion-exchange chromatography have improved. For most analytical applications the strongly acid cation exchangers with functional sulfonic acid groups and the strongly basic anion exchangers with functional quaternary ammonium ions suffice. We should mention here that anion-exchange selectivities are affected by the size of the alkyl groups attached to the quaternary nitrogen, so that special strong-base anion exchangers are sold that carry special R groups in NR_3.

Polystyrene-based exchangers are much used because they have good mechanical strength. They will stand the high-pressure gradients used in high-performance liquid column chromatography. Macroporous styrene polymers (to be described below) are generally used as the starting materials for making chelating resins and other special exchangers, because they have a large internal surface and allow easier access to the reactant molecules and easier exchange of ions in the finished product than do the conventional or gel-type polymers.

B. ACRYLIC POLYMERS

After the polystyrenes these are the most important ion-exchanging polymers. They are softer than the styrene-based polymers, and therefore unsuited to use in closed columns operating under pressure, but they are less hydrophobic than the polystyrenes and better suited to the chromatography of large biological molecules. Polystyrene-based absorbents have aromatic character and therefore absorb nonionic solutes that have aromatic rings or aromatic structures, such as caffeine. This is not so with polyacrylates, unless they are crosslinked with DVB, and then the effect is only slight.

Methyl methacrylate is polymerized in the presence of a crosslinking agent, which may be DVB or another compound that has two ethylenic linkages in it, like divinyl malonate or ethylene glycol dimethacrylate:

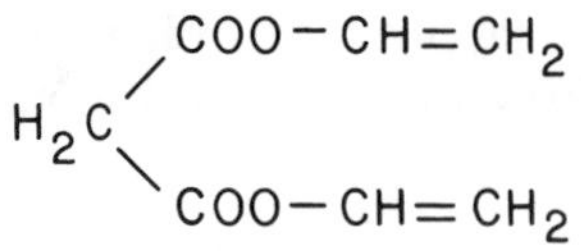

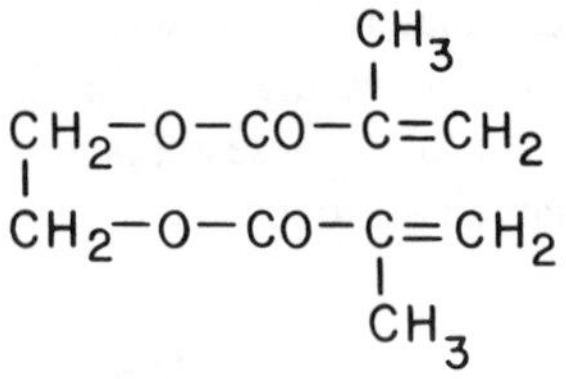

$$H_2C(COO-CH=CH_2)_2 \qquad CH_2-O-CO-C(CH_3)=CH_2 \,|\, CH_2-O-CO-C(CH_3)=CH_2$$

The product is hydrolyzed to give crosslinked polymethacrylic acid, the well-known exchanger called Bio-Rex® 70 or Zeo-Karb® 226:

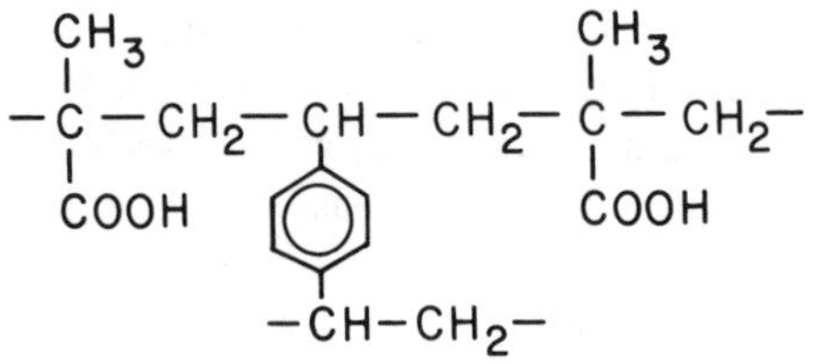

$$-C(CH_3)(COOH)-CH_2-CH(C_6H_4-CH-CH_2-)-CH_2-C(CH_3)(COOH)-CH_2-$$

This product is weakly acidic and thus can differentiate in its binding between bases of different strengths. One can add a number of functional groups to this polymer by appropriate chemistry. Thus, one can form the acid chloride, with the group –COCl. Treating this with phenylhydroxylamine gives a phenyl hydroxamic acid, C–CON(OH)C_6H_5, which is a selective chelating absorbent for Fe(III), U(VI), Cu, Hg, and Pb.[19]

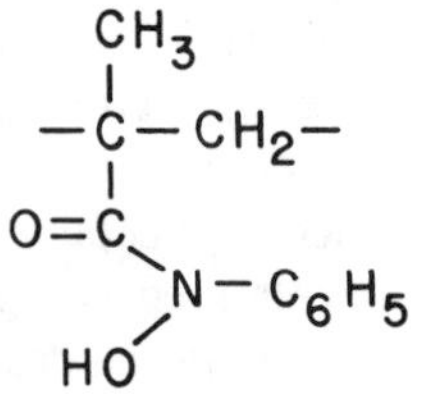

$$-C(CH_3)(-CH_2-)-C(=O)-N(OH)-C_6H_5$$

A special class of acrylate polymers are the Spherons.®[20-23] They are made by copolymerizing 2-hydroxyethyl methacrylate with ethylene glycol dimethacrylate, the latter being the crosslinking agent. Its formula is shown above. Polymerization is done in the presence of an inert solvent to produce a macroporous structure (see below). The beads of polymer are sufficiently rigid that they can be used in high-performance column chromatography under moderate pressures, of the order 5 atm across a column 20 cm long. Prepared as described, the polymers contain no ionic groups. Their use is size-exclusion chromatography. They are hydrophilic and adapted to the chromatography of biopolymers. Ionic functional groups are added by attack on the hydroxyl groups that they contain. Sodium chloroacetate plus sodium hydroxide converts –OH to $-OCH_2COO^-$; chlorsulfonic acid yields the sulfate group, $-OSO_2OH$. Succinic anhydride plus sodium hydroxide converts –OH to $-OCOCH_2CH_2COOH$; oxidation of the 2-hydroxyethyl groups by alkaline permanganate yields the side chain $-COO-CH_2COOH$. To make a weakly basic anion exchanger the monomer $Et_2NCH_2CH_2OCOC(CH_3)C{:}CH_2$ is substituted for a part of the 2-hydroxyethyl methacrylate in the original monomer mix, or the original Spheron® polymer is treated with $ClCH_2CH_2NEt_2HCl$ and sodium hydroxide. These reactions are summarized in Table 1.

C. OTHER POLYMERS

Many other ion-exchanging polymers have been made. Mainly they fall into two classes: hydrophilic polymers suitable for biochemical analysis and polymers with chelating groups

TABLE 1
Modification of Spheron® Polymers[20-23]

$$CH_3-\overset{|}{\underset{|}{C}}-COOC_2H_4OH, \text{ or } (P)C_2H_4OH$$

+ $ClCH_2COOH$	$\xrightarrow{NaOH}$	$(P)C_2H_4OCH_2COOH$
+ $ClSO_3H$	$\longrightarrow$	$(P)C_2H_4SO_3H$
+ $POCl_3$	$\longrightarrow$	$(P)C_2H_4PO(OH)_2$
+ $KMnO_4$	$\xrightarrow{H^+}$	$(P)CH_2COOH$
+ succinic anhydride (CH_2CO–O–CH_2CO ring)	$\xrightarrow{NaOH}$	$(P)C_2H_4OCOC_2H_4COOH$
+ $ClC_2H_4NEt_2$	$\xrightarrow{NaOH}$	$(P)C_2H_4OC_2H_4NEt_2$

Note: A polymer $(P)C_2H_4NEt_2$ can be made by co-polymerizing 2(N,N′-diethylamino)ethyl methacrylate with ethylene glycol dimethacrylate.

that are used to remove heavy metals from water and waste water. Of course chelating groups can be affixed to hydrophilic polymers. Chelating exchangers have important uses in hydrometallurgy. A few of the most important types will be noted here.

1. Polyvinylpyridine, Polybenzimidazole

Vinylpyridine, crosslinked with DVB, gives a weakly basic anion exchanger that has aromatic character and therefore has an attraction for aromatic solutes. The beads are as strong and rigid as crosslinked polystyrene. These resins bind metals as their anionic chloride complexes from hydrochloric acid solutions, and they recover traces of copper ions form sea water.[24] Copolymers of benzimidazole perform similarly.[25]

2. Polyethylene Imine

Ethylene imine and its linear polymer have the formulas:

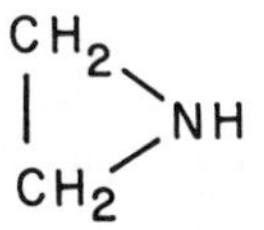

$$\begin{array}{l} CH_2 \\ | \quad \rangle NH \\ CH_2 \end{array} \qquad (-CH_2-CH_2-NH-CH_2-CH_2-NH-)_n$$

Polymers of moderate molecular weight (oligomers) are soluble and are available commercially. To make ion exchangers they must be crosslinked or incorporated into other crosslinked structures. An easy way to crosslink the linear polymers is by reaction in alkaline solution with epichlorhydrin:

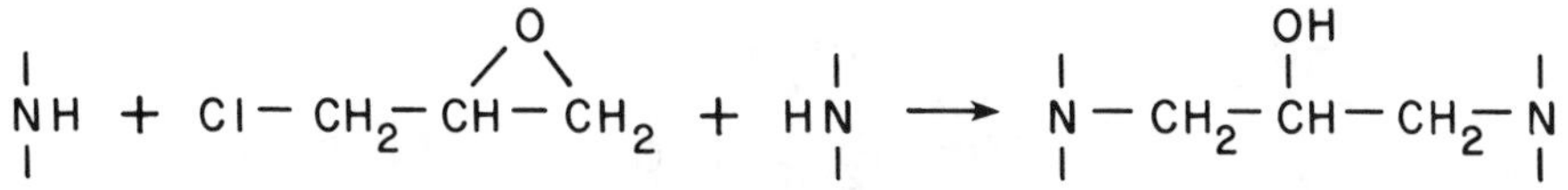

$$\underset{|}{\overset{|}{N}}H + Cl-CH_2-\overset{O}{\overbrace{CH-CH_2}} + H\underset{|}{\overset{|}{N}} \longrightarrow \underset{|}{\overset{|}{N}}-CH_2-\overset{OH}{\overset{|}{CH}}-CH_2-\underset{|}{\overset{|}{N}}$$

Epichlorhydrin forms bridges between secondary amine groups. The result is a weakly basic anion exchanger containing secondary and tertiary amino groups. Co-polymerizing ethylene imine with phenylene diisocyanate or polymethylene polyphenyl isocyanate one obtains a polyamine-polyurea, into which carboxyl and dithiocarbamate groups can be introduced.[26,27] These materials are used to collect trace metals from water and from concentrated salt brines for measurement by inductively coupled plasma emission spectrometry (see Chapter 9). As chelating exchangers they are highly selective.

Polymers and silica-based exchangers that incorporate polyethyleneimine have been developed for chromatography of proteins (Chapter 7).

3. Ion Exchangers from Dextran and Cellulose

Dextran and cellulose are carbohydrates whose structures carry many hydroxyl groups. These react with epichlorhydrin in the same way that secondary amine groups do (see above):

$$\text{(P)OH} + \text{ClCH}_2\text{CH(O)CH}_2 = \text{(P)O–CH}_2\text{–CHOH–CH}_2\text{O(P)}$$

The well-known Sephadex® absorbents are made by crosslinking dextran with epichlorhydrin.[28] They are polar absorbents with no ionic groups. They are made with a range of pore sizes, corresponding to the degree of crosslinking. Ionic functional groups can be introduced by the same reactions that are noted in Table I, to give a range of ion exchangers with strong-acid, weak-acid, strong-base, and weak-base character. One of the most useful of these ion exchangers is DEAE-Sephadex®, made by reaction with $ClCH_2CH_2NEt_2$ in alkaline solution.

The same kinds of materials can be made from cellulose. All of them are soft, and must be used in open columns or in thin-layer or paper chromatography. Their applications are mainly in the biochemical field. They may be used in batch separations; a cellulose-based exchanger called Hyphan which has these functional groups:

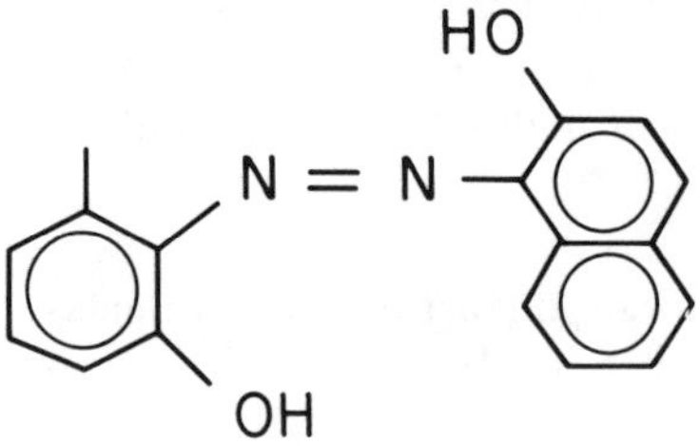

HO

N = N

OH

has been used to adsorb and preconcentrate transition-metal ions, especially Ag, Au, Pd, and Pt, from very dilute solutions.[29]

4. Fluorocarbon Exchangers

In complete contrast to the exchangers based on cellulose and dextran, fluorocarbons are very hydrophobic. The best known ion exchanger is Nafion, whose chemical structure is shown below:

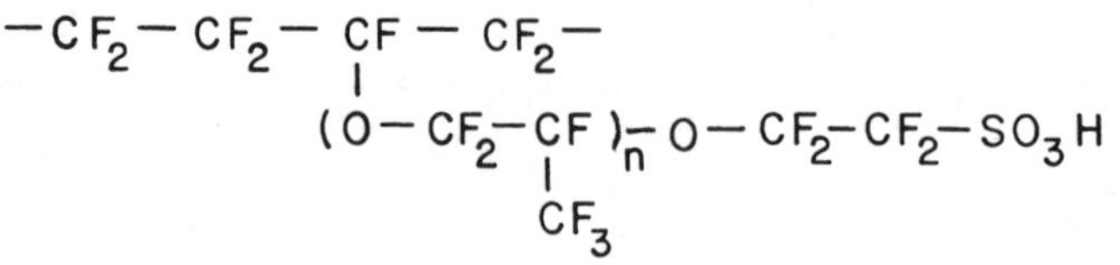

$$-CF_2-CF_2-\underset{\displaystyle (O-CF_2-\underset{\displaystyle CF_3}{\underset{|}{CF}})_n-O-CF_2-CF_2-SO_3H}{\underset{|}{CF}}-CF_2-$$

TABLE 2
Amberlite XAD Resins (Rohm and Haas Co.)

Name	Chemical type	Pore diameter, (nm)	Surface area, (m^2g^{-1})
XAD-1	Styrene-DVB	20	100
XAD-2	Styrene-DVB	9	330
XAD-4	Styrene-DVB	5	750
XAD-7	Acrylate	8	450
XAD-8	Acrylate	25	140

It has a fluorocarbon backbone, $-CF_2-CF_2-$, and sulfonic acid functional groups. Thus, it is a strongly acidic cation exchanger. It is fabricated into thin membranes that are mechanically strong and chemically resistant, and are employed in alkali-chlorine cells, fuel cells, and elsewhere in the electrochemical industry. Ion-exchanging membranes have applications in chemical analysis too. For this reason the ion-exchange selectivity of "Nafion" has been studied.[30,31] There is a big difference between the affinity for sodium ions and that for potassium ions, but the selectivity orders are essentially the same as those for sulfonated polystyrene. Another strong-acid cation exchanger with the $-SO_3H$ functional group has been prepared from the polymer Kel-F.[32]

D. MACROPOROUS POLYMERS

"Macroporous" means having large pores. The word "macroreticular" is synonymous. Macroporous ion exchangers and absorbents have the form of spherical beads, like the microporous or gel-type absorbents that we have described, but to the eye or under a low-power microscope the beads appear opaque and milky, not clear and transparent as do the common gel-type ion exchangers. Viewed with the electron microscope, macroporous ion-exchanger beads are seen to be made up of many much smaller beads, some tens of nm across, .001 the diameter of the visible bead, or less. The primary microbeads are connected loosely together to give an aggregate that has some mechanical strength but not much; the visible or "macrobeads" fall to a powder when one tries to grind them in a mortar. Between the microbeads there are channels that may be up to 500 nm in diameter, far exceeding molecular dimensions. These channels are the macropores (or just "pores") of a macroporous resin. They allow ions and molecules in the solution to move rapidly in and out of the resin, and allow rapid ion exchange compared to the slow diffusion-controlled ion exchange in beads of conventional gel-type exchangers. The macroporous beads are fairly rigid. They do not swell and shrink appreciably when the concentration of the external solution or the nature of the solvent changes.

Macroporous resins without ionic groups are much used as absorbents to remove traces of organic compounds from water. Well known are the Amberlite® XAD resins made by Rohm and Haas Co. Table 2 lists their chemical nature, pore size, and surface area. These materials are also used in size-exclusion chromatography, as are the macroporous methacrylate polymers, the Spherons®, described above. As we have noted in preceding sections, ion-exchanging functional groups can be attached to macroporous polymers by a variety of chemical reactions. In making a macroporous ion exchanger, the usual procedure is to prepare the nonionic polymer first, and then add the functional groups.

Macroporous polymers are a class of *solvent-modified polymers*. To make them, the liquid monomers including the crosslinking agent are mixed with a solvent or diluent, then polymerized by bead polymerization in the manner described above. After the reaction is complete, the beads are washed and dried, and the diluent removed by evaporation. Two classes of diluent may be used; one that is a good solvent for the polymer, the other that is a poor solvent for the polymer. Both kinds, of course, must be good solvents for the monomers. For styrene-divinylbenzene a good solvent is toluene, a poor solvent is hexane, cyclohexane, or carbon tetrachloride.

Consider now the course of polymerization of styrene and DVB in the presence of a solvent. Reaction starts at a number of nuclei and the polymer chains spread outwards from these centers. We recall that a molecule of *m*-DVB adds to styrene faster than a styrene molecule does, and therefore the first portions of polymer to be formed are more highly crosslinked than those formed later. In a good solvent the polymer chains will grow out in a more or less straight fashion, becoming less and less crosslinked until they meet growing chains from another nucleus, when the extended chains will join. In a poor solvent the chains will be very crooked, bunching back on themselves because the polymer segments attract each other more than they do the solvent. Polymer chains become entangled, and crosslinks form between different segments of the same chain. Only a few of the chains will extend far enough to form bridges between two nuclei. We may say that the polymer precipitates out of solution as it is formed.

In the final products remaining after polymerization has stopped and the solvent (diluent) has been removed we may distinguish two kinds of porosity: swelling porosity and permanent porosity. Swelling porosity can be measured by the uptake of toluene, a good solvent that enters among the extended chains, solvates them, and forces them apart. Permanent porosity, or macroporosity, is measured by the uptake of hexane, a poor solvent, which goes into the spaces between the masses of highly crosslinked, highly entangled polymer but cannot separate the individual polymer chains. Macroporous ion-exchange resins have this kind of porosity. Macroporosity is favored by (1) high proportion of DVB or other crosslinking agent, (2) high proportion of diluent, (3) low solvent strength, or low affinity of the diluent for the polymer, and (4) faster reaction of the crosslinking agent with respect to that of the main monomer, e.g., rate of reaction of DVB: rate of reaction of styrene.

Extensive studies have been made of solvent uptake by these materials. The experiments of Millar and collaborators,[33-36] described in a series of four papers are especially instructive. An exhaustive review by Seidl and others[37] described macroporous styrene-DVB co-polymers and the ion exchangers that can be made from them. Macroporous structures are very heterogeneous, and so are macroporous ion exchangers. In ion exchange, selectivity increases with crosslinking. In a macroporous exchanger, the first ions of another kind to displace the ions that are already present are bound very strongly, but as exchange proceeds, this selectivity becomes less and less. The dependence of ion-exchange distribution ratio on the proportion of ions in the exchanger makes for distorted, unsymmetrical peaks in chromatography. Nevertheless, ion exchangers made from macroporous styrene-DVB polymers are used successfully in modern high-performance ion chromatography. The point here is that these exchangers have low ion-exchange capacity, 0.01 meq or less per gram.[38,39] Only the surfaces of the small, highly crosslinked microbeads are sulfonated or aminated, and these surfaces are fairly homogeneous.

We have seen that a great variety of structures can be obtained by polymerizing in the presence of solvents. The term macroporous describes a broad range of materials. Today we are seeing spectacular results in high-performance chromatography with specially made ion-exchanging polymers. These developments are happening in industrial research laboratories, and understandably little is published on the details of polymer synthesis.

III. ION EXCHANGERS FOR CHROMATOGRAPHY

For efficient, high-performance chromatography it is essential to have fast mass transfer between the stationary and the mobile phase. Another requirement is that the particles of the stationary phase be uniform in size. If they are not, the large particles will cause band broadening and the small ones will clog the column or limit the flow rate. The particles must be mechanically strong to stand high-pressure gradients and they should be fairly rigid. They should not undergo large volume changes when the composition of the mobile phase is changed. Small volume changes can be tolerated if the particles are tightly packed and slightly elastic.

The rate of ion exchange is controlled under some circumstances by diffusion to and from the exchanger surface across a liquid film (film diffusion), but more often it is controlled by particle diffusion; the diffusion of ions in and out of the exchanger bead. As we shall see in the next chapter, diffusion coefficients in ion exchangers are very small. To speed up the mass transfer the diffusion paths, the distances that the ions must traverse to move in and out of the resin, must be made small.

An obvious way to accomplish this is to use small particles. Conventional gel-type ion-exchange resins are routinely used in particle diameters of 10 μm and less, sometimes as small as 3 μm. The gel-type resins have the advantage of high ion-exchange capacity (quantity of ions per unit volume) and homogeneity, but the disadvantage to working with small particles is the high back pressure; the pressure gradient needed to maintain a given flow rate is inversely proportional to the square of the particle diameter. It is not easy to produce very uniform resin spheres of small diameter. Another problem may be volume changes that accompany changes in concentration of the mobile phase solution. The volume changes are less, and the mechanical strength is greater, the greater the crosslinking of the gel-type resin. A crosslinking of 8% is usually the minimum that can be used in high-performance columns under pressure, though with great care and maintaining low-pressure gradients it is possible to work with 4% crosslinked sulfonated polystyrene resins. We noted above that the Spheron® acrylate polymers can be used in columns under pressure provided that pressure does not exceed 5 to 10 bar (70 to 140 psi).

The next possibility for faster mass transfer is to use a macroporous ion exchanger. Spheron®, just mentioned, is macroporous. Macroporous styrene-DVB co-polymers (''PRP-1'') are used in reversed-phase liquid chromatography, but fully sulfonated macroporous styrene-DVB beads have limited acceptance because their chromatographic peaks show excessive tailing. This is a sign of nonuniformity of the exchange sites. Lightly sulfonated (in general, lightly functionalized) styrene-DVB polymer beads have been used very successfully by Fritz and others in ion chromatography. Lee[40] made a strong-base anion exchanger by a proprietary process with a capacity of 0.17 meq/g, intermediate between that of the surface-functionalized resins and the fully functionalized resins, and got very good chromatography of inorganic anions, but the loading, or amount of the anions to be analyzed, was small.

In 1967 Horvath et al.[41] described the preparation of pellicular resins. These were small and uniform glass beads, 20 μm in diameter or more, coated with films of ion-exchanging polymers that were no more than 1 to 2 μm thick. The ion-exchange diffusion paths were very short, yet the beads were large enough that they did not impose high back pressures. (In 1967 high back pressures were seen as a disadvantage, because controlled-flow high-pressure pumps were not available). Horvath used pellicular resins for chromatography of nucleic acid derivatives and other compounds of biological interest. Pellicular resins have two disadvantages: their ion-exchange capacity is very low, and the resin film tends to flake away from the glass bead. Their use today is mostly limited to guard columns. The columns are very easy to pack and may be packed dry.

In 1975 there appeared a historic paper by Small et al.[42] that marked the beginning of ''ion chromatography''. These authors prepared surface-functional, rapidly acting ion-exchanger beads by taking beads of styrene-DVB copolymer and sulfonating them on the surface. The beads were dipped into hot sulfuric acid for a limited time, then taken out and washed. The beads had a diameter of 20 μm or more; the active surface layer was only 1 to 2 μm thick, and the thickness could be controlled according to the conditions of sulfonation. Later studies by Sevenich and Fritz,[43] Schmuckler and Goldstein,[44] and others showed that the boundary between sulfonated and unsulfonated polymer was sharp. Hajos and Inczedy[45] showed that the thin sulfonated layer had the same ion-exchanging properties as the bulk gel resin, that is, its ion-exchange selectivity was the same.

It was harder to make a surface-functional anion exchanger by a chemical reaction at the bead's surface. Instead, Small and co-workers[42] made their first anion exchangers by latex coating. They first put a very thin layer of sulfonic acid groups on the polymer beads, packed the lightly sulfonated beads into a column, then pumped through the column a suspension of very finely ground high-capacity, gel-type styrene-DVB strong-base anion exchanger. Electrostatic attraction between the negative sulfonate ions on the bead surface and the positive quaternary ammonium ions on the fine suspended particles gave a very strong bond. The fine particles of anion-exchange resin gave very rapid ion exchange and narrow chromatographic bands.

In later developments the suspension of finely ground ion exchanger was replaced by an ion-exchanging polymer in latex form prepared as described above.[46] The anion-exchanging latex was made by co-polymerizing vinyl benzyl chloride with 5% DVB, then converting the group CH_2Cl to a quaternary ammonium group by reaction with a tertiary amine. The latex particles were some 50 nm across; the polymer beads on to which they were deposited had diameters 10 to 20 μm. Today, smaller beads are used, often 5 μm and cation-exchanging as well as anion-exchanging latex coatings are made. Figure 1, obtained with a scanning electron microscope, shows a latex-coated bead. To get a good picture these latex particles were larger than those commonly used in chromatography.

Latex-based columns are popular because they combine the internal homogeneity and uniformity of gel-type exchangers with fast mass transfer. They are applied not only to inorganic ions, but to the chromatography of peptides and proteins (see Chapter 7).

A new development is to coat a macroporous polymer with a latex whose particles are just too large to enter the polymer pores. The pores are accessible to the mobile phase while the latex particles (which are covalently bound to the polymer beads) can exchange ions. The structure is shown in Figure 2. The result is a stationary phase that can act as a hydrophobic or reversed-phase adsorbent and also as an ion exchanger. One can change from one mode to the other simply by changing the eluent.[47] Furthermore, the polymer is not swollen by acetonitrile and other organic solvents. The polymers normally used in ion chromatography will not tolerate high concentrations of acetonitrile; they swell and may burst the column. Thus, one can separate aromatic hydrocarbons and aromatic acids in the same run. Figure 3 shows such a chromatogram. The new packing can also be used for ion-pair chromatography.

One must recall the conventional styrene-DVB ion-exchange resins have the character of reversed-phase adsorbents too. Polycyclic aromatic hydrocarbons may be separated on a sulfonated polystyrene cation exchanger loaded with Ca(II) or Fe(III) — doubly and triply charged counter-ions give more retention than singly charged ions — using water-acetonitrile mixtures as eluents.[48] No ion exchange is involved, only hydrophobic or reversed-phase effects, strengthened by π-electron interaction. Nonionized water-soluble polar aromatic compounds can also be separated, with water-methanol or water-ethanol eluents. What the new latex-coated resin provides is macroporous character, leading to faster mass transfer and narrower peaks, and also volume stability. There is no drastic expansion with water replacing an organic solvent, as is the case with conventional resins.

Another approach to the problem of getting fast mass transfer in ion exchange is to take porous silica, which is itself a macroporous inorganic polymer, and attach carbon chains to it that carry ionic functional groups. The techniques of making porous silica spheres in controlled particle sizes and controlled porosities for liquid chromatography are well known, and so are the ways to attach organic groups. Bonded silica ion exchangers have the great

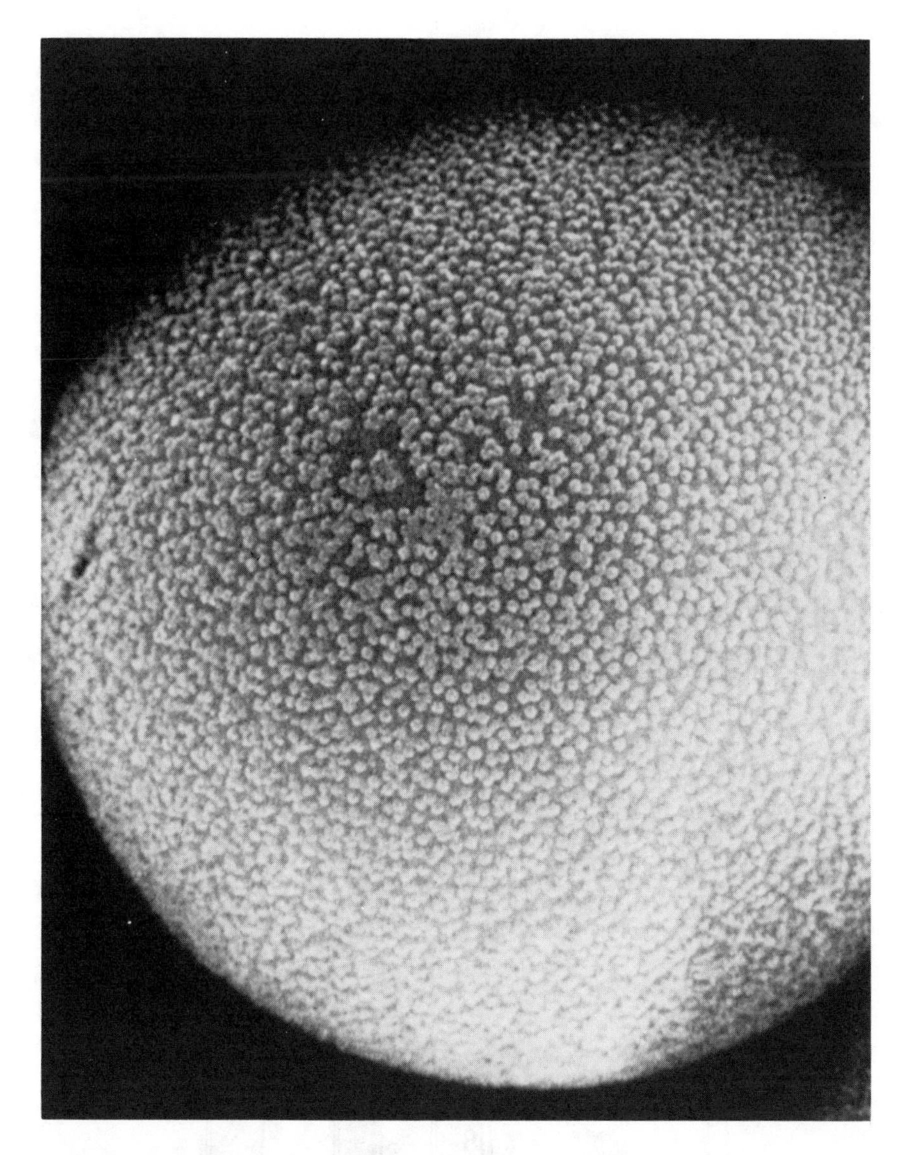

FIGURE 1. Photomicrograph of latex-agglomerated resin bead. Bead is 5 μm in diameter, latex particles are 0.2 μm in diameter. (From Slingsby, R. W. and Pohl, C. A., *J. Chromatogr.*, 458, 241, 1988. With permission. Photograph courtesy of Dionex Corp.)

advantage of rigidity; they will stand high pressures without deforming and they do not swell or shrink; mass transfer between the solid and the mobile phase is fast, and the exchange sites are uniform. A disadvantage to ion exchangers based on silica is that they can only be used in a limited pH range, between (roughly) 3 and 8 in aqueous solution. Below pH 3 the organic carbon chains break away from the silica, and above pH 8 the silica dissolves to form silicate ions. Acidic and alkaline solutions are often used in the chromatography of inorganic ions. Nevertheless, the importance of bonded silica ion exchangers is so great that we shall devote a separate section to them.

A. SILICA-BASED ION EXCHANGERS

Porous silica beads for liquid chromatography are made by the hydrolysis of ethyl orthosilicate in the presence of solvents. Accounts of the preparation are given by Unger et al.[49,50] The beads have a macroporous structure and are ion exchangers; the silanol groups,

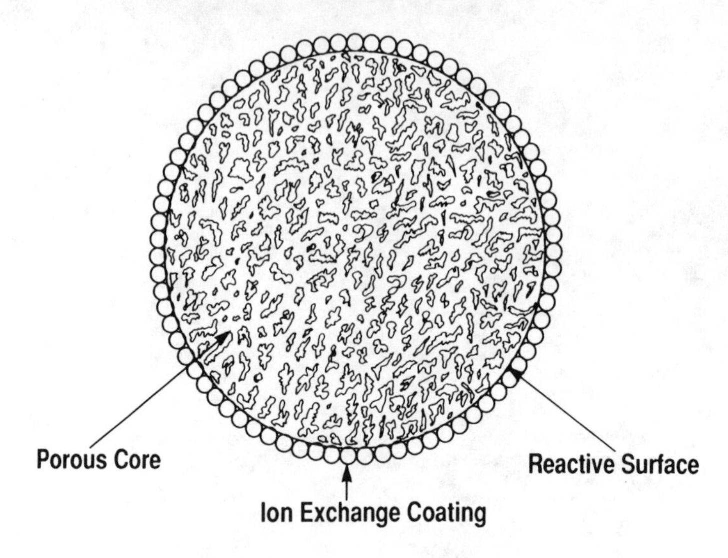

FIGURE 2. Latex-coated macroporous ion-exchange resin: schematic. (Courtesy of Dionex Corp.)

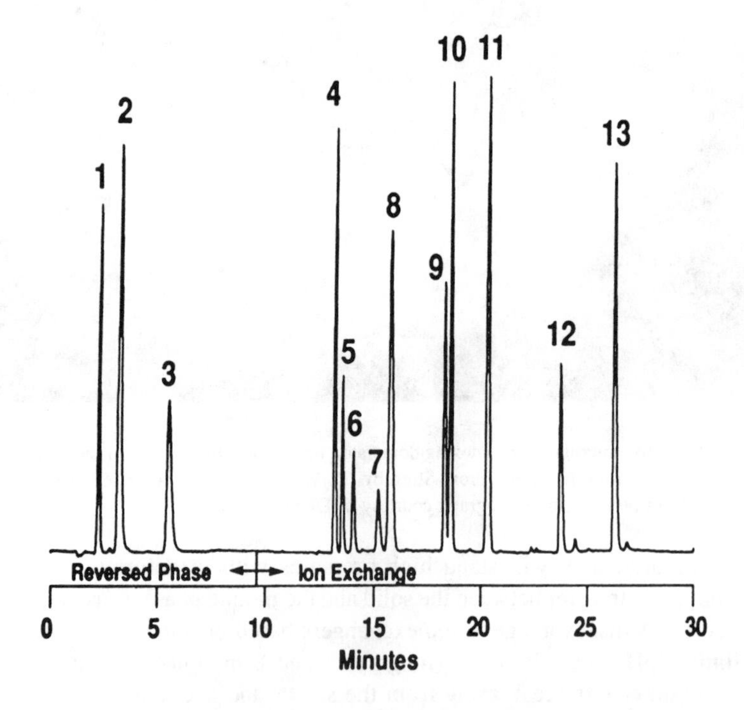

FIGURE 3. Chromatogram of aromatic compounds on the resin shown in Figure 2. Peaks 1, 2, and 3 are benzyl alcohol, diethyltoluamide, and benzene, eluted by 80% acetonitrile. Peaks 4 to 13 are, respectively, benzoic acid, benzenesulfonic acid, *p*-toluenesulfonic acid, *p*-chlorobenzenesulfonic acid, *p*-bromobenzenesulfonic acid, phthalic acid, terephthalic acid, *p*-hydroxybenzenesulfonic acid, trimesic acid, and pyromellitic acid. Eluent was 20% acetonitrile (v/v) with a sodium chloride gradient from zero to 0.4 *M*. Detection was by UV at 254 nm. (Courtesy of Dionex Corp.)

SiOH, on the surface are weakly acidic. They are also the reactive sites that allow one to attach organic groups to silica. To make the well-known reversed-phase packing, porous silica is reacted with a chlorosilane or a methoxy- or ethoxy-silane, $RSiX_3$ or $RSi(CH_3)_2X$, where R is a long-chain alkyl group, commonly $C_{18}H_{37}$, and X is Cl, OCH_3, or OC_2H_5. The reaction is performed in the absence of water by refluxing for several hours with toluene. It may be written thus:

$$-\overset{|}{\underset{|}{Si}}-OH + XSi(CH_3)_2R = -\overset{|}{\underset{|}{Si}}-O-Si(CH_3)_2-R$$

Trichloro- or trimethoxy-silanes are commonly used because they are cheaper and react faster than monochlorosilanes, but they leave unsubstituted Cl atoms or OCH_3 groups attached to the silicon atom:

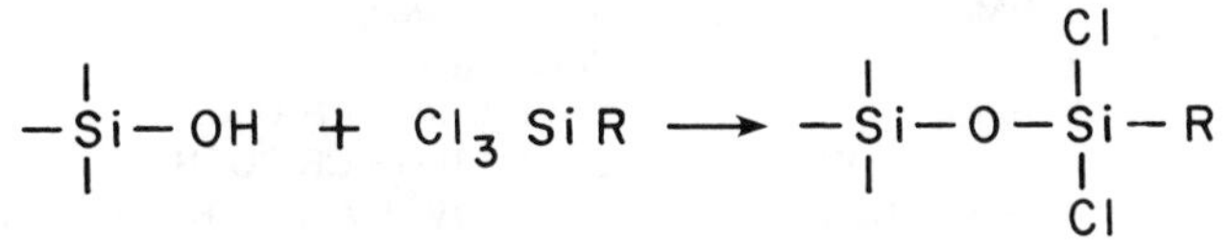

$$-\overset{|}{\underset{|}{Si}}-OH + Cl_3SiR \longrightarrow -\overset{|}{\underset{|}{Si}}-O-\overset{Cl}{\underset{Cl}{Si}}-R$$

Later, these are hydrolyzed and leave silanol groups, SiOH. These either crosslink with silanol groups on the silica surface or they are deactivated with trimethylchlorosilane.

To make a bonded ion exchanger one must use an R group that has a reactive center in it. Such centers are the vinyl group, the amino group, and the epoxy group:

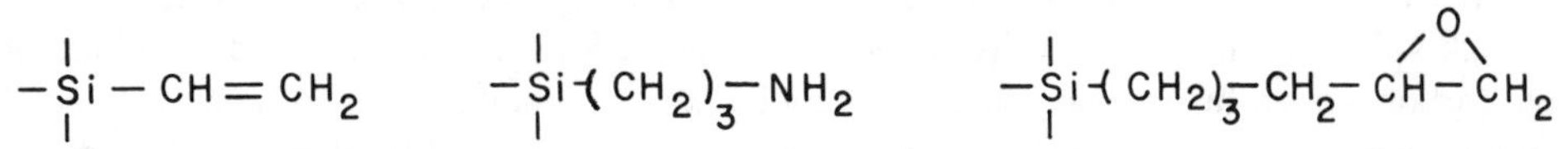

$$-\overset{|}{\underset{|}{Si}}-CH=CH_2 \qquad -\overset{|}{\underset{|}{Si}}-(CH_2)_3-NH_2 \qquad -\overset{|}{\underset{|}{Si}}-(CH_2)_3-CH_2-\overset{O}{CH-CH_2}$$

Attached amino groups make the silica into a weakly basic anion exchanger. Attached epoxy groups can be hydrolyzed to give a bonded diol, with polar character. Some reactions that give rise to bonded ion exchangers are summarized in Table 3. The last group, where one starts with glycidoxytrimethoxysilane or glycidoxypropyltrimethoxysilane, leads to a great variety of products. Chang et al.[51] has used these reactions to make a large family of ion-exchanging absorbents for the chromatography of proteins, the glyco-phases. These will be described in more detail in Chapter 7. The first group, starting with vinyl trichlorosilane or vinyl dimethyl chlorosilane, has been developed by Caude et al.[52-54] These exchangers are distinguished by high ion-exchanging capacity, as high as the capacities of conventional gel-type ion-exchange resins if one compares them per unit volume of the packed column. Of course, to get a large exchange capacity one starts with a porous silica having a large specific surface area, and this implies a small pore diameter and, hence, slow ion exchange.

Silica-based ion exchangers are used in single-column ion chromatography with eluents in the neutral or weakly acid range. We recall that bonded silica packings are attacked by solutions having pH less than 3 or more than 8. Their major uses are in the chromatography of proteins and nucleic acid derivatives, in the collection of trace metals (to be described in Chapter 9) and in certain applications of ligand-exchange chromatography. Silica-based chelating exchangers are used for concentrating trace metals, and have potential use in high-performance chromatography of heavy-metal ions. Here, their innate selectivity makes it unnecessary to use complexing eluents. Control of pH is of course necessary.

The preparation of chelating exchangers and other silica-bonded ion exchangers is summarized in Table 3. A recent synthesis of 8-hydroxyquinoline-bonded silica that does not

TABLE 3
Silica-Bonded Ion Exchange

First product	Treatment	Derived products	Ref.
Si–CH=CH$_2$	+ styrene	Si–CH$_2$CH=CHCH$_2$C$_6$H$_5$	52—54
	then H$_2$So$_4$	Si–CH$_2$CH=CHCH$_2$C$_6$H$_4$SO$_3$H	
	then ClCH$_2$OCH$_3$ and NMe$_3$	Si–CH$_2$CH=CHCH$_2$C$_6$H$_4$CH$_2$NMe$_3$Cl	
	+ methacrylic acid	Si–CH$_2$CH=CHMe · COOH	
Si–CH$_2$C$_6$H$_5$	+ H$_2$SO$_4$	Si–CH$_2$C$_6$H$_4$SO$_3$H	49
Si–(CH$_2$)$_4$C$_6$H$_5$	+ H$_2$SO$_4$	Si–(CH$_2$)$_4$C$_6$H$_4$SO$_3$H	56
Si(CH$_2$)$_4$C$_6$H$_4$CH$_2$Cl	+ NMe$_3$	Si(CH$_2$)$_4$C$_6$H$_4$CH$_2$NMe$_3$Cl	56
Si–(CH$_2$)$_3$NH$_2$	+ ClCOC$_6$H$_4$NO$_2$	Si(CH$_2$)$_3$NHCOC$_6$H$_4$NO$_2$	57, 60
	then reduced, diazotized,	coupled to make chelating exchangers	
	+ ClC$_6$H$_4$NO$_2$	Si(CH$_2$)$_3$C$_6$H$_4$NO$_2$	58, 59
	then reduced, diazotized,	coupled as before	
	+ CS$_2$	Si–(CH$_2$)$_3$NHCSSH	61
Si–(CH$_2$)$_3$Cl	+ NMe$_3$	Si–(CH$_2$)$_3$NMe$_3$Cl	56
Si–(CH$_2$)$_3$SH	+ peracetic acid	Si–(CH$_2$)$_3$SO$_3$H	62
O	+ H$_2$SO$_4$	Si–CH$_2$CH(OH)CH$_2$OSO$_3$H	51
/ \	+ ClCH$_2$COOH	Si–CH$_2$CH(OH)CH$_2$COOH	
Si-CH CH — CH$_2$	+ HOC$_2$H$_4$NEt$_3$	Si–CH$_2$CH(OH)CH$_2$OC$_2$H$_4$NEt$_2$	
Si–C$_6$H$_4$NH$_2$	+ HNO$_2$ then 8-quinolinol (oxine)	Si–C$_6$H$_4$–N=n–oxine (5–position)	63

fit the format of Table 3 is described by Weaver and Harris.[55] A four-step process starts with 8-hydroxyquinoline and allyl bromide and leads to a product Cl(CH$_3$)$_2$Si(CH$_2$)$_3$–Q, where Q is the 8-hydroxyquinoline radical bound in the 7-position; the phenolic –OH is protected with Si(CH$_3$)$_3$; this product is then coupled to dry silica, yielding a final product in which silicon is attached directly to the 7-position of 8-hydroxyquinoline by a bridge of three carbon atoms. In this way the attachment by the azo group is avoided, and the bonded phase is very stable. Coordination with certain metal ions, notably aluminum, produces fluorescence, which enables them to be measured at the same time as they are concentrated from dilute solution. (The azo group suppresses fluorescence.)

B. INORGANIC ION EXCHANGERS

The most abundant inorganic ion exchangers found in nature are the clay minerals. The most common artificial inorganic ion exchanger is glass. The ion-exchanging property of glass surfaces must be borne in mind in the laboratory, especially by chemists who work with trace metal analysis. A simple experiment will demonstrate this property. Take a dilute solution of methylene blue, tetramethyl thionine chloride, a well-known dye that has an intensely blue-colored cation. Place it in a clean glass flask and leave it there for a few minutes, then pour the solution out and rinse the flask several times with distilled water, until the rinsings are colorless. Pour some 10 ml of the rinse water into a small, clean beaker; it will look colorless. Now put 10 ml of concentrated sodium chloride solution into the flask (previously drained) that had held the methylene blue, swirl it around for a minute or two and pour the salt solution into a second small beaker. It will be a very distinct blue. The sodium ions of the salt have displaced the methylene blue cations from the glass surface by ion exchange. This same retention of adsorbable cations can happen with trace heavy metal ions unless the precaution is taken of always adding acid or salt to the solutions.

The pH response of the glass electrode is caused by ion exchange at the glass surface and a few molecular diameters within the surface, and the classic studies of G. Eisenman on ion-exchange selectivity (see Chapter 3) were made in the first instance to interpret the glass-electrode potential response to cations other than hydrogen. The surface of glass carries negatively charged silicate ions or silanol groups, Si–O^- or Si–OH, and also aluminate ions:

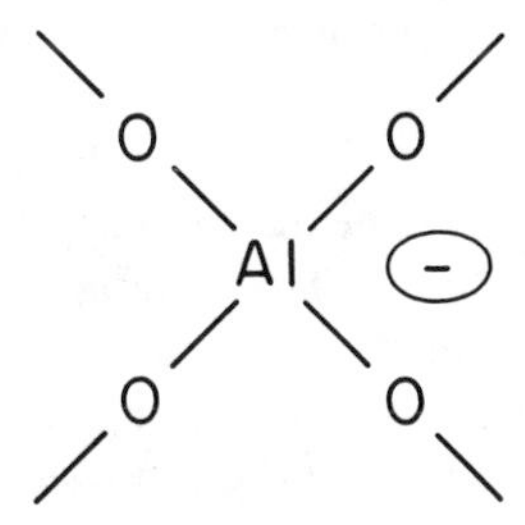

In the silanol ions the negative charge is concentrated in the terminal oxygen atom and the electrostatic field strength around this fixed ion is high. In the aluminate ions the negative charge is spread around four oxygen atoms and the filed strength is much weaker. This difference is the basis of Eisenman's approach to cation-exchange selectivity between the alkali metal ions.

Porous silica is a cation exchanger because of its surface silanol groups, –SiOH. When silica is used in "normal-phase" liquid chromatography the polar character of the silanol groups is what makes the silica effective as a stationary phase. In reversed-phase chromatography the silanol groups are not wanted, and one tries to eliminate them by "end-capping" or conversion to Si–O–$Si(CH_3)_3$. Nevertheless, some silanol groups always remain, and they are present in the silica-bonded ion exchangers discussed in the last section.

One may, however, use porous silica as a cation exchanger. Smith and Pietrzyk[65] used a column of porous silica to separate alkali and alkaline-earth cations by chromatography, using aqueous lithium salts as eluents, and keeping the pH high, up to 7 or 8, to ionize the silanol groups. The order of elution is the same as that found with a sulfonated polystyrene resin, i.e., Na eluted first, then K, Rb, and Cs in that order, a fact that is not consistent with Eisenman's theory. The alkaline-earth metal ions were retained very strongly indeed, and to make them come out of the column in a reasonable time citrate or tartrate was added as a complexing anion.

Porous alumina can be either a cation exchanger or an anion exchanger, depending on the pH. At lower pH values, 5 or less, the surface is protonated and carries a positive charge, and thus functions as an anion exchanger. Schmitt and Pietrzyk[66] used a sodium acetate buffer as eluent and eluted a series of anions in the following order: iodide first, then bromide, nitrate, nitrite, bromate, and chlorite. This is the reverse of the elution order found with conventional strong-base quaternary amine exchangers (Chapter 3).

Other metal oxides have ion-exchanging properties. A classic study was made by Kraus and colleagues.[67] The ion exchange seems to be a surface effect, because it is the amorphous hydrous oxides rather than the crystalline oxides that have the high ion-exchange capacity. An oxide that was effective was zirconium oxide. Like aluminum oxide and other hydrous oxides this can be a cation exchanger at high pH and an anion exchanger at low pH; the isoelectric point is about pH 7, and at this pH, a certain amount of cation and anion exchange occur simultaneously. As an anion exchanger zirconium oxide has a great affinity for fluoride ions. The same is true of tin dioxide.[68] Tin dioxide has an isoelectric point at pH 4. Below this pH it acts as an anion exchanger; above, it acts as a cation exchanger. The stoichiometry of anion exchange with fluoride ions is unusual; one –OH group is released and two F^- ions are adsorbed, along with one alkali-metal cation to maintain electroneutrality. Hydrous antimony pentoxide or HAP is a cation exchanger in 1 *M* acid solutions, as such it has a great affinity for sodium ions. It is used in activation analysis in post-irradiation treatment to remove strongly radioactive sodium-24.[69,70] Crystalline antimonic acid adsorbs alkali-metal cations in the order Li(weakest)-K-Cs-Rb-Na(strongest), and alkaline-earth cations in the order Mg(weakest)-Ba-Ca-Sr(strongest). It was used to separate lithium (weakly adsorbed) from large amounts of sodium and other ions in the analysis of rocks.[82] Hydrous manganese dioxide has interesting ion-exchange properties, including a tenfold preference for strontium over calcium.[71]

A large family of cation exchangers is prepared by combining oxides of elements of groups 4, 5, and 6 of the periodic table. The most important of these is zirconium phosphate, which can be represented as $ZrO_2 \cdot P_2O_5$.[64] The material is amorphous or microcrystalline, and the ratio P:Zr is variable. The higher this ratio is, the greater is the cation-exchange capacity. The limiting formula $Zr(HPO_4)_2$ has been suggested. The great practical advantage of zirconium phosphate is its high selectivity; the differences in affinity among the alkali and alkaline-earth metal cations is much greater than in sulfonated polystyrene resins. Cesium is held much more strongly than rubidium, and rubidium much more strongly than potassium. The selectivity is related to the spacings between layers of atoms in the crystal lattice. An important use for zirconium phosphate in the nuclear power industry is the selective adsorption of radioactive cesium-137. It also has a high selectivity for strontium over calcium. The inorganic exchangers have high resistance to radiation, whereas organic exchangers are rapidly destroyed by high-energy gamma-radiation.

Zirconium molybdate and zirconium tungstate have ion-exchanging properties like those of zirconium phosphate.

Many oxide combinations have been made, especially by Rawat and Singh[72] and Qureshi et al.,[73] and each shows distinctive selectivity patterns. These combinations include Zr-Sb or zirconium antimonate; Fe(III)-As(V); Sn(IV)-Sb(V); Bi(III)-W(VI); Sn(IV)-V(V)-As(V), and many more. Some of them have inconveniently high solubility in water, and all suffer from poor physical stability that makes them difficult to use in columns. Some are formed as precipitates, others as gels that are washed, dried, and broken up before use. The precipitates are generally refluxed before filtering to promote crystal growth.

Another class of inorganic ion exchangers is the salts of heteropoly acids. The familiar yellow precipitate of ammonium phosphomolybdate, formed in the gravimetric determination of phosphate, is a cation exchanger which is highly selective for Cs over Rb, Rb over K. Phosphotungstate, arsenomolybdate, and so forth have similar properties. Ferrocyanides like Prussian blue show selective cation exchange; the insoluble potassium silver ferrocyanide is extremely selective for cesium over rubidium. Potassium hexacyanocobalt(II) ferrate has been made on the commercial scale.[74] Information on these materials may be found in the literature, particularly in the book by Amphlett.[75]

At one time these exchangers offered great hope for selective, high-resolution ion-exchange chromatography. The pioneer paper of Kraus, already cited, shows very impressive chromatograms of alkali and alkaline-earth cations obtained using radioactive tracers, but macro amounts of cations, not tracer amounts. But these materials are all amorphous and degrade to fine powders. They are suitable to batch operations but not to high-performance chromatography. From the analytical point of view they have been superseded by ion chromatography (Chapter 4). They have found uses in trace collection, particularly of cesium. Traces of radioactive Cs-137 may be removed from sea water after adding inactive cesium as a carrier.[73]

Another class of inorganic ion exchangers that should be mentioned are the molecular sieves, synthetic minerals, notably zeolites and faujasites, that have characteristic open crystal structures that allow water and ions to move in and out. Their ion-exchanging properties have been carefully studied by Walton and Sherry.[77] Carrying exchangeable lanthanide or transition-metal ions, these materials have an enormously important use as catalysts in the petroleum industry, but as ion exchangers in the analytical laboratory they have no role.

In closing, we note the aluminosilicate gels, which were the only ion exchangers manufactured in the years before Adams and Holmes' discovery of ion-exchange resins. They were made by rapidly mixing solutions of sodium aluminate and sodium silicate. A few minutes after mixing the solution set to a stiff milky-white gel. The gel was broken up and washed with water, then left to dry, ground, and screened to the appropriate size range. The product was used for water softening. It adsorbed calcium and magnesium ions from hard water, exchanging them for sodium ions. When the bed was exhausted it was regenerated with a concentrated sodium chloride brine. In due course this product was displaced, first by sulfonated bituminous coal, then by sulfonated polystyrene resins, which are used today. The weakness of the aluminosilicate gel exchangers was their instability in acid solutions, and also in strongly alkaline solutions. Nevertheless, one analytical use should be recorded; the method of Folin for measuring ammonia and ammonium salts in urine. The aluminosilicate gel, treated carefully with dilute acetic acid to introduce replaceable hydrogen ions, then washed and finely ground, was added to the urine sample and shaken, then left to settle. The liquid was decanted off, and the solid exchanger was washed two or three times with distilled water, again by decantation. Then the solid exchanger, which retained most of the ammonium ions, separating them from unwanted organic compounds, wa stirred with dilute sodium hydroxide; this extracted the ammonium ions and converted them to ammonia. Then Nessler's reagent (essentially K_2HgI_4) was added to the solution, producing a yellow or brown color which was matched against standards.

Since then, sodium aluminosilicate gel has found some limited use in thin-layer chromatography, but the wider use of ion exchange in analytical chemistry had to await the organic exchangers and other exchangers that were resistant to acid and alkali.

C. LIQUID ION EXCHANGERS

Liquid ion exchangers are insoluble in water, but have ions that can be exchanged for other ions in aqueous solutions. Their molecules have ionic groups attached to long hydrocarbon chains, as do soaps and surfactants, but the hydrocarbon part of the molecule is so large that these substances are insoluble in water. They dissolve in water-immiscible solvents like kerosene and trichloroethylene. Typical liquid ion exchangers are dinonylnaphthalene sulfonic acid, di-(2-ethylhexyl)phosphoric acid, tri-*n*-octylamine, tridodecyl methyl ammonium chloride. The functional groups of cation exchangers include sulfate, sulfonate, carboxylate, and phosphonate; those of anion exchangers include primary, secondary, and tertiary amines and quaternary ammonium ions.

They are more often used as solutions in a nonaqueous solvent than as pure liquids. They are used in batch extractions, and many chemists prefer to call them extracting solvents rather than ion exchangers. They can be used for chromatography in columns if they are coated as liquid films on inert porous media, but this is seldom done today. A special recent application is in dynamically coated and permanently coated ion exchangers for high-performance chromatography. The exchanging liquid is deposited on alkyl bonded silica from a solution in acetonitrile or methanol, and it remains in place when the solvent is changed to water or a mixed solvent that is low in organic modifier. This application will be described in more detail in Chapter 5. It has enabled very fine chromatographic separations of lanthanide ions and transition metal ions. In this context, liquid ion exchangers are seen as extreme types of ion-pairing agents.

Thin layers or films of liquid ion exchangers are used for the selective transport of ionic species across membranes, and for ion-selective membrane electrodes. The most common example of such a use is in the calcium ion-selective electrode. Here, the active layer is dioctyl or dodecyl phosphate (which has one free P–OH group in the molecule) in which the hydrogen ions are replaced by calcium ions, dissolved in dioctyl phenyl phosphate, and perhaps incorporated into a poly(vinyl chloride) matrix. A combination of ion-exchange selectivity with differences in transport rates makes these electrodes selective for particular ionic species. This electrode responds more sensitively to calcium ions than it does to magnesium, strontium, or barium ions, but these ions do interfere and modify the electrode response. (Ion-selective electrodes, by the way, supplement ion chromatography for the rapid and approximate measurement of cations and anion concentrations in aqueous solutions.)

An interesting feature of liquid ion exchangers is that the concentration of active groups can be varied at will by diluting with a solvent, without causing any radical change in the environment, as would be the case in a crosslinked polymer. In an important study of liquid anion exchangers, Schindewolf,[78] using radioactive tracers, measured the partition of bromide, zinc, and silver ions between water and methyl di-*n*-octylamine at various concentrations dissolved in trichloroethylene. The tertiary amine was converted to its hydrochloride salt. To the water was added hydrochloric acid, lithium chloride, or cesium chloride over a range of concentrations. From the results, Schindewolf drew these conclusions: Donnan electrolyte invasion (see Chapter 3) in the liquid ion exchanger phase was negligible; silver and zinc were taken into the exchanger as $AgCl_3^{2-}$ and $ZnCl_4^{2-}$, respectively; and the distribution of tracer zinc between water and liquid ion exchanger exactly paralleled that between water and a strong-base anion-exchange resin, showing that it was the immediate interaction between the $ZnCl_4^{2-}$ complex ion and the amine cations that determined the equilibrium, not the interaction with the resin polymer matrix. Schindewolf's study will be mentioned again in Chapter 8.

Many procedures for selective extraction of metal salts with liquid ion exchangers have been worked out, and are tabulated by Inczedy[79] and Marcus and Kertes.[80] The physical chemistry of metal-complex extraction is treated by Högfeldt.[81]

REFERENCES

1. **Adams, B. A. and Holmes, E. L.,** Adsorptive properties of synthetic resins, *J. Soc. Chem. Ind. London,* 54, 1T, 1935.
2. **Kunin, R. and Myers, R. J.,** *Ion Exchange Resins,* John Wiley & Sons, New York, 1950.
3. **Vernon, F. and Nyo, K. M.,** Synthesis optimization and properties of 8-hydroxyquinoline ion exchange resins, *Anal. Chim. Acta,* 93, 203, 1977.
4. **Parrish, J. R. and Stevenson, R.,** Chelating resins from 8-hydroxyquinoline, *Anal. Chim. Acta,* 70, 189, 1974.
5. **Vernon, F. and Eccles, H.,** Chelating ion exchangers containing salicylic acid, *Anal. Chim. Acta,* 72, 331, 1974.
6. **D'Alelio, G. F.,** Ion Exchangers, U. S. Patents 2,366,007 and 2,366,008, 1944.
7. **Storey, B. T.,** Copolymerization of styrene and *p*-divinylbenzene, *J. Polym. Sci.,* A3, 265, 1965.
8. **Wiley, R. H. and Sale, E. E.,** Monomer reactivity ratios in copolymerization with divinyl monomers, *J. Polym. Sci.,* 42, 479, 1960.
9. **Barron, R. E. and Fritz, J. S.,** Reproducible preparation of low-capacity anion-exchange resins, *React. Polym.,* 1, 215, 1983.
10. **Sahni, S. K. and Reedijk, J.,** Coordination chemistry of chelating resins and ion exchangers, *Coord. Chem. Rev.,* 59, 1, 1984.
11. **Griessbach, M. and Lieser, K. H.,** Eigenschaften von chelatbildenden Ionenaustauscher auf Basis von Polystyrol, *Fresenius Z. Anal. Chem.,* 302, 109, 1980.
12. **Vernon, F. and Eccles, H.,** Chelating ion-exchange resins containing 8-hydroxyquinoline as the functional group, *Anal. Chim. Acta,* 63, 403, 1973.
13. **Hirsch, R. F., Gancher, E., and Russo, F. R.,** Macroreticular chelating ion exchangers, *Talanta,* 17, 483, 1970.
14. **Fritz, J. S. and Moyers, E. M.,** Concentration and separation of trace metals with an arsonic acid resin, *Talanta,* 23, 590, 1976.
15. **Moyers, E. M. and Fritz, J. S.,** Preparation and applications of a propylenediaminetetraacetic acid resin, *Anal. Chem.,* 49, 418, 1977.

16. **Barron, R. E. and Fritz, J. S.,** Effect of functional group structure on selectivity for monovalent anions, *J. Chromatogr.*, 284, 13, 1984.
17. **Koster, G. and Schmuckler, G.,** Separation of noble metals from base metals by means of a new chelating resin, *Anal. Chim. Acta,* 38, 179, 1967.
18. **Warshawsky, A., Frieberg, M. M. B., Mihalik, P., Murphy, T. G., and Rao, Y. B.,** The separation of platinum group metals in chloride media by isothiouronium resins, *Sep. Purif. Methods,* 9, 209, 1980.
19. **Vernon, F. and Eccles, H.,** Chelating ion exchangers containing N-substituted hydroxylamine groups, *Anal. Chim. Acta,* 79, 229, 1975.
20. **Mikes, O., Strop, P., and Coupek, J.,** Ion exchange derivatives of Spheron. I. Characterization of polymer support, *J. Chromatogr.*, 153, 23, 1978.
21. **Mikes, O., Strop, P., Zbrozek, J., and Coupek, J.,** Chromatography of biopolymers on ion-exchange derivatives of Spheron, *J. Chromatogr.*, 119, 339, 1976.
22. **Mikes, O., Strop, P., Smrz, M., and Coupek, J.,** Ion exchange derivatives of Spheron. III. Carboxylic cation exchangers, *J. Chromatogr.*, 192, 159, 1980.
23. **Mikes, O., Strop, P., Zbrozek, J., and Coupek, J.,** Ion exchange derivatives of Spheron. II. Diethylaminoethyl Spherson, *J. Chromatogr.*, 180, 17, 1979.
24. **Sugii, A., Ogawa, N., Iinuma, Y., and Yamamura, H.,** Selective metal sorption on crosslinked poly(vinylpyridine) resins, *Talanta,* 28, 551, 1981.
25. **Chanda, M., O'Driscoll, K. F., and Rempel, G. L.,** Sorption of phenolics and carboxylic acids on polybenzimidazole, *React. Polym.*, 4, 39, 1985.
26. **Dingman, J., Siggia, S., Barton, C., and Hiscock, K. B.,** Concentration and separation of trace metals by complexation on a polyamine-polyurea resin, *Anal. Chem.*, 44, 1351, 1972.
27. **Dingman, J., Gloss, K. M., Milano, E. A., and Siggia, S.,** Concentration of heavy metals by complexation on a dithiocarbamate resin, *Anal. Chem.*, 46, 774, 1974.
28. **Janson, J. C.,** On the history of development of Sephadex, *Chromatographia,* 23, 361, 1987.
29. **Kenawy, I. M., Khalifa, M. E., and El-Defrawy, M. M.,** Preconcentration and AAS determination of silver, gold, platinum and palladium using a cellulose ion exchanger, *Analusis,* 15, 314, 1987.
30. **Szentirmay, M. N. and Martin, C. R.,** Ion exchange selectivity of Nafion films on electrode surfaces, *Anal. Chem.*, 56, 1898, 1984.
31. **Steck, A. and Yeager, H. L.,** Water sorption and cation-exchange selectivity of a perfluorosulfonate ion exchanger, *Anal. Chem.*, 52, 1215, 1980.
32. **Siergiej, M. N. and Danielson, N. D.,** Preparation and characterization of Kel-F ion-exchanging HPLC packing, *J. Chromatogr. Sci.*, 21, 362, 1983.
33. **Millar, J. R., Smith, D. G., Marr, W. E., and Kressman, T. R. E.,** Solvent-modified polymer networks, part I, *J. Chem. Soc.*, 218, 1963.
34. **Millar, J. R., Smith, D. G., Marr, W. E., and Kressman, T. R. E.,** Solvent-modified polymer networks, part II, *J. Chem. Soc.*, 2779, 1963.
35. **Millar, J. R., Smith, D. G., Marr, W. E., and Kressman, T. R. E.,** Solvent-modified polymer networks, part III. Cation-exchange equilibria with some univalent ions, *J. Chem. Soc.*, 2740, 1964.
36. **Millar, J. R., Smith, D. G., and Kressman, T. R. E.,** Solvent-modified polymer networks, part IV. Styrene-DVB copolymers made in the presence of nonsolvating diluents, *J. Chem. Soc.*, 304, 1965.
37. **Seidl, J., Malinsky, J., Dusek, K., and Heitz, W.,** Makroporöse Styrol-DVB-Copolymere and ihre Verwendung in der Chromatographie und zur Darstellung von Ionenaustauschern, *Adv. Polym. Sci. (Fortschr. Hochpolym. Forsch.*, 5, 113, 1967.
38. **Fritz, J. S. and Story, J. N.,** Selective behaviour of low-capacity, partially sulfonated macroporous resin beads, *J. Chromatogr.*, 90, 267, 1984.
39. **Gjerde, D. T. and Fritz, J. S.,** Effect of capacity on behaviour of anion-exchange resins, *J. Chromatogr.*, 176, 199, 1979.
40. **Lee, D. P.,** A new anion-exchange phase for liquid chromatography, *J. Chromatogr. Sci.*, 22, 327, 1984.
41. **Horvath, Cs., Preiss, B., and Lipsky, S. R.,** Fast liquid chromatography: an investigation into operating parameters and the separation of nucleotides on pellicular ion exchangers, *Anal. Chem.*, 39, 1422, 1967.
42. **Small, H., Stevens, T. S., and Bauman, W. C.,** Novel ion exchange chromatographic method using conductimetric detection, *Anal. Chem.*, 47, 1801, 1975.
43. **Sevenich, G. J. and Fritz, J. S.,** Preparation of sulfonated gel resins for use in ion chromatography, *React. Polym.*, 4, 195, 1986.
44. **Schmuckler, G. and Goldstein, S.,** Interface mass transfer rates of chemical reactions with crosslinked polystyrene, *Ion Exch. Solvent Extr.* 7, 1, 1977; Marinsky, J. and Marcus, Y., Eds.; Marcel Dekker, New York, 1977.
45. **Hajos, P. and Inczedy, J.,** Preparation, examination and parameter optimization of cation exchangers in ion chromatography, in *Ion Exchange Technology,* Nadan, D. and Streat, M., Eds., Ellis Horwood, Chichester, 1984, 450.

46. **Stevens, T. S. and Langhorst, M. A.,** Agglomerated pellicular anion-exchange columns for ion chromatography, *Anal. Chem.*, 54, 950, 1982.
47. **Pohl, C. A.,** Private communication.
48. **Ordemann, D. M. and Walton, H. F.,** Liquid chromatography of aromatic hydrocarbons on ion-exchange resins, *Anal. Chem.*, 48, 1728, 1976.
49. **Unger, K. K.,** *Porous Silica,* Elsevier, Amsterdam, 1979.
50. **Unger, K. K., Schick-Kalb, J., and Krebs, K.,** Preparation of porous silica spheres for column liquid chromatography, *J. Chromatogr.*, 83, 5, 1973.
51. **Chang, S. H., Gooding, K. M., and Regnier, F. E.,** Use of oxiranes in the preparation of bonded phase supports, *J. Chromatogr.*, 120, 321, 1976.
52. **Lefèvre, J. P., Divry, A., Caude, M., and Rosset, R.,** Etude des propriétés de silices échangeuses de cations sulfonate et carboxylate, *Analusis,* 3, 533, 1975.
53. **Gareil, P., Hériter, A., Caude, M., and Rosset, R.,** Etude des propriétés de silices échangeuses d'anions, *Analusis,* 4, 71, 1976.
54. **Caude, M. and Rosset, R.,** Comparison of new high-capacity ion-exchanging silicas of the Spherosil type and normal microparticulate ion exchangers of the polystyrene type in HPLC, *J. Chromatogr. Sci.*, 15, 405, 1977.
55. **Weaver, M. R. and Harris, J. M.,** In situ fluorescence studies of aluminum ion complexation by 8-hydroxyquinoline covalently bound to silica, *Anal. Chem.*, 61, 1001, 1989.
56. **Asmus, P. A., Low, C. E., and Novotny, M.,** Preparation and chromatographic evaluation of chemically bonded ion-exchange stationary phases, *J. Chromatogr.*, 123, 109, 1976.
57. **Fulcher, C., Crowell, M. A., Bayliss, R., Holland, K. B., and Jezorek, J. R.,** Synthetic aspects of the characterization of some silica-bound complexing agents, *Anal. Chim. Acta,* 129, 29, 1981.
58. **Hill, J. M.,** Silica gel as an insoluble carrier for the preparation of selective chromatographic adsorbents: the preparation of 8-hydroxyquinoline silica gel, *J. Chromatogr.*, 76, 455, 1973.
59. **Jezorek, J. R. and Freiser, H.,** Metal-ion chelation chromatography on silica-immobilized 8-hydroxyquinoline, *Anal. Chem.*, 51, 336, 373, 1979.
60. **Faltynski, K. H. and Jezorek, J. R.,** Liquid chromatographic separation of metal ions on several silica-bonded chelating stationary phases, *Chromatographia,* 22, 5, 1986.
61. **Chow, F. K. and Grushka, E.,** High-performance liquid chromatography with metal-solute complexes, *Anal. Chem.*, 50, 1346, 1978.
62. **Fazio, S. D., Crowther, J. B., and Hartwick, R. A.,** Synthesis and characterization of aliphatic sulfonic acid cation exchangers for HPLC, *Chromatographia,* 18, 216, 1984.
63. **Marshall, M. A. and Mottola, H. A.,** Synthesis of silica-immobilized 8-quinolinol with (aminophenol)trimethoxy-silane, *Anal. Chem.*, 55, 2089, 1983.
64. **Clearfield, A., Nancollas, G. H., and Blessing, R. H.,** New inorganic ion exchangers. *Ion Exch. Solvent Extr.*, Marinsky, J. A. and Marcus, Y., Eds., vol. 5, Marcel Dekker, New York, 1973, chap. 1
65. **Smith, R. L. and Pietrzyk, D. J.,** Liquid chromatographic separation of metal ions on a silica column, *Anal. Chem.*, 5, 610, 1984.
66. **Schmitt, G. L. and Pietrzyk, D. J.,** Liquid chromatographic separation of inorganic anions on an alumina column, *Anal. Chem.*, 57, 2247, 1985.
67. **Kraus, K. A., Phillips, H. O., Carlson, T. A., and Johnson, J. S.,** Ion exchange properties of hydrous oxides, Vol. 28, Proc. 2nd U. N. Conf. Peaceful Uses Atomic Energy, Geneva, 1958, 3.
68. **Jaffrezic-Renault, N.,** Study of the retention of fluoride ion on tin dioxide, *Radiochem. Radioanal. Lett.*, 29, 47, 1977.
69. **Girardi, F., Petra, R., and Sabbioni, E.,** Radiochemical separations by retention on inorganic precipitates, Special Publ. No. 312, National Bureau of Standards, U.S. Government Printing Office, Washington, D.C., 1969, p. 639.
70. **Girardi, F. and Sabbioni, E.,** Radiochemical separations for activation analysis, *J. Radioanal. Chem.*, 1, 169, 1968.
71. **Tsuji, M. and Abe, M.,** Synthetic inorganic ion-exchange materials. Acid-base properties of a cryptomelane-type hydrous Mn(IV) oxide, *Bull. Chem. Soc. Jpn.*, 58, 1109, 1985.
72. **Rawat, J. P. and Singh, D. K.,** Synthesis, ion-exchange properties and analytical applications of iron (III) antimonate, *Anal. Chim. Acta,* 87, 157, 1976.
73. **Qureshi, M., Kumar, R., and Sharma, V.,** Synthesis of reproducible tin(IV)-based ion exchangers, *Anal. Chem.*, 46, 1855, 1974.
74. **Boni, A. L.,** Rapid ion-exchange analysis of radiocesium in milk, urine, sea water and environmental samples, *Anal. Chem.*, 38, 89, 1966.
75. **Amphlett, C. B.,** *Inorganic Ion Exchangers,* Elsevier, Amsterdam, 1964.
76. **Mason, W. J.,** Rapid method for the separation of cesium-137 from a large volume of sea water, *Radiochem. Radioanal. Lett.*, 16, 237, 1974.

77. **Walton, H. F. and Sherry, H. S.,** The ion-exchange properties of zeolites. II. Ion exchange in the synthetic zeolite Linde 4A, *J. Phys. Chem.,* 71, 1475, 1967.
78. **Schindewolf, U.,** Flüssige Anionenaustauscher, *Z. Elektrochem.,* 162, 335, 1958.
79. **Inczedy, J.,** *Analytical Application of Ion Exchangers,* translated by Pall, A. and Williams, M., Pergamon Press, Oxford, 1966, chap. 10.
80. **Marcus, Y. and Kertes, A. S.,** *Ion Exchange and Solvent Extraction of Metal Complexes,* Wiley-Interscience, New York, 1969.
81. **Högfeldt, E.,** *Liquid Ion Exchangers. Ion Exchange; A Series of Advances,* Vol. 1, Marinsky, J. A., Ed. Marcel Dekker, New York, 1966, chap. 4.
82. **Abe, M., Ichsan, E. A., and Hayashi, K.,** Ion-exchange separation of lithium from large amounts of sodium, calcium and other elements by a double column of Dowex 50W-x8 and crystalline antimonic(V) acid, *Anal. Chem.,* 52, 524, 1980.

Chapter 3

PROPERTIES OF ION-EXCHANGE RESINS

When swollen in water, beads of gel-type, microporous ion exchangers derived from crosslinked polystyrene resemble drops of a concentrated electrolyte solution. Internally they are fairly homogeneous. The inside of the bead can therefore be treated as a distinct phase whose properties depend on the crosslinking and the concentration of fixed ions. Styrene-divinylbenzene polymers with functional sulfonate ions, i.e., strong-acid cation-exchange resins, are especially amenable to study because they are chemically stable and unlikely to contain functional groups other than sulfonate ions. Furthermore, they are strong electrolytes; there is little or no ion association. Therefore, their physical chemistry has been studied intensively, mainly in the decade 1960 to 1970. Today there is less incentive for such studies, because, in laboratory applications at any rate, interest has shifted to composite materials like surface-functional and macroporous ion exchangers, but we shall understand the behavior of these materials better if we understand the physical chemistry of the conventional microporous or gel-type resins. An excellent account of the physical chemistry of ion exchange is given by Helfferich in a book that has become a classic.[1]

I. IONIC CAPACITY

A fundamental property of an ion exchanger is its capacity, the quantity of exchangeable ions per unit quantity of exchanger. Commonly the capacity is expressed in milliequivalents of replaceable ions per gram of exchanger on a dry basis. The ionic form of the exchanger must be specified.

To measure the capacity of a sulfonated polystyrene, strong-acid cation exchanger one first converts the exchanger completely to its hydrogen form. One places the exchanger in a column (if possible under gravity flow) and passes enough dilute hydrochloric or nitric acid to displace all the other cations. Then one rinses well with water and dries the exchanger in the air at room temperature. In the air-dry form it contains some 20 to 50% of water, depending on the atmospheric humidity. Some of this water is present as the hydronium ion, H_3O^+, some as water of hydration. To find the moisture content one heats a small, weighed sample in a drying oven at 110 to 130°C. In the course of the drying some slight decomposition takes place with loss of sulfuric acid. Thus, the sample taken for the moisture determination is not used for further experiments. A new sample of the air-dried resin is weighed and transferred to a flask for titration. Half a gram or so of sodium or potassium chloride or nitrate is added, plus a convenient quantity of water; an indicator is added (methyl red is suitable) and the resin and salt solution together are titrated with standard alkali. The resin gives up all its replaceable hydrogen ions.

Another and very useful way to express the capacity is on a volume basis, the number of milliequivalents of replaceable ions per milliliter of the packed column. It is also useful to know the total amount of exchangeable ions in a particular column. One would first wash the column with excess acid to convert the exchanger to its hydrogen form, then wash with water, then pass an excess of a salt solution (such as KCl) and titrate the acid in the effluent.

It is easy to define the capacity of a strong-acid cation-exchange resin, because the internal solution is completely ionized, and the quantity of exchangeable ions equals the quantity of fixed ions. It is not as easy to define the capacity of a strong-base anion-exchange resin, because weakly basic groups, primary, secondary, and tertiary amines, are always present. They are formed by decomposition of the quaternary ammonium hydroxide functional groups. The decomposition is slow at room temperature, but nevertheless, strong-base anion-exchange resins should never be stored in the hydroxide form. (The chloride and other salt forms are more stable.) The presence of weakly basic groups in the resin can be established by treating a sample of resin with alkali, then washing, transferring it to a salt solution, and titrating with standard acid, following the pH with a glass electrode. One sees two inflections, the first from neutralization of the strong-base quaternary ammonium groups, the second from protonation of weakly basic amino groups.

The capacity of these resins for exchanging ions other than hydroxyl ions depends on pH. The same is true of weak-acid cation exchangers; at low pH undissociated acid groups like –COOH are present, and the ability to exchange ions other than hydrogen is reduced. In other words the effective exchange capacity of weak-acid and weak-base exchangers depends on pH.

Another ambiguity in defining ion-exchange capacities arises with inorganic exchangers. Here, too, capacity depends on pH, because the materials are weakly acidic or basic (see Chapter 2). Also, larger cations such as Cs^+ and Ba^{2+} cannot enter the narrower spaces of the crystal lattice; this is true of natural and synthetic zeolites, which have an intricate internal framework structure.

II. SWELLING

When dry beads of an ion-exchange resin are placed in water they take up water and swell, generating considerable pressure as they do so. Therefore, one must never pack a column with dry beads. If one does pack a glass column with dry resin beads and then pours water into the column, the beads will swell and shatter the glass tube. Before packing a column one must stir the resin beads with water in an open vessel, and a minute or two later, when swelling is complete, pour the resin slurry into the column. To pack a steel column that is to be run under pressure one pours the slurry into a reservoir tube connected to the column and transfers the resin by pumping.

Swelling and movement of water into the resin beads continues until it is constrained by the crosslinks. The more the degree of crosslinking, the less the resin swells. Swelling continues until the chemical potential of the water is the same inside the resin as out. Now the chemical potential of a substance, which is the same as its partial molal free energy, depends on the pressure, as follows:

$$\frac{\partial \overline{G}}{\partial P} = \overline{V} \tag{1}$$

Here, $\overline{G}$ is the partial molal free energy, P the pressure, and $\overline{V}$ the partial molal volume. The partial molal volume of water, here, is virtually equal to its molal volume under atmospheric pressure.

To use Equation 1 to find the internal pressure we need the experimental data shown in Figure 1. This figure is a simplified presentation of the data of Boyd and Soldano.[2] The ordinates are the moles of water in the resin per mole of fixed ions; the abscissas are the relative humidity, the partial pressure of water vapor in equilibrium with the resin divided by the vapor pressure of pure water at the same temperature. The measurements were made by the isopiestic method, i.e., weighted quantities of resin were brought to equilibrium with the water vapor above a series of salt solutions whose vapor pressures were known.

The numbers on the curves show the degrees of crosslinking. The upper curve is for a specially prepared resin having 0.25% divinylbenzene (DVB), just enough to keep the sulfonated polymer from dissolving. The internal pressure inside this resin can be considered to equal atmospheric, i.e., the swelling pressure may be considered zero. The interaction of water with this resin is then due entirely to ionic hydration and contact with the polymer

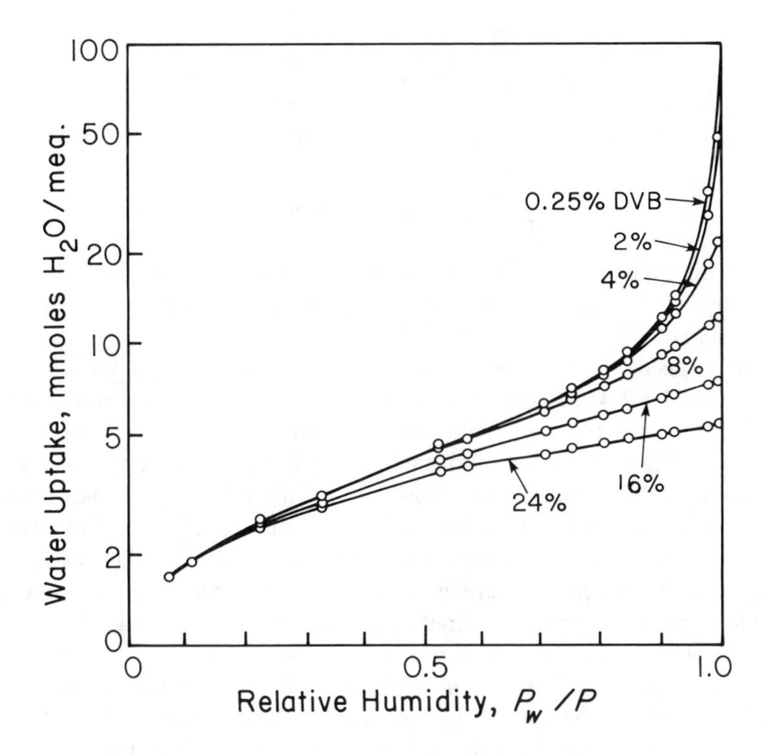

FIGURE 1. Water-vapor sorption isotherms of hydrogen-form sulfonated polystyrene ion-exchange resins. Percent crosslinking is indicated. (From Boyd, G. E. and Soldano, B. A., *Z. Elektrochem.*, 57, 162, 1953. With permission.)

matrix. We now make the assumption that these interactions are the same in all resins, regardless of crosslinking.

The graph tells us that the 0.25% crosslinked resin at a relative humidity of 90% has the same water content as the fully swollen 8% crosslinked resin at relative humidity 100%. (Incidentally this is 11.8 mol of water per mole of fixed ions, or 4.7 mol of fixed ions per kg water, that is, 4.7 *m*). The vapor pressure of the water in the higher crosslinked resin is 100/90 = 1.11 times the vapor pressure in the weakly crosslinked resin, even though the water contents are the same. The higher vapor pressure in the higher crosslinked resin is caused by the internal or swelling pressure. The vapor pressure measures the chemical potential or partial molal free energy, $\overline{G}$, according to this equation:

$$\overline{G} = \overline{G}^0 + RTlnp \tag{2}$$

where p is the partial pressure of water vapor. It is considered that the vapor is a perfect gas. $\overline{G}^0$ is the energy at a ''standard state'', usually chosen, in this case, to be pure water at the same temperature.

Comparing the free energy of water in the 8% crosslinked resin (subscript 2) with the free energy in the 0.25% crosslinked resin (subscript 1),

$$\overline{G}_2 - \overline{G}_1 = RTln\frac{p_2}{p_1} = RTln\frac{100}{90}$$

and

$$\frac{\overline{G}_2 - \overline{G}_1}{P_2 - P_1} = \overline{V} = 0.018 \text{ l/mol} \tag{3}$$

Putting the gas constant R = 0.082 l/mol and T = 298 K, we find:

$$P_2 - P_1 = 141 \text{ atmospheres}$$

This is the swelling pressure. A similar calculation for the 16% crosslinked resin, where this resin, fully swollen at 100% humidity, has the same water content as the 0.25% crosslinked resin at 78% humidity, gives 337 atm.

The solution inside the 8% crosslinked resin, we noted above, is 4.7 *m*. The osmotic pressure of an ideal 4.7 *M* solution at 25°C would be 115 atm. It seems that only ions of one charge sign, presumably the mobile ions, contribute to the osmotic pressure.

The graphs of Figure 1 are for sulfonated polystyrene resins with hydrogen ions as the mobile counter-ions. With other counter-ions the swelling is less. In the alkali metal series the swelling decreases in the order Li-Na-K-Rb-Cs. This is the order of hydration of the cations in solution. Lithium ions are hydrated the most, cesium ions the least. Likewise in the alkaline-earth metal series, swelling decreases in the order Mg-Ca-Sr-Ba, with the magnesium ions, the most strongly hydrated, causing the most swelling.

When we look at the data, however, we find that the resin swelling changes less rapidly form one ion to another than does the ionic hydration. We also find that the amount of water in the resin is greater than the water of hydration alone. Now, "water of hydration" of ions in solution is hard to define. The lithium ion, for example, has its inner hydration shell consisting of four water molecules coordinated with the lithium ion through the electrons of their oxygen atoms, but the interaction of the positive charge with the water dipoles extends out into the solution, dying away with distance, and it is impossible to define objectively a radius or distance from the central ion within which the water molecules are bound by hydration, and outside of which they are not. The definitions of hydration number and radius of the hydrated ion are arbitrary. One useful definition for the radius comes from the Debye-Hueckel distance of closest approach of positive and negative ions in solution.[3] Another comes from the partial molal volume of water; it is considered that the molecules of water of hydration are oriented towards the central ion and therefore occupy less volume than the unbound water.[4] The two approaches give very similar estimates of the water of hydration, the bound water. Only the cations are considered to be hydrated. There is ample evidence that the fixed ions, the large sulfonate ions are unhydrated because of their low electrostatic field strength.

Thus, one can make a distinction between "bound" and "free" water within an ion-exchange resin. The total amount of water inside the resin beads is found by centrifuging and weighing, making a correction for the external water that always sticks between the beads. Table 1 shows some estimates of free and bound water.[5] The distinction helps to interpret the sorption of nonionic organic compounds by these resins. We must bear in mind that the polymer backbone of an ion-exchange resin gives the resin the character of an organic solvent or a reversed-phase packing, as well as its ion-exchanging character. Thus, caffeine, phenacetin (4-ethoxyacetanilide), and several other polar organic compounds are adsorbed by polystyrene-based ion exchangers, and they are adsorbed increasingly as the cation counter-ions are changed from K to Na to Li. Though the total water in the resin increases in this sequence, the free water decreases, and the resin becomes more like a nonpolar organic solvent as one changes from K to Na to Li. Hydroxylic solutes like phenols, on the other hand, become less strongly adsorbed as one goes form K to Na to Li. Here hydrogen bonding is important, and one presumes that it is the free water that takes part in hydrogen bonding. The rates of diffusion of nonionic organic compounds in cation-exchange resins are also determined by the free water, rather than the total water content.[6]

TABLE 1
Water Uptake by Ion-Exchange Resins

Counter-ion	Crosslinking (%)	Millimol water per meq resin	Bound water mmol/meq of cation	Free water mmol/meq of cation
H	8	12.0	—	—
Li	8	11.9	6.5	5.4
	4	19.8	6.5	13.3
Na	8	10.4	3.5	6.9
	4	17.4	3.5	13.9
K	8	8.9	1.9	7.0
	4	15.4	1.9	13.5
Mg	8	10.3	6.95	3.35
	4	15.9	6.95	8.95
Ca	8	9.1	5.95	3.15
	4	14.0	5.95	8.05

Note: Bound water is water of hydration of the ions, estimated from Debye-Hueckel distance of closest approach in aqueous solutions.

From Dieter, D. W. and Walton, H. F., *Anal. Chem.*, 55, 2109, 1983. With permission.

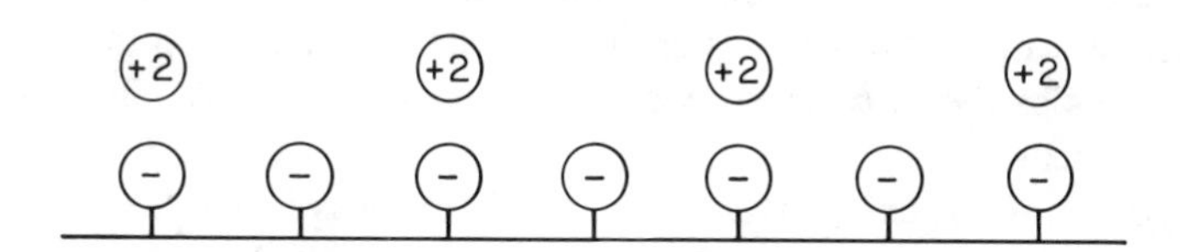

FIGURE 2. Divalent counter-ions on a polyelectrolyte chain: schematic.

Table 1 shows the important fact that doubly charged cations cause less swelling than singly charged ions. Of course there are only half as many doubly charged ions as there are singly charged ions per mole of fixed ions, but the ions Mg^{+2} and Ca^{+2} are much more hydrated than Na^+ and K^+, and still they cause less swelling. The quantity of free water is considerably less for doubly charged ions than for singly charged ions. A possible interpretation of this effect is that the doubly charged ions cause electrostatic crosslinking between neighboring polymer chains. We must bear in mind that a doubly charged ion on an isolated polyelectrolyte chain or string of single negative charges has its position of maximum stability near *one* of the fixed negative charges, not halfway between them (see Figure 2). This is a simple consequence of Coulomb's law. Thus, one fixed negative charge has a +2 cation sitting on it while the neighboring fixed negative charge is bare. This condition would promote mutual attraction of two separate polyelectrolyte chains.

Whatever the reason, the resins generally shrink when one doubly charged ion is substituted for two singly charged ions. The shrinkage is more with transition-metal or heavy-metal ions like Zn^{2+} or Cu^{2+} than with Ca^{2+}, probably because of ion association. Triply charged ions cause even more shrinkage.

When a swollen ion-exchange resin bead is transferred from water to a salt solution the bead shrinks, even if no ion exchange takes place. The swelling pressure is caused by osmosis and results from the difference in osmotic pressure between the solution inside the bead and outside. Thus, the swelling is greatest in pure water and decreases as the concentration of the external solution rises.

Another factor that affects the swelling is the nature of the solvent. In a nonpolar solvent like hexane the beads do not swell at all. Polar solvents like ethyl or methyl alcohol will enter the resin and cause it to swell; their dielectric constants are high enough that some separation of positive and negative ions takes place. In mixtures of polar solvents like water and methanol the resin swells, though not as much as it would in pure water. There is now a difference in solvent composition between the liquid inside the resin and that outside. In general, the proportion of water is greater inside the resin than outside, though the difference depends on the ionic composition of the resin. This is true of anion-exchange resins as well as cation-exchange resins.

The first definitive studies of swelling in mixed solvents were made by Rueckert and Samuelson[7] in 1957. They measured the partition of ethyl alcohol and water between solutions and polystyrene-based anion and cation exchangers in different ionic forms. Two of their graphs are reproduced in Figure 3, and some of their data are presented in different form in Figure 4, which is taken from a comprehensive review by Rodriguez and Poitrenaud.[8] Two experimental quantities are of interest: the total mass of solvents absorbed, and the proportions of organic solvent to water in the resin and the external phase. These generalizations can be made concerning the solvent by cation-exchange resins:

1. The quantity of liquid absorbed depends on the nature of the resin cation. It is greater, the more hydrated the cation; i.e., it increases (for the alkali metals) in the sequence Cs, Rb, K, Na, Li, being greatest with Li.
2. The quantity of liquid absorbed is greater, the less the crosslinking, and it decreases as the proportion of organic solvent increases. (There are one or two exceptions to this statement; in mixtures of water and dioxane the total sorption on a mass basis first increases as the dioxane proportion increases, then passes through a maximum and then decreases.)
3. The proportion of organic solvent is nearly always less inside the resin than outside. The disparity is greater, the greater the crosslinking. It depends on the ionic form, and in a somewhat complex manner. Thus, Figure 3a shows that the ethanol to water ratio in the resin is greater for Li^+ than it is for Na^+ or K^+. One might have expected the opposite, since Li^+ is the most hydrated. However, the *total* solvent uptake is greatest with Li^+, which makes the discrimination less.
4. Comparing different organic solvents, one finds that the more polar the solvent is, the more readily it enters the resin. Methanol, for example, is more polar than ethanol, and the proportion of methanol to water inside the resin is not much less than it is outside.

These generalizations and others are discussed critically by Rodriguez and Poitrenaud.[8] Samuelson,[9] who made the first studies of this kind, went on to use the preference of the resin phase for water, rather than ethanol, to develop a system for analyzing mixtures of sugars by partition chromatography. The mobile phases were ethanol-water mixtures containing 85 to 95% ethanol, and the stationary phases were cation- and anion-exchange polystryene-based resins.[9] The resins absorbed sugars because they contained proportionately more water than the external mobile phase.

The use of the past tense here means that this method has been superseded, for detection and measurement, by other, faster chromatographic methods to be described later (Chapters 6 and 7). However, Samuelson's method is still useful for separating and collecting rare sugars and related carbohydrates.

III. SWELLING AND CHROMATOGRAPHY

For high-performance chromatography in packed columns any volume changes in the stationary phase are bad. However, beads of gel-type ion-exchange resins are elastic and

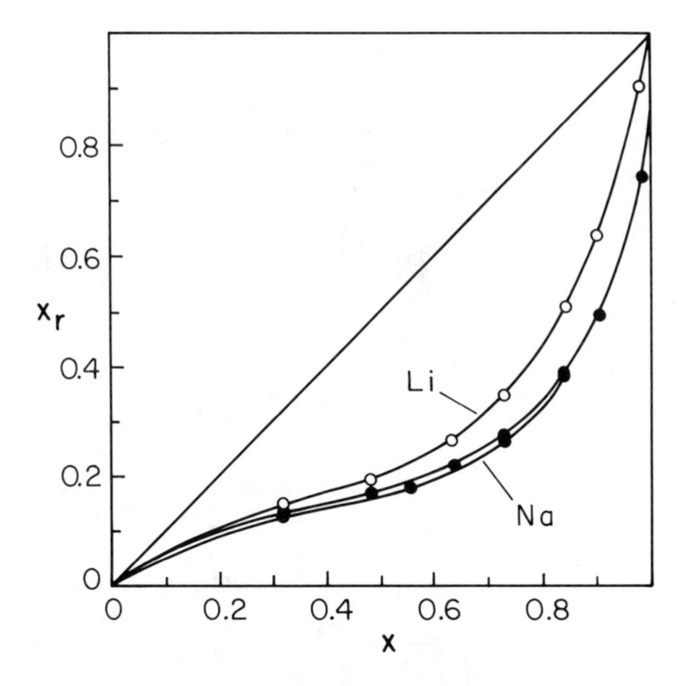

FIGURE 3. Partition of ethyl alcohol between water and Dowex®-50 sulfonated polystyrene cation-exchange resin in three ionic forms. The curve for the potassium form is just above that for the sodium form. X is the mole fraction of alcohol in the external solution; X_r is the mole fraction of alcohol inside the resin. (From Rueckert, H. and Samuelson, O., *Acta Chem. Scand.*, 11, 303, 1957. With permission.)

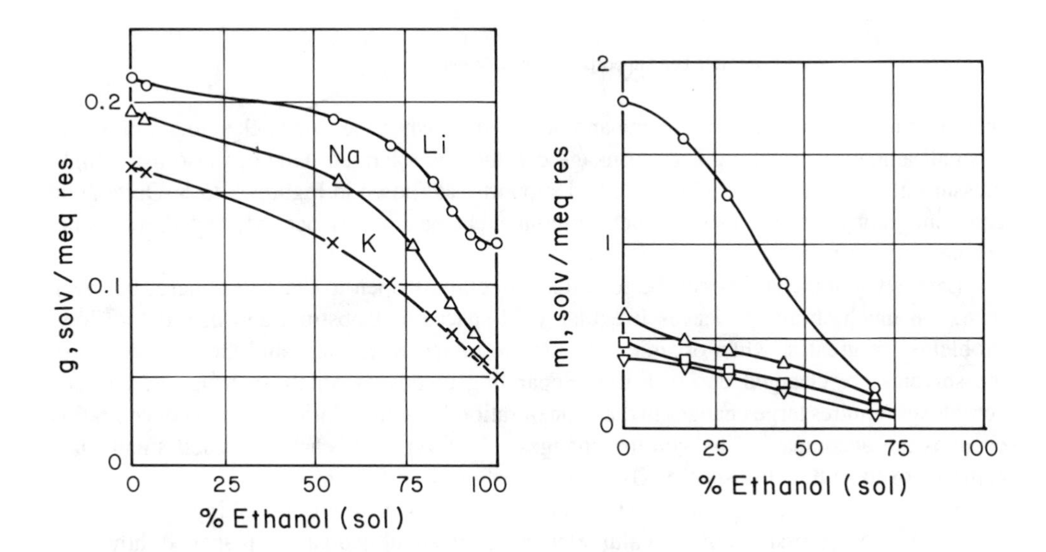

FIGURE 4. Total solvent uptake by ion-exchange resins from water-alcohol mixtures. (Left figure) Dowex®-50 × 8 cation-exchange resin in the ionic forms indicated. (Right figure) Dowex®-50 with different crosslinking: lowest curve, × 16; next, × 8; next, × 4; highest curve, × 1. (From Rodriguez, A. R. and Poitrenaud, C., *J. Chromatogr.*, 127, 29, 1976. With permission.)

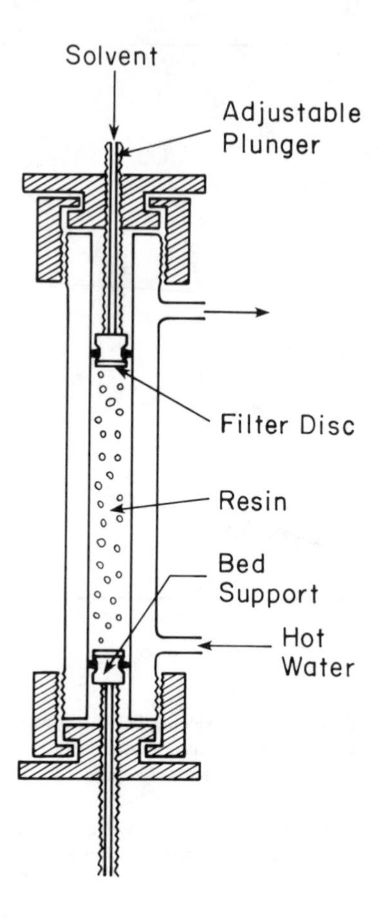

FIGURE 5. Glass column with heating jacket and adjustable bed supports. (Glenco Scientific).

press into one another, deforming one another slightly, when they are packed under pressure. A small amount of swelling and shrinkage can then be tolerated. One can also use a high-pressure glass column with adjustable bed supports, as shown in Figures 3 to 5. Quite good chromatography can be done in such columns, but patience is needed, and they are not adapted to routine use.

Larger volume changes can be tolerated in columns open to the atmosphere, but even here, too much shrinkage leaves irregular void spaces that obstruct and distort the flow. Problems are greatest when organic solvents are mixed with water and the proportions of the solvents are changed during a run. Separating metals by anion exchange of chloride complexes requires large changes in the concentration of hydrochloric acid, and concentration changes are accompanied by volume changes. However, it is easy to repack small glass columns open to the atmosphere. Often backwashing is sufficient.

Swelling and shrinkage are most serious with resins of low crosslinking, say 4% and below. With 8% crosslinking the volume changes are small and can be managed; however, volume changes can be avoided altogether by using silica-based ion exchangers or surface-functional polymer beads. Modern macroporous polymers, like the Spheron® and PRP resins, are rigid without being brittle, and ion exchangers derived from them can be packed well and show minimal volume changes.

IV. SALT UPTAKE

A. DONNAN EQUILIBRIUM

If a dry gel-type ion-exchange resin is added to a dilute salt solution that has an ion in common with the replaceable counter-ions of the exchanger, e.g., if a cation exchanger carrying replaceable sodium ions is placed in a dilute sodium chloride solution, the resin will swell and absorb water in the manner we have described, but it will take up relatively little salt. The concentration of the sodium chloride solution surrounding the resin will be found to have increased.

Thermodynamics requires that the chemical potential of the salt be the same inside the resin as outside, once equilibrium is reached. That is, the activity product, $a_{Na} \times a_{Cl}$, must be the same in both phases. If we make the usual approximation and equate activities with molal concentrations, we write:

$$[Na^+]_{in} \times [Cl^-]_{in} = [Na^+]_{out} \times [Cl^-]_{out} \tag{4}$$

We recall that the interior of a gel-type ion-exchange resin is like a drop of concentrated electrolyte solution. The 8% crosslinked sulfonated polystyrene cation exchanger described in Figure 1 above had, when fully swollen, an internal solution that was 4.7 *M* in hydrogen ions. The concentration of counter-ions in the exchanger is high, and therefore the concentration of the co-ions, the ions having the same charge as the fixed ions, must be low, compared with the concentration of the external solution.

Let the concentration of salt in the external solution be c, the concentration of fixed ions in the exchanger be X, and the concentration of co-ions (chloride ions in this instance) be x. Then Equation 4 may be written:

$$(X + x)x = c^2 \tag{5}$$

The quantity $(X + x)$ is, of course, the concentration of counter-ions, sodium ions in our example. Let X be 5.0 *M*. If $c = 0.1$, $x = 0.002$; if $c = 0.01$, $x = 0.00002$. A tenfold change in the external concentration makes a 100-fold change in the internal co-ion concentration, as long as the internal fixed-ion concentration is large compared with the external concentration. Crudely we may say that if the external solution is dilute compared with the internal, the co-ions are excluded from the exchanger, and salt is excluded. If the concentrations of the internal and external solutions are nearly the same, considerable salt invasion takes place.

Equation 5 refers to salts whose ions have equal charges. If the charges are not equal, as in a uni-divalent salt like Na_2SO_4, Equation 5 becomes

$$(X + 0.5x)^2x = 4c^3 \tag{1b}$$

Again putting $X = 5$ and $c = 0.1$, we find $x = 0.00016$. That is to say, the exclusion of divalent and polyvalent co-ions is very efficient.

The Donnan equilibrium was first enunciated by the English biochemist F. G. Donnan in 1924.[10] Ion-exchange resins were unknown at that time. What Donnan was studying was the distribution of salts across membranes that were impermeable to large ions of biological interest, namely proteins. A cell membrane would have protein on one side of it and not on the other, and the presence of the (ionized) protein would affect the salt distribution.

There have been few experimental tests of the Donnan relation in ion-exchange resin beads at low concentrations, because the experiments are hard to do. The very low electrolyte uptake is overshadowed by the salt in the solution adhering to the beads. Glueckauf and

Watts[11] found an ingenious way to overcome this difficulty, and their results showed that the salt uptake, while very small, was considerably larger than the Donnan relation predicted. They attributed the difference to inhomogeneities within the resin. Small, submicroscopic, local zones of lower crosslinking than average, hence lower fixed-ion concentration, would accept more co-ions than the average. In the light of what we now know about polymer growth, that DVB molecules add on to growing polymer chains faster than styrene molecules do (Reference 8, Chapter 2), this interpretation is reasonable.

An extreme case of inhomogeneity is that of macroporous resins, those prepared in the presence of precipitating solvents. These resins show negligible salt exclusion, because the macropores do not exclude salt. Polymers with extended chains, prepared in the presence of good solvents for the polymers, show intermediate salt exclusion (Reference 36, Chapter 2).

Electrolyte exclusion is applied to chemical analysis in many ways. The topic of "ion-exclusion chromatography" will be discussed at length in a later chapter. Ion-exchanging membranes that are permeable to ions of one sign but not the other have important uses in the electrochemical industry and are the basis of the "membrane suppressor" used in ion chromatography (Chapter 4). They all depend on the Donnan membrane equilibrium.

V. EQUILIBRIUM AND SELECTIVITY

Ion exchanges are reversible, and all reversible reactions proceed towards an equilibrium state in which all reactants and products are present. Their concentrations are related through an equilibrium quotient, or selectivity coefficient that to a first approximation is constant. It would really be constant if the solution and the exchanger were ideal solution. The simplest ion exchange is that between two singly charged ions, A and B. The ion-exchange reaction and the equilibrium quotient can be expressed thus:

$$\mathrm{A + BRes = B + ARes}$$

and

$$\frac{[\mathrm{B}][\mathrm{ARes}]}{[\mathrm{A}][\mathrm{BRes}]} = \mathrm{K}$$

where Res stands for the resin or exchanger. Often a quantity in the exchanger phase is denoted by a bar; thus, the concentration of ions A in the exchanger could be written $[\overline{\mathrm{A}}]$.

To make K in Equation 6 a true constant, one must correct for nonideal behavior by inserting activity coefficient. For ionic solutions the standard reference state is the ideal 1 *m* solution. An ideal solution is one in which the partial molal free energy, G, of a solute changes with concentration, c, according to the relation

$$\partial\overline{\mathrm{G}} = \mathrm{RT}\partial\ln\mathrm{c}$$

As c approaches zero this equation becomes more and more nearly true. Thus, measurements made in very dilute solutions can be extrapolated to define activity coefficients in higher concentrations. For the exchanger phase the standard state usually chosen is the homoionic resin, the exchanger in which all the replaceable ions are of one kind, A or B.

Now one can define a thermodynamic equilibrium constant,

$$\mathrm{K_a} = \frac{[\mathrm{B}][\mathrm{ARes}]}{[\mathrm{A}][\mathrm{BRes}]}\frac{\gamma_\mathrm{B}\overline{\gamma}_\mathrm{A}}{\gamma_\mathrm{A}\overline{\gamma}_\mathrm{B}} \tag{7}$$

The ratio of activity coefficients in solution γ_B/γ_A, can be evaluated exactly by electromotive force measurements in some cases, or it can be found to a very close approximation by taking the ratio of the mean activity coefficients of two uni-univalent salts having a common counter-ion. Thus,

$$\frac{\gamma_H}{\gamma_{Na}} = \left(\frac{\gamma_\pm \text{ for HCl}}{\gamma_\pm \text{ for NaCl}}\right)^2$$

The activity coefficients are taken for the same ionic strength as the solution in equilibrium with the resin.

A quantity K′ may be defined in which the quotient K is corrected for activity coefficients in the solution but not in the exchanger. By measuring this quantity over the whole range of resin composition, from nearly pure ARes to nearly pure BRes, and integrating over the whole composition range, one can evaluate K_a by this equation:

$$K_a = \int_0^1 \ln K' dN_b \tag{8}$$

Here, N_b is the equivalent fraction of ion B in the exchanger. This is a simplified version of an equation derived by Gaines and Thomas.[12] Comparing K_a with the values of K′ at different resin compositions one can evaluate the activity coefficients in the resin. In uni-valent-divalent exchanges these coefficients can be quite far from unity, ranging from 0.5 to 2.

So far we have considered exchanges of ions of equal valence. More complex equations result when one ion of one kind displaces two or more ions of another kind. In Equations 6 and 7 above, it does not matter what units are chosen to express concentrations, as long as the same units are chosen for both ions. If the ratio of the concentrations of ions A and B in the solution remains constant, the concentrations in the exchanger remain the same, no matter what the individual concentrations are in the solution. In other words, diluting the external solution does not affect the proportion of ions in the exchanger, apart from the small change in the ratio of activity coefficients. This is not true for exchanges between singly charged and doubly charged ions, like the exchange of one calcium ion for two sodium ions. Here the equation for the reaction is

$$Ca^{2+} + 2NaRes = 2Na^+ + CaRes_2$$

and the corresponding equilibrium quotient is

$$\frac{[CaRes_2][Na^+]^2}{[NaRes]^2[Ca^{2+}]} = K \tag{9}$$

Here it does matter what units are used to express concentrations. It is common, but not universal, to express concentrations in solution as gram-equivalents per liter and concentrations in the resin as equivalent fractions.

The important point to grasp is that the ratio of calcium to sodium ions in the exchanger (in general, divalent to univalent ions) does depend on the total concentration in solution. Adding water to a mixture of solution and exchanger causes calcium ions to move from the solution into the exchanger. In ion-exchange chromatography, doubly charged ions are retained more strongly with respect to singly charged ions, the more dilute is the eluent. In water softening, which is the biggest industrial application of ion exchange, calcium ions are more efficiently removed from water the lower their concentration, and they are more

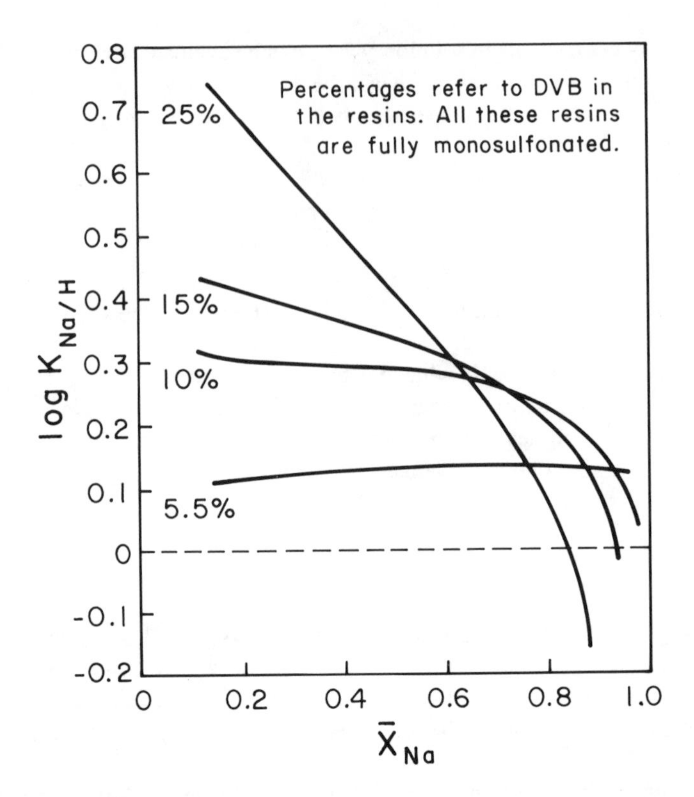

FIGURE 6. Sodium-hydrogen ion-exchange distributions on crosslinked sulfonated polystyrene resins. Abscissas are equivalent fractions of sodium ions; K is the equilibrium quotient for the exchange Na^+ + HRes. (From Reichenberg, D. and McCauley, D. J., *J. Chem. Soc. London*, 2741, 1955 and Reichenberg, D., *Ion Exchange: A Series of Advances*, Vol. 1 Marcel Dekker, New York, 1966, chap. 7. Courtesy of Marcel Dekker, Inc.)

efficiently removed from the exchanger in the regeneration step, the more concentrated is the sodium chloride brine used in this step.

Returning to univalent exchanges and looking at experimental data, Figure 6, taken from a classic paper by Reichenberg and McCauley,[13] shows how the equilibrium quotient for the sodium-hydrogen exchange in sulfonated polystyrene resins of different crosslinking depends on the equivalent fractions of the ions in the exchanger. We note that at low crosslinking, say 8% or less, the quotient is almost independent of resin composition and can be regarded as a true constant, but that at higher crosslinking, the preference of the resin for sodium ions falls as the concentration of these ions increases. At very high crosslinking and high proportions of sodium ions there is actually a selectivity reversal; hydrogen ions become more strongly bound than sodium ions. Several sets of graphs like Figure 6 are shown in the references cited.[14]

For analytical ion-exchange chromatography one works at low loading, that is to say the proportion of the ions being separated, in both resin and solution phases, is generally small. Moreover, in ion chromatography the concentration of the eluent is small. Thus, the equilibrium quotient or selectivity coefficient can be regarded as constant, and the effects of resin composition shown in Figure 6 can be ignored. It is enough to note that the selectivity increases with increasing crosslinking. For preparative chromatography, on the other hand, one does have to take into account the changes of selectivity coefficient with resin composition.

VI. ELUTION VOLUME IN ANALYTICAL COLUMN CHROMATOGRAPHY

In analytical column chromatography one introduces into the eluent stream a sample containing a small amount of the substance sought (the analyte) and measures the time taken to move the analyte from the column inlet to the exit, or, what is more significant, the volume of the mobile phase that passes. This time or volume is larger, the greater the proportion of analyte in the stationary phase. In ion-exchange chromatography, analyte ions are driven through the column by a much larger concentration of eluent ions. The ion-exchange reaction is:

$$E\ An^{A} + A\ El_s^{E} = E\ An_s^{A} + A\ El^{E}$$

Here, An is the analyte ion whose charge is A, and El is the eluent ion whose charge is E. The subscript s refers to the stationary phase. The thermodynamic equilibrium constant of this reaction is:

$$K_a = \frac{[An]_s^E[El]^A\gamma_{An(s)}^E\gamma_{El}^A}{[An]^E[El]_s^A\gamma_{An}^E\gamma_{El(s)}^A} \tag{10}$$

Here, the γ are molal activity coefficients. Equation 10 is the same as Equations 7 and 9, written for the general case of ions with any charge.

In chromatography the retention of an analyte is best measured by the capacity factor, k′, which is a dimensionless value equal to the total mass of analyte in the stationary phase divided by the total mass of analyte in the mobile phase:

$$k' = \frac{V_R - V_m}{V_m} = \frac{V_s[An]_s}{V_m[An]} \tag{11}$$

Here, V_R is the retention volume of the analyte, and V_m and V_s are the volume of mobile phase in the column (i.e., the void volume) and the volume of the stationary phase. The dependence of an analyte's retention on eluent concentration is described by Equation 12, which is derived by substituting Equation 10 into Equation 11

$$k' = \frac{V_s}{V_m} K_a^{1/E}[El]_s^{A/E}[El]^{-A/E}\gamma_{An}\gamma_{El}^{-A/E}\gamma_{An(s)}^{-1}\gamma_{El(s)}^{A/E} \tag{12}$$

$[El]_s$ is the concentration of the eluent in the stationary phase. It equals Q, the total ion-exchange capacity of the column, divided by the volume of the stationary phase, V_s. This is true for low eluent concentrations (below 0.1 *M*) since excess eluent ions are excluded from the resin by the Donnan effect. For the same reason the activity coefficients in the stationary phase do not depend on the eluent concentration.

If activity coefficients in the mobile phase are not included, all of the terms in Equation 12 except for the eluent concentration can be combined into a single isocratic constant, C_i. Then,

$$k' = C_i[El]^{-A/E} \tag{13}$$

This assumption is valid for low eluent concentrations. In logarithmic form, Equation 13 reads:

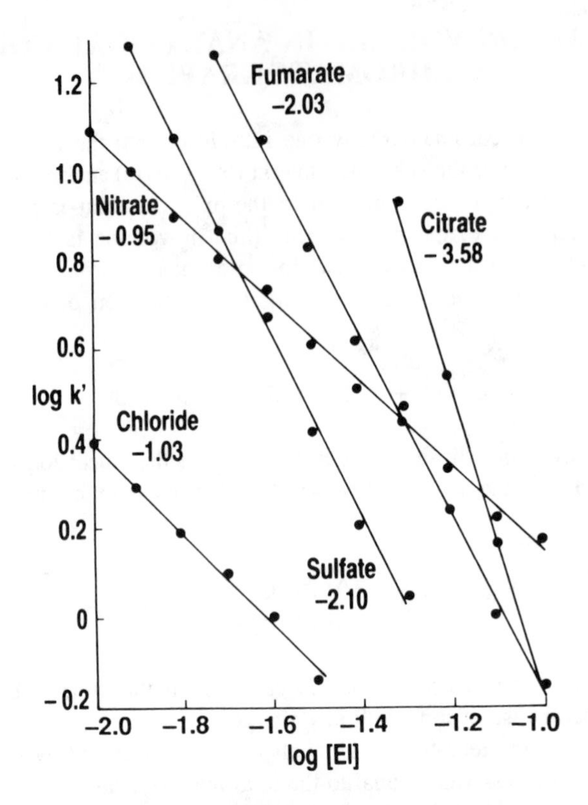

FIGURE 7. Log k′ vs. log (eluent concentration) for isocratic separation of anions, with the slopes of the lines listed. A low capacity exchanger is used, with sodium hydroxide eluent. Eluent concentrations are moles per liter. (From Rocklin, R. D., Pohl, C. A., and Schibler, J. A., *J. Chromatogr.*, 411, 107, 1987. With permission.)

$$\log k' = -\frac{A}{E}\log [El] + \log C_i \qquad (14)$$

Equation 14 predicts that a plot of log k′ vs. log [El] will be a straight line for each analyte with a negative slope equal to the charge of the analyte divided by the charge of the eluent ion. If [El] is measured in millimolars then $C_i = k'$ for an eluent concentration of 1 m*M*.

This prediction is born out by plots shown in Figure 7.[15] The slopes of the log-log plots are in reasonable agreement with the values predicted from theory. One exception is the greater than predicted slope for citrate. The most likely cause of this deviation is the dependence of the activity coefficients of multiply charged ions on the eluent concentration.

VII. GRADIENT ELUTION

During gradient elution, k′ changes with the eluent concentration. For this reason, retention will be described in terms of $(V_R - V_m)/V_m$ instead of k′. (In isocratic elution these two expressions are equal.) For a linear gradient with eluent concentration beginning at zero and increasing with time, the instantaneous eluent concentration at the column inlet is:

$$[El] = RV \qquad (15)$$

where V is the volume of eluent pumped since the beginning of the run, and R is the gradient ramp slope, for example in m*M*/ml. R is the change in eluent concentration with time divided by the flow rate. The instantaneous capacity factor at the band maximum, k'_a, can be found by substituting Equation 15 into Equation 13:

$$k'_a = C_i(RV)^{-A/E} \tag{16}$$

k'_a is the capacity factor that would result if the instantaneous eluent concentration were held constant from injection to elution. To calculate $(V_R - V_m)/V_m$ the effective capacity factor for gradient elution, k'_a is integrated from injection to elution. This is done by considering k'_a as dV/dx, where a fractional volume of eluent dV passes over the band maximum and moves the band dx down the column. The x is expressed as the column void volume, rather than as a length. Therefore, from Equation 16,

$$k'_a = \frac{dV}{dx} = C_i(RV)^{-A/E}$$

which can be rearranged and integrated thus:

$$\int_0^{V_R - V_m} C_i^{-1}(RV)^{A/E}\, dV = \int_0^{V_m} dx$$

and

$$\frac{V_R - V_m}{V_m} = \left(\frac{E}{A + E}\right)^{-E/(A+E)} V_m^{-A/(A+E)} C_i^{E/(A+E)} R^{-A/(A+E)}$$

The three constants in front of R can be combined into one gradient constant, C_g, thus:

$$\frac{V_R - V_m}{V_m} = C_g R^{-A/(A+E)} \tag{17}$$

Taking the log of both sides gives:

$$\log \frac{V_R - V_m}{V_m} = \frac{-A}{A + E} \log R + \log C_g \tag{18}$$

If R is measured in millimoler per milliliter, then C_g equals $(V_R - V_m)/V_m$ when R equals 1, i.e., when the eluent concentration increases at a rate of 1 m*M*/ml of eluent pumped.

Equation 18 for gradient elution is very similar to Equation 14 for isocratic elution. It predicts that a plot of log $(V_R - V_m)/V_m$ against log R will produce a straight line for each analyte. With a singly charge eluent ion, the slopes for mono-, di-, and trivalent analytes will be $-1/2$, $-2/3$, and $-3/4$, respectively. Again this prediction is born out by experiment. As shown in Figure 8, the slopes of the log-log plots for gradient elution agree with the predicted values.

VIII. ISOCRATIC ELUTION AND STATIONARY-PHASE CONCENTRATION

The fundamental equation of elution chromatography, where a small amount of an analyte, An, is displaced by a large excess of an eluent, El, is

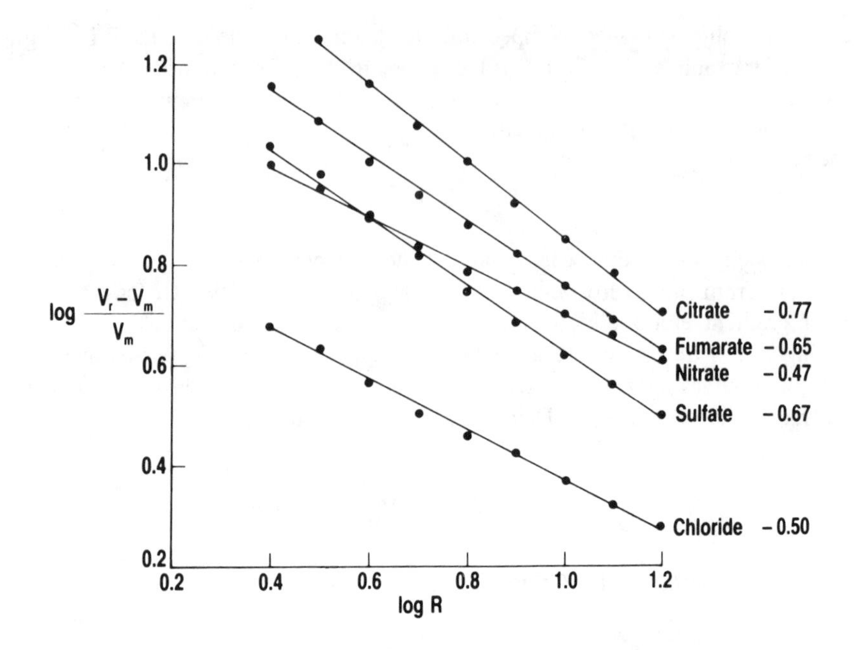

FIGURE 8. Log $(V_R - V_m)/V_m$ vs. log R for gradient elution, with the slopes of the lines listed. R is the rate of change of the eluent concentration with volume passed; see text. (From Rocklin, R. D., Pohl, C. A., and Schibler, J. A., *J. Chromatogr.*, 411, 107, 1987. With permission.)

$$V_R = V_m + k'V_m$$

This is a restatement of Equation 11 above. V_R and V_m are, respectively, the retention volume of the analyte and the void volume of the column, and k' is the capacity factor.

For the simplest kind of ion exchange, the displacement of one singly charged ion by another, and assuming that activity coefficients cancel (which they nearly do), the equation reads:

$$V_R = V_m + \frac{K \cdot V_s \cdot [El]_s}{[El]} \tag{19a}$$

K is the ion-exchange equilibrium quotient for the displacement:

$$An + El_s = An_s + El$$

The ionic capacity of the column, Q, is the product $V_8[El]_8$. Since the eluent ions are in great excess, we may equate $[El]_8$ with ionic capacity of the resin, as defined at the beginning of this chapter. (Note the difference between the capacity of the column measured in equivalents, and the capacity of the resin measured in equivalents per unit volume.) Then the corrected retention volume, $V_R - V_m$, is inversely proportional to the eluent concentration, as shown in Figure 7, and directly proportional to Q, the column's ion-exchange capacity.

If the analyte ion, An, is doubly charged, the equation takes the form:

$$V_R = V_m + \frac{KV_s[El]_s^2}{[El]^2} \tag{19b}$$

Now the corrected retention volume is inversely proportional to the square of the eluent concentration, in agreement with the experimental values shown in Figure 7. The relation to the capacity of the resin is not so clear. First, it is not as easy to change the resin capacity at will as it is to change the eluent concentration. Second, it makes a difference whether the solid exchanger is homogeneous or not. The easiest way to change the resin capacity is to use a heterogeneous, surface-modified exchanger of the types used in ion chromatography (see Chapter 4, Figure 3, and Chapter 2, Section III). One can vary the depth of sulfonation in surface-sulfonated styrene-DVB polymer beads, and one can vary the size of the latex particles in latex-coated beads (what matters here is the ratio of the radii of latex particles to the underlying support beads.) In both cases one produces two phases, one fully functionalized, the other not. (As noted in Chapter 2, in surface-sulfonated beads there is a sharp boundary between the sulfonated layer and the underlying non-sulfonated polymer.) Thus, in Equations 19a and 19b V_8 changes, but $[El]_8$ remains the same if this quantity is defined as the concentration of fixed ions in the active functionalized phase that contains these ions. One can also consider that the equilibrium constant for ion exchange, K, as defined in Equation 10, remains constant as the depth of sulfonation changes. Then the corrected retention volume is proportional to the first power of the overall capacity, or the total column capacity, Q, and this is true for divalent ions as well as for univalent ions.

Another way to vary the resin capacity is to keep the volume V_8 the same, or nearly so, and to change the density of ionic charges within this volume. One way to do this is to take fully sulfonated polystyrene beads and partially desulfonate them by heating in water under pressure to 180°C. Homogeneous partially sulfonated resins with differing capacities are thus obtained. Boyd et al.[16] measured ion-exchange equilibrium constants for Na–H exchanges in these resins and also measured self-diffusion rates. The equilibrium constants (really quotients) dropped with increasing proportion of sodium ions, and what is more important for chromatography, they were approximately proportional to the capacities. Boyd et al. did not measure chromatography retentions, but since raising the capacity raised K in Equation 19a as well as $[El]_8$, the corrected retention volume would have increased as the square of the capacity.

It is possible to change the capacity at will for dynamically coated ion exchangers, which will be described in Chapter 5. These exchangers have a unimolecular layer of an ionic surfactant, a compound with an ionic group at the end of a long hydrocarbon tail, adsorbed on the surface of porous polystyrene or C18-bonded silica. Changing the capacity through changing the surfactant loading changes retention and separations of heavy-metal and transition-metal ions can be optimized in this way. However, the effect remains empirical. No systematic study has been made of the effect of the capacity of dynamically coated exchangers on chromatographic retention.

IX. DISTRIBUTION COEFFICIENTS

For many analytical purposes one does not need to know equilibrium constants or quotients; it is enough to know a distribution coefficient, a ratio of solute concentrations in the exchanger and solution phase. The distribution coefficient depends on the composition of the solution and its concentration, but is independent of the concentration of the solute. Distribution coefficients are especially useful for low concentrations of dissolved metal ions in solutions that form complexes with the metal ions, like concentrated hydrochloric acid (which forms anionic complexes) or citrate buffers. They are usually measured by shaking a known weight of exchanger and a known volume of the solution together, adding a small amount of the metal salt, perhaps a radiotracer, and noting the distribution of the metal between the two phases. They are usually quoted in milliliters per gram. Calling the distribution coefficient D, and the mass of resin in a column m, the equation for the elution volume becomes:

$$V_R = Dm + V_m$$

Extensive tables of distribution coefficients of metals in both cation and anion exchange are to be found in the literature, especially in the works of Strelow (see Chapter 8).[17]

X. SELECTIVITY ORDERS

The orders of selectivity or strengths of binding of cations and anions to exchangers are well known for the alkali and alkaline-earth metal ions and for simple inorganic anions on strong-acid and strong-base polystyrene-type resins. For the cations they are:

$$Li < Na < K < Rb < Cs; \qquad Mg < Ca < Sr < Ba$$

Cesium is held the most strongly of the alkali-metal ions, with the ammonium ion between sodium and potassium. Barium is held the most strongly of the alkaline-earth metal ions. The doubly charged alkaline-earth metal ions are held considerably more strongly than the singly charged alkali-metal ions. Of course one must qualify this statement; the equilibrium between singly and doubly charged ions depends on the concentration. The more dilute the eluent, the farther apart are the distribution coefficients for singly and doubly charged ions. High-performance ion chromatography uses dilute eluents, and it is difficult to elute and measure sodium, potassium, magnesium, and calcium ions in the same run.

The orders of elution are the orders of decreasing hydration. Lithium ion is the most strongly hydrated of the alkali metal ions, and it is the most weakly held. An early theory of ion-exchange selectivity, due to Gregor,[18] attributed selectivity to the difference of swelling of the resins. The free energy change on replacing sodium ions by lithium ions was set equal to the work done against the swelling pressure. We have seen that the swelling pressure is considerable; it is greater, the greater the crosslinking, and so is the ion-exchange selectivity, but calculations have shown that the work of swelling is not in itself sufficient to account for the observed equilibrium constants.

As to the stronger binding of doubly charged ions with respect to singly charged, we have already noted that the resin shrinks when one Mg^{2+} is substituted for two Na^+, in spite of the fact that in solution, Mg^{2+} is much more hydrated than Na^+. Perhaps the stronger binding of the divalent ion is due to the electrostatic crosslinking effect.

Among the doubly charged transition-metal cations there are no marked differences in selectivity and no regular pattern. To separate these ions by chromatography on sulfonated polystyrene resins one must rely on complexing ions in the eluent.

Anion-exchange selectivity among the halide ions follows the order:

$$F < Cl < Br < I$$

There is much more difference in binding strength among the halide anions than there is among the alkali cations. On an 8% crosslinked sulfonated polystyrene cation exchanger there is about a factor of four difference between Li and Cs. Between F and I on an 8% crosslinked quaternary base anion exchanger there is a factor of 100. Hydration of the ions does not seem to play a part. Anions are hydrated very little in solution if at all. The ionic size, that is the size of the unhydrated ion in the crystal lattice, is more significant. A very interesting relation is found with singly charged oxy-anions. If the central atom remains the same and the number of oxygen atoms around it increases, so that the ion gets bigger, the affinity to the exchanger increases with the ionic size. Thus, NO_3^- is bound more strongly than NO_2^-, ClO_4^- is bound more strongly than ClO_3^-, which is bound more strongly than ClO_2^-, which is bound more strongly than ClO^-. Sulfate is bound more strongly than sulfite,

TABLE 2
Ion-Exchange Selectivities

Cations		Anions	Type I	Type II
H^+	1.0	OH^-	1.0	1.0
Li^+	0.85	F^-	1.6	0.3
Na^+	1.5	Ci^-	22	2.3
NH_4^+	1.95	Br^-	50	6
K^+	2.5	I^-	175	17
Rb^+	2.6			
Cs^+	2.7			
Ag^+	7.6	ClO_3^-	74	12
		BrO_3^-	27	3
		IO_3^-	5.5	0.5
Mg^{2+}	2.5	NO_2^-	24	3
Ca^{2+}	3.9	NO_3^-	65	8
Sr^{2+}	5.0	HCO_3^-	6	1.2
Ba^{2+}	8.7	HSO_3^-	27	3
		HSO_4^-	85	15
Fe^{2+}	2.5	$H_2PO_4^-$	5	0.5
Zn^{2+}	2.7			
Co^{2+}	2.8	Acetate	3.2	0.5
Cu^{2+}	2.9	Phenate	110	27
Ni^{2+}	3.0	Salicylate	450	65
Pb^{2+}	7.5	Benzene-sulfonate	500	75

Note: These are relative values for 8% crosslinked strong-acid and strong-base resins. Type I anion exchangers have functional groups $-N(CH_3)_3$; Type II have $-N(CH_3)_2CH_2CH_2OH^+$.

Adapted from Price List K, Bio-Rad Laboratories, Richmond, California. With permission.

though here the difference is not so definite, and on certain surface-functional anion exchangers there is little difference.

An explanation for these effects was proposed by Chu et al.[19] It invokes the hydrogen-bonded structure of liquid water. An ion can interact with water in two ways. If the ion is small and has a high charge, it will attract water dipoles around it and the free energy will be lowered. If it large and has only a single charge, the main effect on the water will be to break hydrogen bonds, which will require energy and cause an increase in free energy. A large ion will require more energy to break hydrogen bonds and will have more difficulty going into the water phase outside the resin than a smaller ion. This argument seems convincing until one compares the series ClO_3^-, BrO_3^-, IO_3^-. Here the affinity to the exchanger decreases as the size of the central atom increases, and iodate ions are held far more weakly than iodide ions.

Table 2 gives some selectivity data for the most common anion- and cation-exchangers of the crosslinked polystyrene type.

Selectivities and selectivity orders depend on the nature of the fixed ionic group. In strong-base anion exchangers the size of the quaternary ammonium group affects selectivity; comparing $-N(C_6H_{13})^3_+$ with $-N(CH_3)_3^+$, Barron and Fritz[20] found that the larger fixed ion discriminated better between bromide and nitrate, the nitrate ion being retained longer, while it retained the divalent anions sulfate and thiosulfate less than did the smaller fixed ion. It seems that ion pairing is a factor in anion-exchange selectivity. In cation exchangers with

TABLE 3
Diffusion Coefficients at 25°C

CO_2 in air	1.5×10^{-1}
NaCl in water	1.5×10^{-5}
Ions in 10% crosslinked sulfonated polystyrene cation exchanger (self-diffusion coefficients):	
Na^+	2.5×10^{-7}
Zn^{2+}	3×10^{-8}
Y^{3+}	3×10^{-9}
Br^-	5×10^{-6}

Note: units, $cm^2\ sec^{-1}$

From B. A. Soldano (1953), quoted by F. Helfferich, Reference 1, Chapter 6.

functional carboxylate ions sodium ions are held more strongly than potassium ions, which is the opposite to what happens in sulfonate resins. Selectivity orders in cation exchange have been studied exhaustively by Eisenman,[21] who began his studies by looking at the effects of alkali metal cations on glass electrodes. Glass is a cation exchanger with two kinds of fixed-ion sites: silanol ions, ($Si{-}O^-$) and aluminate ions. The proportions of these sites vary from one glass to another. In glasses very high in aluminate, the order of ion-exchange binding of the alkali metal ions was Li < Na < K < Rb < Cs, as in sulfonated polystyrene resins and in glasses very low in aluminate, virtually pure silica, this order was reversed. Eisenman's interpretation, which applies to organic fixed negative ions as well, refers to the electrostatic field strength around the fixed ions. If the field strength is very high, as it is with the small oxide ion, the electrostatic force can strip away the water of hydration of the alkali-metal ion and bind it directly, as an ion pair, to the fixed ion. Lithium ion is the largest hydrated ion, but it is the smallest bare, nonhydrated ion; it is therefore held the most strongly by a fixed ion of high field strength. If the field strength is very low, as it is with the large aluminate ion (and larger sulfonate ion), the alkali-metal ions retain their water of hydration, and the familiar sequence, Li < Na < K < Rb < Cs, results. With intermediate field strengths, intermediate selectivity sequences result.

To return to the organic ion exchangers, the carboxylate ion, $-COO^-$, is smaller than the sulfonate ion $-SO_3^-$, and exerts a higher field strength; it binds Na^+ more strongly than K^+, presumably because it can strip some of the water of hydration from the sodium ion.

XI. KINETICS OF ION EXCHANGE

The rate of ion exchange is governed by diffusion, either "film diffusion" in a layer of solution next to the surface of the exchanger, or more usually, "particle diffusion" within the exchanger bead itself. Self-diffusion rates and diffusion coefficients have been measured in ion exchangers using radioactive tracers. They are presented in Table 3.

We see that diffusion of counter-ions in gel-type cation exchangers is extremely slow, factors of ten slower than in aqueous solutions. It appears that a counter-ion must jump from one position of lowest potential energy next to a fixed ion, to another position of lowest potential energy next to another fixed ion, and to make this jump it needs activation energy. This energy is greater, the greater the charge on the ion. Co-ions, on the other hand, stay away from the fixed ions and move in the spaces or channels between them; they diffuse nearly as fast in the ion exchanger as they do in water.

During ion exchange, one ion is moving into the bead while another is moving out. If

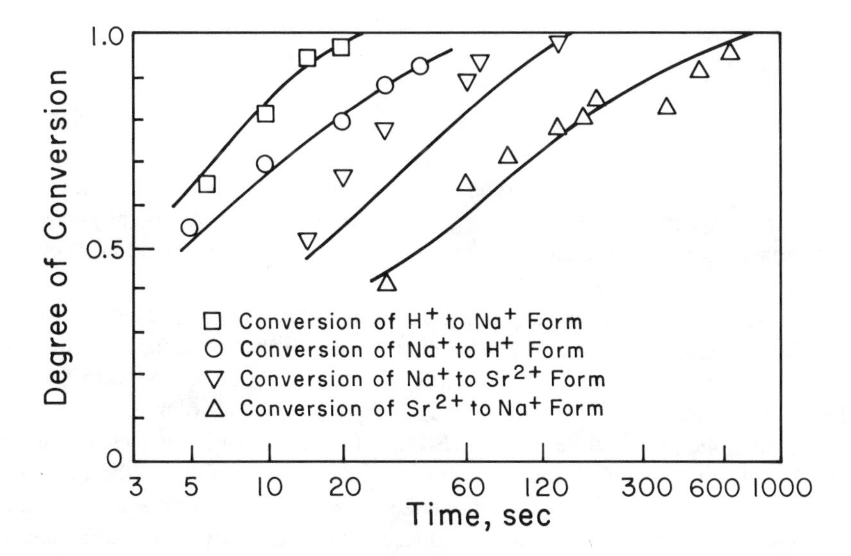

FIGURE 9. Comparison of forward and reverse ion-exchange rates. The points are experimental values; the lines are theoretical curves. (From Helfferich, F. G., *J. Phys. Chem.*, 66, 39, 1962. Copyright, American Chemical Society, 1962. With permission.)

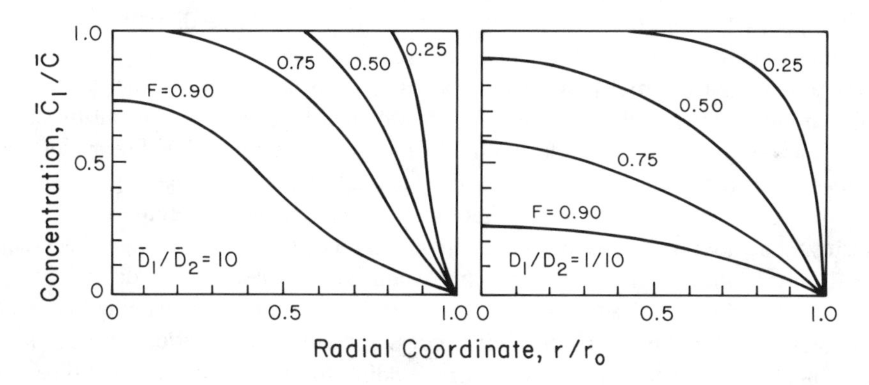

FIGURE 10. Radial concentration profiles in a resin bead. F is the fraction converted; D_1 and D_2 are diffusion coefficients within the resin. Ion 1 is in the exchanger originally, and is displaced by ion 2. (From Helfferich, F. and Plesset, M. S., *J. Chem. Phys.*, 28, 418, 1958. With permission.)

their diffusion rates are unequal, one ion will tend to move faster than the other; yet electrical neutrality must be maintained. What happens is that an electrical potential gradient is set up. The faster-diffusing ion moves against the potential gradient, i.e., it is held back, while the other moves with the gradient, i.e., it is speeded up. The nonlinear differential equations that describe the motion were solved by Helfferich and Plesset.[22] An important conclusion from these calculations, which has been confirmed by experiment, is that the forward and reverse exchanges in a bead proceed at different rates. Consider the displacement of hydrogen ions (faster-diffusing) from a resin bead by sodium ions (slower-diffusion) coming in from outside. It is nearly twice as fast as the reverse reaction, which is the displacement of sodium ions in the bead by hydrogen ions form the surrounding solution[23] (see Figure 9). The exchange is faster when the faster-moving ion is leaving the resin. An extreme example of this effect is the lanthanum-hydrogen ion exchange. Lanthanum ions enter the resin rapidly, displacing hydrogen ions, but getting the lanthanum ions out again is exasperatingly slow.

Figure 10 shows concentration profiled within an exchanger bead as one ion (ion 1) is displaced by another of equal charge (ion 2), and how these profiles depend on the ratio of their diffusion coefficients.

XII. PROPERTIES OF SURFACE-FUNCTIONAL RESINS

The slow diffusion of ions in gel-type resins has prompted the use of surface-functional exchangers in high-performance chromatography. To what extent do the physico-chemical studies that have been made on gel resins serve as a guide to the understanding of composite, surface-active ion exchangers?

Consider first the elution volumes in a chromatographic column packed with a surface-functional resin. Equations 19a and 19b include the ion-exchange capacity of the column. This is low for surface-active exchangers, 0.01 to 0.1 meq/g compared with 5 meq/g for fully sulfonated styrene-DVB exchangers. To keep the elution volume, V_R, from being excessively small the eluent concentration, [El], must be small also. The question then arises, does the thin film of sulfonated or aminated polymer at the surface of the polymer bead have the same properties as a fully functionalized bead? Early experiments of Stevens and Small[24] suggested that it did not; indirect evidence showed that diffusion coefficients of ions in the film were much smaller than in the fully functionalized gels, partly counteracting the advantage of the short diffusion paths. Anchored firmly to the underlying polystyrene bead, the film carrying the fixed ions is not as free to swell as in the gel. However, another test suggested that the functionalized layer does have the ion-exchange properties of the gel. From Equations 19a and b the corrected elution volume should be directly proportional to the exchanger capacity in this case. Hajos and Inczedy[25] prepared exchangers of capacities up to 0.165 meq/g by partially sulfonating beads of crosslinked polystyrene, and compared them with polystyrene resins that were fully sulfonated or nearly so. For all these resins, elution volumes for alkali-metal ions with dilute nitric acid eluents gave linear plots against the capacity. The surface-sulfonated and the bulk-sulfonated exchangers gave points that fell on the same straight line. The equilibrium quotients, K, were therefore the same in all the resins, showing that the sulfonated layers all had the same ion-exchange properties and differed only in their depth. The same conclusion was reached by Gjerde and Fritz[26] for macroporous styrene-DVB adsorbents carrying strong-base anion-exchanging layers of different thickness. Certainly the selectivity orders for anions and cations are the same in surface-functional exchangers as in the corresponding fully functionalized gel-type resins.

If the surface-functional layer is one molecule thick or less, it can no longer be regarded as a distinct phase, and its ion-exchanging properties are best understood in terms of the electrical double layers.[27]

We have mentioned dynamic ion exchange, a technique in which ions having long hydrophobic carbon chains are spread along the surface of a hydrophobic adsorbent, octadecyl silica or a nonionic macroporous styrene-DVB polymer. An advantage of this technique is that the concentration of ionic groups in the stationary phase, hence its exchange capacity, can be changed at will. By so doing, elution orders and elution volumes of metal ions can be manipulated to give effective chromatography. This effect was put to good use by Cassidy (Chapter 5). The systems are complex and only partly understood, but the effect of ionic charge is apparent.

XIII. TEMPERATURE AND ION EXCHANGE

Temperature has little effect on simple ion-exchange equilibria. The temperature dependence of an equilibrium constant is given by this equation:

$$\frac{d\ln K}{d(1/T)} = -\frac{\Delta H}{R}$$

TABLE 4
Free Energy and Enthalpy of Ion Exchange

Cation	ΔG°	ΔH°	ΔS° (cal. deg^{-1})
Li^+	360	1460	3.6
H^+	230	1180	3.0
K^+	−260	−550	−1.0
Ag^+	−830	110	3.1
Mg^{2+}	−40	1500	5.2
Ca^{2+}	−350	1410	5.9
Sr^{2+}	−440	1090	5.1
Zn^{2+}	−410	1430	6.2

Note: In calories per gram-equivalent, for the displacement of sodium ions from 8% crosslinked polystyrene sulfonic acid at 25°C.

From Boyd, G. E., *Chem. Abstr.*, 74, 57618q, 1971. With permission.

where ΔH is the enthalpy change, or heat absorbed at constant temperature and pressure. The enthalpies, free energies, and entropies of ion-exchange reactions were measured by several investigators in the years before 1970, and a table from a summary paper by Boyd[28] is shown here (Table 4). These enthalpies were measured calorimetrically. We note that none of them are much over 1 kcal/g equivalent. From the equation above, we can calculate that an enthalpy gain of 1000 cal implies an increase in the ion-exchange equilibrium constant of only 6% for a 10° rise in temperature.

Table 4 has some interesting features, like the evidence that the preference of the exchangers for divalent over univalent cations is caused by the entropy gain when the divalent cation releases part of its water of hydration on entering the resin.

Similar data are available for anion exchanges; here again, the enthalpy changes are of the order of 1 or 2 kcal only. The heat quantities are greater, the greater the crosslinking, both for cation and for anion exchange.

When temperature affects ion-exchange elution volumes significantly it is usually because of complex-forming and acid-base reactions in solution. Thus, the elution volumes of lanthanide ions and transition-metal ions in anion-exchange chromatography fall with rising temperature because the complexes formed with eluent ions in solution become less stable. The pH of buffer solutions depends on temperature because of the effect of temperature on the ionization constants; however, the ionization of carboxylic acids, where a separation of electric charges accompanies ionization, changes very little in the room-temperature region. The ionization of the ammonium ion, where the positive charge is simply transferred to the water molecule and there is no separation of charges, increases markedly with rise in temperature. The same is true of the "tris" cation ("tris" = tris(2-hydroxyethyl)aminomethane); the pH of "tris" buffers falls with rising temperature some 20 times as fast as the pH of phthalate and citrate buffers. Elution volumes of amino acids change enough with temperature that careful temperature control is needed to separate their peaks; this change is due to the ionization of the amino acids themselves as well as to that of the buffers (see Chapter 7).

Aside from the effect on equilibrium, which is generally small, rising temperature increases diffusion rates and makes the bands sharper. If raising the temperature brings the bands closer together, however, there may be no gain in resolution.

REFERENCES

1. **Helfferich, F.,** *Ion Exchange,* McGraw-Hill, New York, 1962.
2. **Boyd, G. E. and Soldano, B. A.,** Osmotic free energies of ion exchangers, *Z. Elektrochem.,* 57, 162, 1953.
3. **Harned, H. S. and Owen, B. B.,** *Physical Chemistry of Electrolyte Solutions,* 3rd ed., Reinhold, New York, 1958.
4. **Pepper, K. W., Reichemberg, D., and Hale, D. K.,** Properties of ion-exchange resins in relation to their structure, *J. Chem. Soc.,* 3129, 1952.
5. **Dieter, D. W. and Walton, H. F.,** Counter-ion effects in ion-exchange partition chromatography, *Anal. Chem.,* 55, 2109, 1983.
6. **Gregor, H. P., Collins, F. C., and Pope, M.,** Studies on ion-exchange resins. III. Diffusion of neutral molecules in a sulfonic acid cation-exchange resin, *J. Colloid Sci.,* 6, 304, 1951.
7. **Rueckert, H. and Samuelson, O.,** Die Verteilung von Aethylalkohol und Wasser bei Ionenaustausch auf Harzbasis, *Acta Chem. Scand.,* 11, 303, 1957.
8. **Rodriguez, A. R. and Poitrenaud, C.,** Gonflement des resines echangeuses de cations par des melanges eau-solvant organique, *J. Chromatogr.,* 127, 29, 1976.
9. **Samuelson, O.,** Partition chromatography of sugars, sugar alcohols and sugar derivatives, in *Ion Exchange; A series of Advances,* Vol. 2, Marcel Dekker, New York, 1969, chap. 5.
10. **Donnan, F. G.,** A theory of membrane equilibria, *Chem. Rev.,* 1, 73, 1924.
11. **Glueckauf, E. and Watts, R. E.,** The Donnan law and its application to ion-exchange polymers, *Proc. R. Soc. London,* A268, 350, 1962.
12. **Gaines, G. L. and Thomas, H. C.,** Adsorption studies on the clay minerals: formulation of thermodynamics of exchange adsorption, *J. Chem. Phys.,* 21, 714, 1953.
13. **Reichenberg, D. and McCauley, D. J.,** Properties of ion-exchange resins in relation to their structure. VII. Cation-exchange equilibria on sulfonated polystyrene resins of different crosslinking, *J. Chem. Soc. London,* 2741, 1955.
14. **Reichenberg, D.,** Ion-exchange selectivity, in *Ion Exchange: A Series of Advances,* Vol. 1. Marcel Dekker, New York, 1966, chap. 7.
15. **Rocklin, R. D., Pohl, C. A., and Schibler, J. A.,** Gradient elution in ion chromatography, *J. Chromatogr.,* 411, 107, 1987.
16. **Boyd, G. E., Soldano, B. A., and Bonner, O. D.,** Ionic equilbria and self-diffusion rates in desulfonated cation exchangers, *J. Phys. Chem.,* 58, 456, 1954.
17. **Strelow, F. W. E.,** Application of ion exchange to element separation and analysis, in *Ion Exchange and Solvent Extraction,* Vol. 5, Marcel Dekker, New York, 1973, chap. 2.
18. **Gregor, H. P.,** Gibbs-Donnan equilibria in ion-exchange systems. *J. Am. Chem. Soc.,* 73, 642, 1951.
19. **Chu, B., Whitney, D. C., and Diamond, R. M.,** On anion-exchange resin selectivities, *J. Inorg. Nucl. Chem.,* 24, 1405, 1962.
20. **Barron, R. E. and Fritz, J. S.,** Effect of functional group and ion-exchange capacity on the selectivity of anion exchangers, *J. Chromatogr.,* 316, 201, 1984.
21. **Eisenman, G.,** Electrochemistry of cation-sensitive glass electrodes, in *Advances in Analytical Chemistry and Instrumentation,* Vol. 4, John Wiley & Sons, New York, 1965, 213.
22. **Helfferich F. and Plesset, M. S.,** Ion-exchange kinetics; a nonlinear diffusion problem, *J. Chem. Phys.,* 28, 418, 1958.
23. **Helfferich, F.,** Ion-exchange kinetics: experimental test of the theory of particle-diffusion control, *J. Phys. Chem.,* 66, 39, 1962.
24. **Stevens, T. S. and Small, H.,** Surface-sulfonated polystyrene: optimization of performance in ion chromatography, *J. Liquid Chromatogr.,* 1, 123, 1978.
25. **Hajos, P. and Inczedy, J.,** Preparation, examination and parameter optimization of cation exchangers in ion chromatography, in *Ion Exchange Technology* Naden, D. and Streat, M., Eds., Society of Chemical Industry, London, 1984, 540.
26. **Gjerde, D. T. and Fritz, J. S.,** Effect of capacity on the behaviour of ion-exchange resins, *J. Chromatogr.,* 176, 199, 1979.
27. **Hux, R. A. and Cantwell, F. F.,** Surface adsorption and ion exchange in chromatographic retention of ions on low-capacity cation exchangers, *Anal. Chem.,* 56, 1258, 1984.
28. **Boyd, G. E.,** Thermal effects in ion-exchange reactions with organic exchangers: enthalpy and heat capacity changes, *Chem. Abstr.,* 74, 57618q, 1971.

Chapter 4

ION CHROMATOGRAPHY

In 1950 Beukenkamp and Rieman[1] separated sodium and potassium ions as their chlorides by chromatography on a cation-exchange resin column 2 × 59 cm with gravity flow. The eluent was 0.7 *M* hydrochloric acid. Fractions of 8 ml each were collected, then evaporated to remove hydrochloric acid, and weighed. The run time was 5 h. Results of the analyses are shown in Figure 1.

Compare this separation with the modern chromatography shown in Figure 2. Here, all of the alkali metals through cesium plus ammonium are separated in less than 3.5 min using continuous electrical conductivity detection. This is an example of high-performance ion-exchange chromatography, commonly called ion chromatography. Originally, the term implied the use of conductivity detection. Today, all of the detection methods commonly used in high performance liquid chromatography (HPLC) are used in ion chromatography. This includes electrochemical as well as optical methods such as UV absorbance. Because conductivity detection is the natural form of detection for ionic species, it is the most commonly used method in ion chromatography. Therefore, we begin this chapter with a discussion of ion chromatography with conductivity detection.

I. ION CHROMATOGRAPHY WITH CONDUCTIVITY DETECTION

To get fast chromatographic analysis with good resolution one needs small and uniform particles and, especially in ion exchange, short diffusion paths. We have noted in Chapter 2 how these conditions are achieved. Figure 3 shows the types of small, rapidly acting ion exchangers now used: gel-type resins in very small particle sizes, macroporous ion exchangers, surface-functionalized polymer beads, and latex-coated polymer beads. Latex-coated beads, also called "latex agglomerated" beads, are popular in ion chromatography because they combine fast mass transfer with the uniformity of gel-type exchangers. Also, their ion-exchange capacity can be controlled by choosing the latex particle size. Macroporous polymers can be surface functionalized and so can porous silica.

Along with small particles and short diffusion paths one needs a means of continuous monitoring or detection. Electrical conductivity is the obvious property to use for this purpose, but as long as eluents like 0.7 *M* hydrochloric acid had to be used, electrical conductivity could not be considered. The conductivity of the eluent was so high that analyte ions could not be detected.

Both these problems, fast mass transfer and continuous detection by electrical conductivity, were solved together in masterly fashion by Small, Stevens, and Bauman of Dow Chemical Co. and described in their paper in 1975 (Reference 42, Chapter 2). This paper marks the birth of ion chromatography.

First, these authors attacked the mass transfer problem by taking 20 μm beads of 2% crosslinked styrene-DVB polymer and sulfonating them to a depth of less than 1 μm. To make an anion exchanger, they took the same polymer beads and sulfonated them very lightly, to give as near a unimolecular layer as possible, then coated these beads with fine particles of a conventional strong-base anion-exchange resin, less than a micron in diameter (the latex). The electrostatic force between the sulfonate anions of the polymer bead and the quaternary ammonium cations of these fine resin particles was so great that once coated, the particles stayed in place. This process was described above (Chapter 2). Today, exchangers are made from polymer beads having diameters from 25 down to 5 μm and the size of the latex particles can be varied, as noted in Chapter 2.

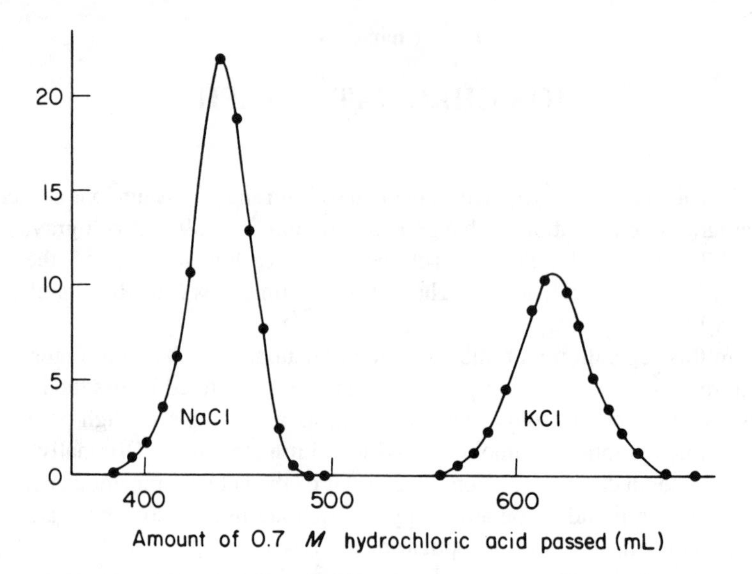

FIGURE 1. Chromatographic separation of sodium and potassium chlorides. The 20 × 590 mm column was packed with sulfonated polystyrene cation-exchange resin. The cations were eluted with 0.7 *M* HCl using gravity flow. Fractions were taken every 8 ml and analyzed by evaporating off the HCl and weighing the remaining salt. (From Beukenkamp, J. and Rieman, W., *Anal. Chem.*, 22, 582, 1950. Copyright, American Chemical Society, 1950. With permission.)

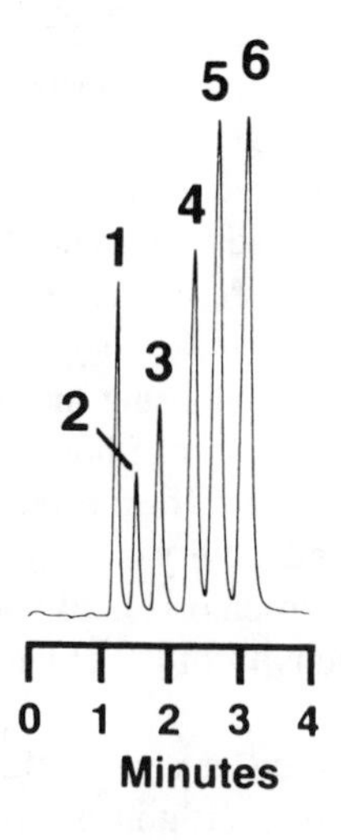

FIGURE 2. Ion chromatography of alkali metals using suppressed conductivity detection. Peaks: (1) Li^+ 0.8 ppm; (2) Na^+ 2 ppm; (3) NH_4^+ 3 ppm; (4) K^+ 6 ppm; (5) Rb^+ 20 ppm; and (6) Cs^+ 30 ppm. Column: Dionex Fast Cation I, 4 × 250 mm, 13 μm latex-coated PS/DVB resin. Eluent: 20 m*M* HCl, 0.2 m*M* DAP·HCl at 2.0 ml/min.

The ion-exchange capacity of the first surface-sulfonated beads was about 0.02 meq/g, compared to 5 meq/g for fully sulfonated beads in the hydrogen form. This would correspond to a sulfonation depth of 0.02 μm in a bead of 10 μm radius. Equation 12, Chapter 3, relates the elution volume (as described by k′) in chromatography to the capacity of the column and the concentration of the displacing ion (the eluent). If the capacity is divided

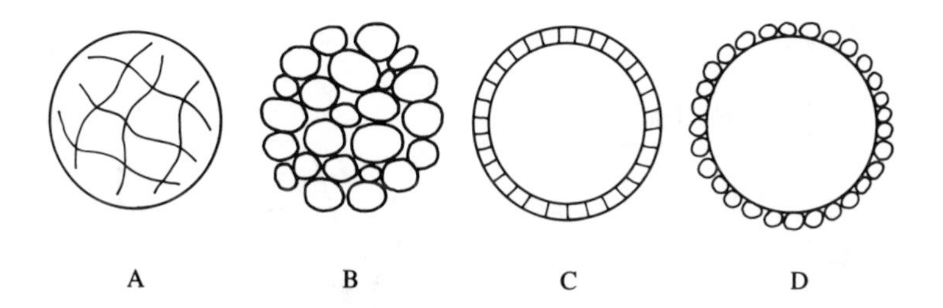

FIGURE 3. Forms of ion exchangers: (A) microporous or gel-type; (B) macroporous; (C) surface functional; (D) latex coated.

by 250 (5/0.02 = 250), the eluent concentration must also be divided by 250 to keep a similar elution volume. Instead of the 0.7 *M* HCl used in Rieman's work, we must now use about 2.8 m*M* (0.7/250 = 0.0028 *M*) acid. Small et al. used 10 m*M*.

Using a dilute eluent makes conductivity detection more feasible than before. The background conductivity of the eluent is smaller, and there is more hope of seeing the differences in conductivity that accompany the elution of ions of interest. There are now two possibilities: to accept the background conductivity and choose the nature and concentration of the eluent so as to make these differences as large as possible, or to remove or neutralize as much of the excess eluent as possible by chemical means before measuring the conductivity. Small et al. chose the second path, using a device called a suppressor. Both techniques are in use today. The first is called ''single-column'' or ''nonsuppressed'' ion chromatography; the second is called ''suppressed'' ion chromatography. The distinction between the two methods is in the way to present the column effluent for detection by electrical conductivity. If a different detection method is used, such as UV absorbance or amperometry, suppression is unnecessary and the distinction between the two methods disappears. (Single-column ion chromatography should really only be called ''nonsuppressed'', since today only a single column is used in suppressed ion chromatography. The suppressor is no longer a column.)

A. NONSUPPRESSED (SINGLE-COLUMN) ION CHROMATOGRAPHY

Dionex Corporation holds an exclusive license to the original Dow Chemical Co. patent. Therefore, chemists wishing to practice suppressed ion chromatography must purchase the chromatographic equipment from Dionex. Nonsuppressed ion chromatography was developed to allow owners of standard HPLC equipment to perform ion chromatographic analyses simply by the addition of a good conductivity detector. The detector must be very stable and sensitive, for it must measure small conductivity changes against a large background, perhaps a thousand times larger than the signals observed. Temperature control is very important, for the conductivity of ionic solutions is greatly affected by temperature; a change of 1°C changes the conductivity by about 2%.

1. Principles and Cation-Exchange Eluents

At this point let us bear in mind that the total ionic concentrations in each phase, mobile and stationary, remain constant when ions are exchanged. As one ion enters the solid exchanger from the solution, another ion takes its place, moving from the exchanger into the solution. The concentration of ions in the solution does not change. The conductivity of the solution will change, however, if the ions entering the solution have a different equivalent conductivity from those that leave. (For a definition of equivalent conductivity, see the section on conductivity detection later in this chapter.) Figure 4a shows the situation. Here we show cation exchange, with high conductivity hydrogen ions displacing lower conductivity sodium ions from the exchanger. While the sodium ions are eluting the con-

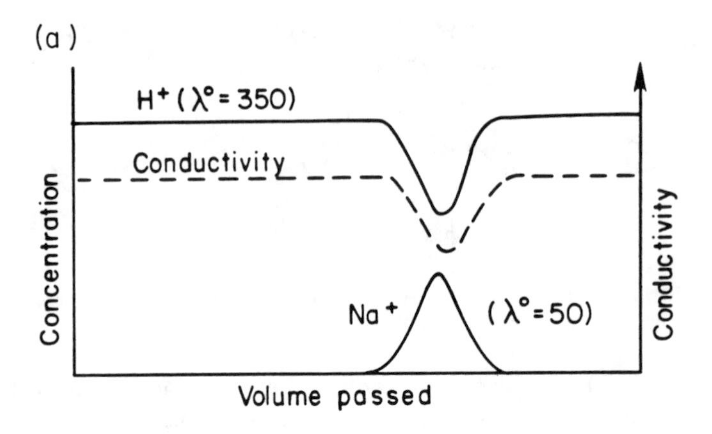

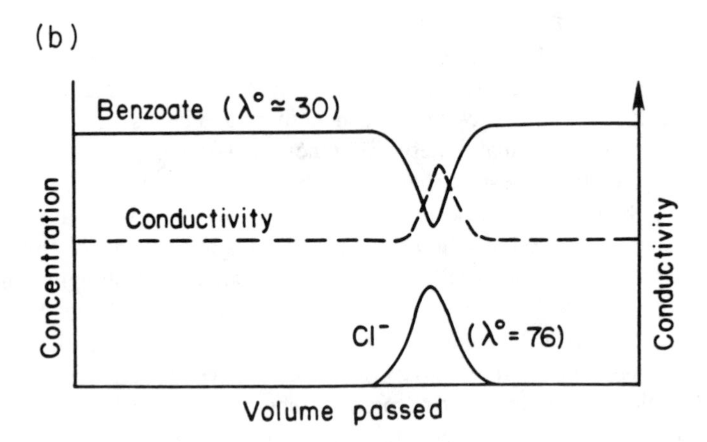

FIGURE 4. Conductivity and concentrations of eluent and analyte during (a) cation exchange and (b) anion exchange. After the column void volume passes, the total ionic strength of eluent and analyte (sum of solid lines) eluting from the end of the column is constant. (From Walton, H. F., ACS Audio Course C-93, American Chemical Society, Washington, D. C., 1987. With permission.)

ductivity will fall, to rise again and resume its original value after all the sodium ions have eluted. When dilute strong acids are used as eluents in cation exchange, the detector produces dips instead of peaks as cation analytes elute. Of course, one could displace sodium ions by an eluent cation with lower conductivity, say large organic cations like tetraethylammonium ions, anilinium ions, or benzylammonium ions.[2] In that case the conductivity would rise as the sodium ions elute, producing peaks. The situation for anion exchange is shown in Figure 4b.

In addition to having favorable eluting strength, eluents for nonsuppressed ion chromatography are chosen so as to maximize the difference between the equivalent conductivities of the eluent and analyte ions. Sensitivity is directly proportional to this difference. Table 1 shows the equivalent ionic conductivities of common anions and cations. Large organic cations and anions appropriate for use in eluents have conductivities of 30 or less.

Alkali metal cations are usually eluted with a dilute acid, generally nitric or sulfuric acid. (Hydrochloric acid attacks stainless steel and can only be used on nonmetallic chromatographs.) To elute the doubly charged alkaline-earth ions of magnesium and calcium the acid must be much more concentrated because these ions are much more strongly held than singly charged ions. This would cause the background conductivity to be far too high.

TABLE 1
Equivalent Conductivities of Ions at 25°C ($S \cdot cm^2 \cdot equiv^{-1}$)

Cations		Anions	
H^+	350	OH^-	198
Li^+	39	F^-	54
Na^+	50	Cl^-	76
K^+	74	NO_3^-	71
Mg^{2+}	53	HCO_3^-	45
Ca^{2+}	60	Acetate$^-$	41
Cu^{2+}	55	Benzoate$^-$	32
NEt_4^+	33	SO_4^{2-}	80

Therefore, a stronger eluent must be used. A common choice is the doubly charged ion of ethylenediamine, $C_2H_4(NH_3)_2{}^{2+}$, which is much more strongly bound than the hydrogen ion, and can therefore be used in much lower concentration. Its equivalent conductivity is lower than that of calcium, strontium, and barium ions; therefore, the conductivity rises as the alkaline-earth cations elute.[3] The doubly charged cation of m-phenylenediamine can also be used. Here, the aromatic nature of the cation causes it to be held strongly by the polystyrene "backbone" of the ion exchanger.

2. Anion-Exchange Eluents

The chromatography of anions is of more importance than the chromatography of simple inorganic cations. Most cations come from metals, and metallic elements can be determined spectroscopically. Atomic absorption spectroscopy is easy and precise. The determination of simple inorganic anions, on the other hand, was very difficult until the development of ion chromatography.

For the chromatography of anions an obvious displacing ion is hydroxide. For nonsuppressed ion chromatography, hydroxide is not a good choice because it is weakly held by the strong-base anion exchangers used. The high concentration necessary to elute common anions produces a background conductivity which is too high. (It is used, however, in the chromatography of anions of very weak acids, such as cyanide, borate, sulfide, and silicate.[4]) More commonly used in nonsuppressed anion-exchange chromatography are benzoate and phthalate ions.[5] These ions have low conductivities (Table 1), so most analyte ions produce positive peaks. They are also strongly held by the anion exchanger on account of their aromatic character. Both acids are strong enough that the pH need not exceed 7 to 8 to have them fully ionized; thus, bonded silica exchangers may be used without fear of dissolving the silica. (This is not the case for hydroxide eluents.) For benzoic acid, $pKa = 4.2$; for phthalic acid, $pK_1 = 2.9$, $pK_2 = 5.5$. The eluent strength is adjusted by adjusting the pH. The pH controls the ratio of benzoate ions to benzoic acid, and the ratio of singly to doubly charged phthalate ions. The doubly charged phthalate anion is a much stronger eluent than the singly charged hydrogen-phthalate anion.[6]

Gjerde et al[7] (the original developers of nonsuppressed ion chromatography) used 0.2 m*M* benzoate and 0.1 m*M* phthalate solution at pH 6.25, where 85% of the phthalate and nearly all the benzoate are in the fully ionized forms. The exchanger was a surface-aminated macroporous styrene-divinylbenzene polymer (Amberlite XAD-1) having an ion-exchange capacity of 0.0007 meq/g. It is characteristic of nonsuppressed ion chromatography that to keep the background conductivity low, the eluent concentrations must be very low, and so must the capacities of the ion exchangers. This is a limitation of the nonsuppressed technique.

A popular eluent system introduced in 1986 is composed of a mix of boric acid, sodium

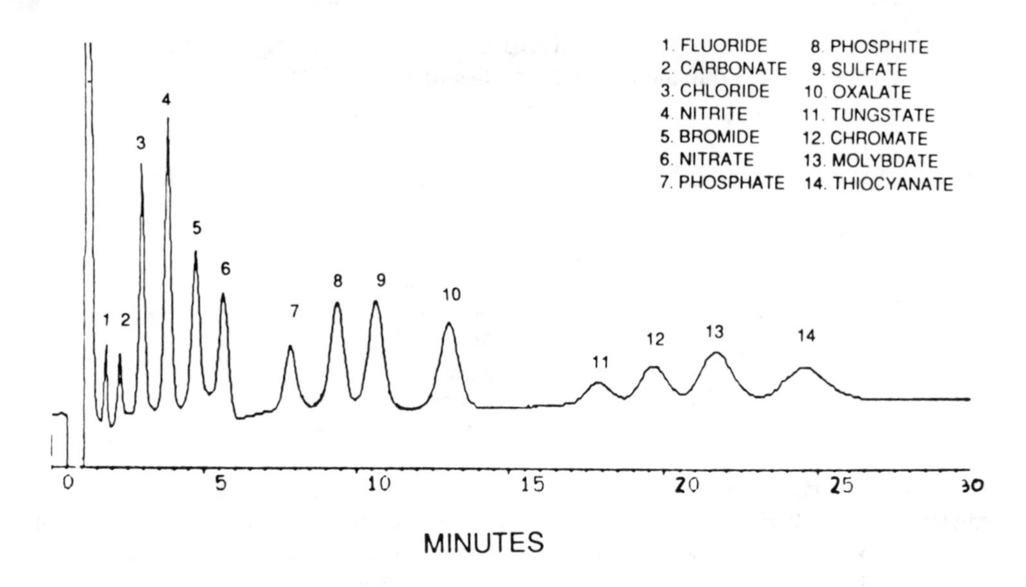

FIGURE 5. Single column ion chromatography of anions with borate/gluconate eluent. (1)F^- 1 ppm; (2) HCO_3^- 2 ppm; (3) Cl^- 2 ppm; (4) NO_2^- 4 ppm; (5) Br^- 4 ppm; (6) NO_3^- 4 ppm; (7) HPO_4^{2-} 6 ppm; (8) SO_4^{2-} 4 ppm; (9) $C_2O_4^{2-}$ 4 ppm; (10) CrO_4^{2-} 10 ppm; (11) MoO_4^{2-} 10 ppm. Sample volume, 0.100 ml. Column, Waters IC-Pak anion, 4.6 × 50 mm, 10 μm surface-functionalized polyacrylate-based resin. Eluent: 11 m*M* boric acid, 1.48 m*M* gluconic acid, 3.49 m*M* KOH, 0.65 m*M* glycerin, 12% CH_3CN. (From Jones, W. R., Jandik, P. and Heckenberg, A. L., *Anal. Chem.*, 60, 1977, 1988. Copyright, American Chemical Society, 1988. With permission.)

tetraborate, and gluconic acid.[8] The borate ion combines with polyhydroxy molecules like gluconic acid to form a much stronger acid than either of the separate components. Borate/gluconate eluents are useful because of their low background conductivity, relatively strong eluting power, and higher pH. The pH of 8.5 ensures quantitative dissociation of carboxylic acids, maximizing their conductivity. Borate/gluconate eluents are also useful for determining strong acid anions as well (Figure 5). Unlike the aromatic eluent systems, borate/gluconate eluents can only be used on polyacrylate-based resins or on silica-based packings. They can not be used on PS/DVB resins.

Other eluents are used for nonsuppressed ion chromatography. Tartrate and malate, being aliphatic and hence weakly adsorbed by polystyrene-based exchangers, are good eluents for weakly bound anions like fluoride.[9] Salicylate and 4-hydroxybenzoate are strong eluents, and 2,4-dihydroxybenzoate is stronger yet.[10]

B. SUPPRESSED ION CHROMATOGRAPHY

Rather than accept the necessity of measuring small conductivity changes against a large background, Small et al. chose to remove the excess eluent, or at least to convert it to a form that would have low background conductivity. For this they used a second ion-exchanging column, which they called a "suppressor column". It was placed between the "separator column" and the conductivity cell. The suppressor column was packed with high capacity ion-exchange resin of opposite type as the separator column. For example, suppressors for cation exchange used anion-exchange resin in the hydroxide form. The dilute nitric acid eluent was neutralized by hydroxide in the resin, virtually eliminating background conductivity. Cationic analytes had their nitrate counter-ions replaced with hydroxide, thus increasing their conductivity. In this manner, the signal-to-noise ratio was greatly increased. After a while, the suppressor resin would be expended and would have to be regenerated by taking the suppressor column out of line and pumping sodium hydroxide through it. Anion exchange was performed by switching all the resin, eluent, and regenerant types.

Today, suppressor columns are no longer used. Instead, membrane suppressors allow

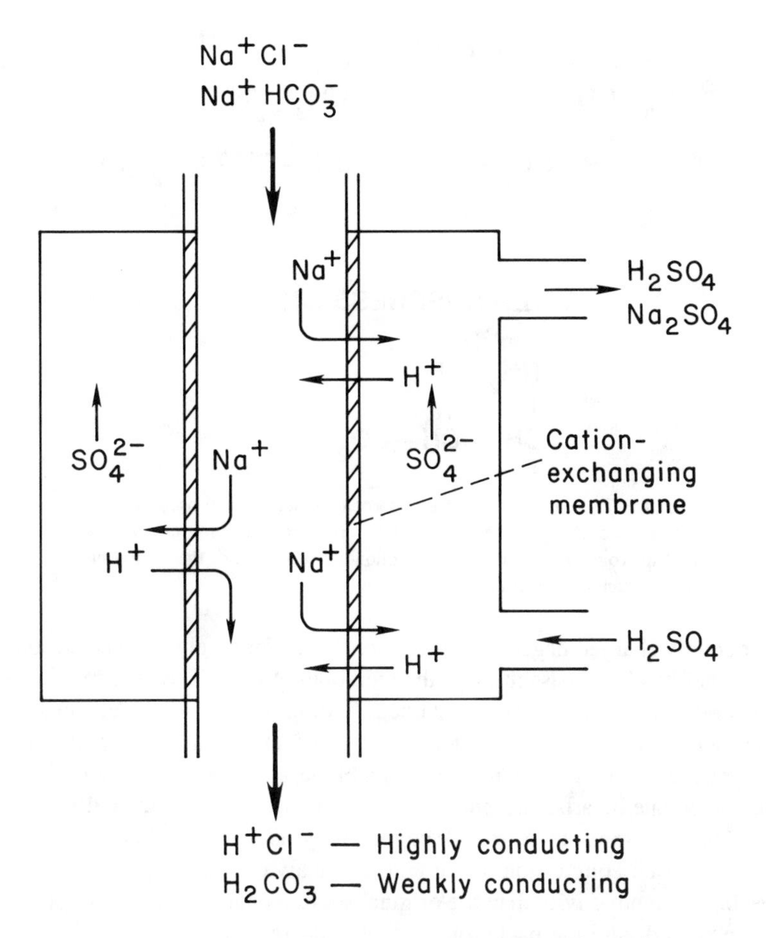

FIGURE 6. Operation of membrane suppressor for anion-exchange chromatography. Sodium counterions from chloride analyte and bicarbonate/carbonate eluent exchange with hydrogen ions across the cation-exchange membrane. The bicarbonate/carbonate eluent conductivity is lowered by the formation of the neutral acid, and the chloride conductivity is enhanced by the hydrogen counter-ion.

continuous ''regeneration'', so analyses do not need to be interupted for suppressor regeneration. In fact, it is even possible to regenerate the regenerant by continuously pumping it through a very large ion-exchange column. In this manner, 500 ml of regenerant can be continuously recycled through both the membrane suppressor and the large column for several weeks before it must be replaced. The large ion-exchange column will last for months in normal use. These advances have made the use of a suppressor far more convenient.

1. How Suppression Works

Ion-exchanging membranes take advantage of the Donnan equilibrium (Chapter 3). A diagram of how this is put into practice is shown in Figure 6, which is a schematic diagram of the membrane suppressor used following anion-exchange separation.[11] A useful eluent in suppressed ion chromatography is a sodium carbonate/bicarbonate buffer. Analyte anions (such as chloride) exit the separator column with sodium counter-ions. The column effluent enters the membrane suppressor between two membranes made of a cation-exchanging polymer. On the opposite side of these membranes a dilute solution of sulfuric acid (the regenerant) flows in the opposite direction, as shown. The membranes have fixed sulfonic acid negative ions and therefore exclude anions to a degree determined by the Donnan equilibrium. Dilute sulfuric acid is the preferred regenerant because exclusion is more

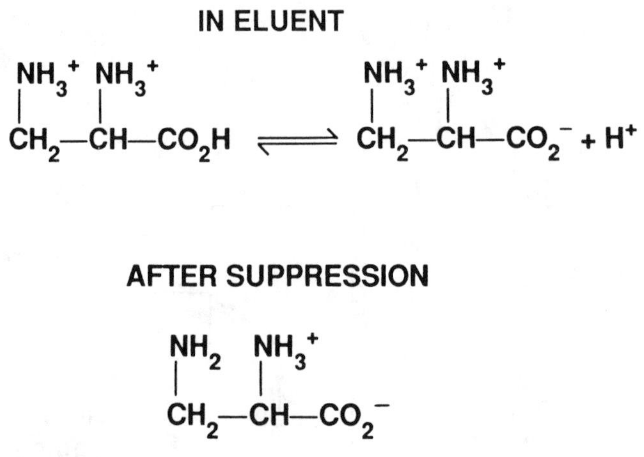

IN ELUENT

$$\underset{CH_2}{\overset{NH_3^+}{|}}\text{—}\underset{CH}{\overset{NH_3^+}{|}}\text{—}CO_2H \rightleftharpoons \underset{CH_2}{\overset{NH_3^+}{|}}\text{—}\underset{CH}{\overset{NH_3^+}{|}}\text{—}CO_2^- + H^+$$

AFTER SUPPRESSION

$$\underset{CH_2}{\overset{NH_2}{|}}\text{—}\underset{CH}{\overset{NH_3^+}{|}}\text{—}CO_2^-$$

FIGURE 7. Forms of 2,3-diaminopropionic acid (DAP) used as eluent for cation exchange. The 2+ and 1+ forms are in equilibrium at 0.047 m*M* hydrogen ion concentration. Following suppression, the neutral and nonconductive zwitterion is formed.

effective for doubly charged negative ions (co-ions) than for singly charged co-ions. Positive ions, on the other hand can pass through the membrane freely. The only requirement is that electrical neutrality must always be maintained. Sodium ions leave the eluent side and flow through the membranes into the regenerant side, and an equal number of hydrogen ions leave the regenerant side, flow across the membrane and enter the eluent side, thus neutralizing the carbonate/bicarbonate buffer to carbonic acid. Carbonic acid is largely undissociated, so the resulting background conductivity is considerably lower than before suppression. Also, the counter-ion to the chloride analyte is now hydrogen ion, which has seven times higher conductivity than the original sodium counter-ion. Therefore, the analyte response is increased, and the background level is decreased.

The effluent from a cation-exchange column is treated similarly. In this case the membranes are made of an anion-exchange material. They allow anions to pass freely but exclude cations. Sodium analytes are eluted from the separator column by hydrochloric acid, and exit the separator column with chloride counter-ions. The sodium analytes are converted in the suppressor to sodium hydroxide, the acid is neutralized by hydroxide, while the chloride ions pass through the membranes and out to waste. The regenerant solution is generally a quaternary ammonium hydroxide solution. The quaternary ammonium cation is chosen because its size and slow diffusion rate minimizes its transport across the membranes.

2. Eluents for Cation Exchange

The eluents used in suppressed ion chromatography must be such that they are removed or converted to weakly conducting compounds by the suppressor. For cation chromatography, a dilute acid such as 10 m*M* hydrochloric acid is a good choice for the elution of alkali metals. However, hydrogen ion is too weakly held by sulfonic acid-type ion exchangers and it is too weak a displacing ion for alkaline-earth ions. The more strongly retained divalent alkaline earths are eluted using 2,3-diaminopropionic acid (DAP, shown in Figure 7).[12] The ionization constants of the acid are 1.3, 6.7, and 9.4. Used along with 47 m*M* HCl (pH 1.3), one-half of this acid is in the fully protonated divalent form, and one-half in the partially protonated monovalent form. The divalent form has much stronger displacing power, making this a good eluent component for the elution of divalent cations. The membrane suppressor converts it to its zwitterionic form, which has no conductivity.

Eluents strong enough to elute alkaline earths from the same column used for alkali

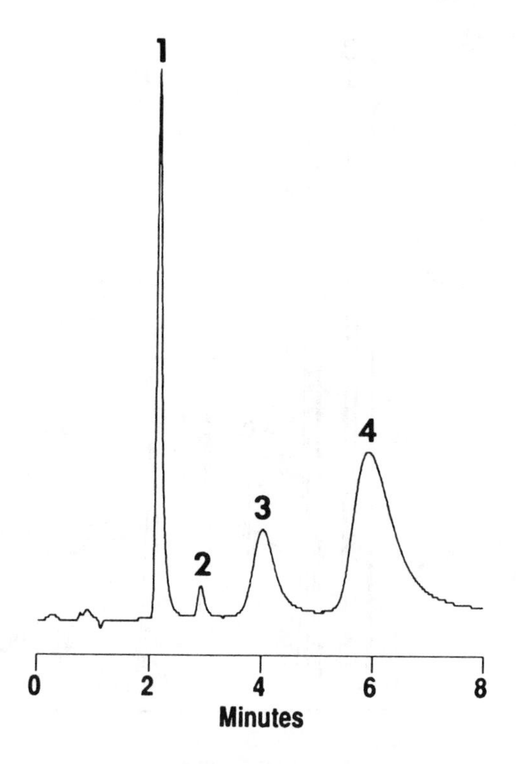

FIGURE 8. Determination of alkali metal and alkaline-earth cations in drinking water (diluted 1/10) using column switching and suppressed conductivity detection. Peaks with concentrations before dilution: (1) Na^+ 21 ppm; (2) K^+ 2.3 ppm; (3) Mg^{2+} 12 ppm; (4) Ca^{2+} 57 ppm. Column 1: Dionex Fast Cation I, 4 × 250 mm; column 2: Fast Cation II, 4 × 50 mm. Eluent: 17 m*M* HCl, 0.26 m*M* DAP·HCl at 2.0 ml/min. (Courtesy of Dionex.)

metals cause the alkali metals to elute in the void volume. Both alkali metals and alkaline earths can be eluted in a single injection using either gradient elution (described below) or a column-switching technique. For this method, a 4 × 50 mm cation-exchange column is placed in front of a 4 × 250 mm column. Following injection, the monovalent alkali metals quickly elute from the short column and onto the long column. Using a valve, the short column is switched to the position after the 250 mm column, effectively ''leapfrogging'' the divalent cations ahead of the monovalents. The monovalents then elute from the long column and cross over the divalents, eluting first. They are followed by the divalent alkaline earths. In this manner, both alkali metals and alkaline earths can be eluted in a single run. An example chromatogram using this method is shown in Figure 8.

3. Eluents for anion exchange

To separted anions, suppressable eluents are salts of weak acids with pKa greater than 6. This includes sodium hydroxide, the salt of water (pKa = 14). Sodium hydroxide is a very useful eluent, especially in gradient elution, since it is suppressed to water regardless of its concentration. However, for isocratic elution, sodium carbonate/bicarbonate buffers have the advantage of being more powerful, allowing more dilute eluents to be used. They are also safer to use. The ionization constants of carbonic acid are 6.4 and 10.3. A commonly used eluent is nearly a 1:1 buffer of carbonate and bicarbonate, so the eluent has a pH of 10.3. An example chromatogram of eight common anions is shown in Figure 9.

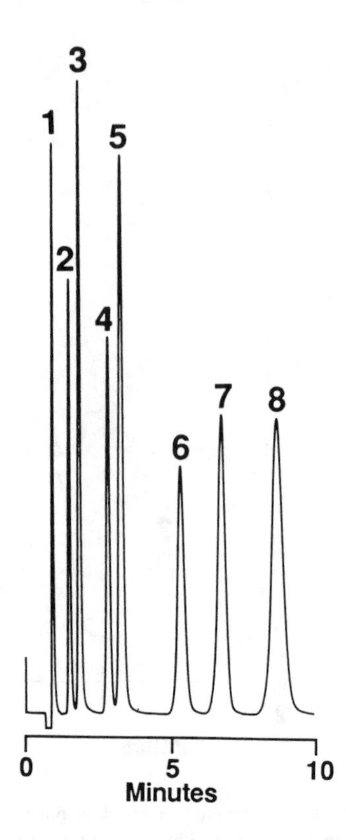

FIGURE 9. Anion exchange and suppressed conductivity detection of anions. Peaks: (1) F^- 1.5 ppm; (2) Cl^- 2.8 ppm; (3) NO_2^- 7.5 ppm; (4) Br^- 7.5 ppm; (5) NO_3^- 10 ppm; (6) HPO_4^{2-} 15 ppm; (7) SO_4^{2-} 10 ppm; (8) $C_2O_4^{2-}$ 15 ppm. Column: Dionex AS4A 4 × 250 mm, 15 μm latex-coated resin. Eluent: 1.8 m*M* Na_2CO_3, 1.7 m*M* $NaHCO_3$ at 2 ml/min.

Analyte elution time is controlled by the carbonate concentration, since it is a considerably more powerful displacer ion than bicarbonate. The retention of most ions is not affected by eluent pH, except to the extent that the carbonate concentration is affected. One exception is phosphate ion. The ionization constants of orthophosphoric acid, H_3PO_4, are 2.1, 7.2, and 12.6. If the pH of the carbonate/bicarbonate eluent is increased and the carbonate concentration held constant, the elution time of phosphate increases, as the proportion of trivalent to divalent phosphate increases.

Other eluents which may be used in suppressed anion chromatography are the sodium salts of boric acid (pKa = 9.2) and p-cyanophenol (pKa = 8.0). Boric acid is a very weak displacer, and is effective at separating very weakly retained anions such as carboxylic acids.[13] It also forms weak bonds with hydroxy organic acids, producing changes in selectivity as compared to hydroxide. Because of its high pKa, background conductivity following suppression is very low, making borate eluents useful for gradient elution. p-Cyanophenol is a powerful monovalent displacer, useful for eluting strongly retained hydrophobic monovalent anions such as iodide and thiocyanante.

C. COMPARISON OF SUPPRESSED TO NONSUPPRESSED ION CHROMATOGRAPHY

Suppressed ion chromatography requires the addition of a suppressor and a reservoir of regenerant to the analytical instrumentation. Although this increases the complexity of the instrumentation somewhat, continuous recycling of the regenerant through a large ion-exchange column allows the suppression system to be used unattended for weeks. Suppressed ion chromatography has the advantages of a signal-to-noise ratio about one order of magnitude better than in nonsuppressed ion chromatography. In addition to the obvious advantages of lower detection limits, this also allows dirty samples to be diluted more, extending column life and minimizing time spent on sample clean-up. Other advantages include wider dynamic range, the ability to handle larger sample volumes and more concentrated eluents, faster equilibration time, the ability to perform gradient elution (described below), elimination of interference from counter ions, and elimination of system peaks (except for a dip at the column void volume).

The main advantage of nonsuppressed ion chromatography is that the instrumentation is simpler, since no suppressor or regenerant reservoir is required. With the addition of a conductivity detector, any standard HPLC can perform ion chromatography. There are several analytical advantages to nonsuppressed ion chromatography. As described below, calibration curves from borate/gluconate and other eluents used in nonsuppressed ion chromatography produce more linear calibration curves than those from the carbonate/bicarbonate eluents commonly used in suppressed ion chromatography. (However, hydroxide- or borate-based eluents for suppressed ion chromatography will also produce more linear calibration curves than carbonate/bicarbonate eluents.) Also, calibration curves form certain weak acids (e.g., fluoride) are more linear in nonsuppressed ion chromatography. Finally, very weak acids (borate, cyanide, sulfide, silicate) can be determined by anion exchange and conductivity detection only without suppression.[4]

D. SENSITIVITY

Ion chromatography is a very sensitive analytical technique. Minimum detection limits (m.d.l.'s) from 50 μl samples range from 5 to 50 ppb (μg/l), with early eluting ions having lower m.d.l.'s than later eluters. Thus, the mass sensitivity is below 1 ng. The concentration sensitivity can be lowered by preconcentration.[14] For this technique, a large volume of sample (5 to 500 ml) is pumped though a short (4 × 50 ml) ion-exchange column. Sample ions are concentrated on the front of the column. Then, the short concentrator column is placed by a valve in the eluent stream directly in front of the separator (analytical) column. If the sample is nearly pure water, it is possible to inject a large volume of sample directly onto the analytical column. The sample ions are concentrated near the entrance to the column and start to move only when the eluent front reaches them. However, column reequilibration produces a less stable baseline as compared to the use of a concentrator column. A classic case of the use of preconcentration is the analysis of Antarctic snow and ice.[15] Volumes of 5 to 10 ml samples were introduced through a sample loop. Average concentrations of some common anions in ppb were sodium 20, ammonium 30, sulfate 120, chloride 35, nitrate 40. The authors note that the amount of meltwater required is an order of magnitude less than for neutron activation, and two orders of magnitude less than for atomic absorption.

Using a membrane suppressor, rainwater was analyzed by direct injection of 100 μl samples.[16] Detection limits for a 100 μl sample injected directly without preconcentration were reported to be 0.05 to 0.15 μ*M* for alkali metals, and 0.2 to 0.6 μ*M* for alkaline earths. These authors, too, note that for flame atomic absorption, 1 ml samples are needed, and for inductively coupled plasma emission spectroscopy, 5 to 10 ml.

E. CONDUCTIVITY AND LINEARITY

The conductivity of a strong electrolyte in dilute solution is directly proportional to its concentration. This would imply that ion chromatography calibration curves based on peak height or area would be linear. However, this is not always the case. There are three causes of nonlinear calibration curves.

1. Overloaded Column

Column overload will cause peak efficiency to decrease and peak symmetry to degrade, resulting in a decrease in the height to area ratio. As long as the analyte concentration at the peak maximum is below approximately 1 m*M*, column overload will not cause nonlinear calibration curves when peak area is used. Calibration curves based on height will not be linear. The slope of the curve will begin to decrease as the column is overloaded.

2. Weak Electrolytes

Unlike strong electrolytes, the conductivity of weak electrolytes is not proportional to concentration except in very dilute solution. If the analyte is the anion of a weak acid and the counter-ion (following suppression) is H^+, then the formation of the neutral free acid is increasingly favored as analyte concentration increases. Similarly, the conversion of ammonium to neutral ammonia following suppression with hydroxide regenerant is increasingly favored as analyte concentration increases. The result is a calibration curve that is linear at low concentration; however, the slope slowly decreases as concentration increases. The concentration at which this curvature begins is dependent on the pK of the analyte, with stronger acids and bases producing linear calibration curves to higher concentrations. This is less of a problem in single column than in suppressed ion chromatography. Single column eluents are either buffered or are at much higher concentration when entering the detector cell, so the eluent pH changes very little as analyte ions elute; i.e., a smaller change in eluent pH produces a smaller change in analyte dissociation.

3. Suppression of Eluent Dissociation by the Analyte

When a weak acid is used as the eluent and the dissociation of the acid is not totally suppressed by the suppressor, then nonlinear calibration curves result.[17] This usually occurs only when carbonate/bicarbonate buffers are used as eluents with suppressed conductivity detection, although it can also be a problem in single column ion chromatography at high analyte concentrations. The cause of nonlinearity is that the elution of analyte anions of a strong acid results in an increase in hydrogen ions in the eluting volume. This causes the dissociation of the weak acid eluent to decrease, lowering its contribution to the background. The result is a calibration curve which can be divided into three sections: a linear section at low concentration with an intercept of zero; a linear section at high concentration with a slope greater than the first section and a negative intercept; and a curved portion connecting the two. The problem of nonlinearity can be overcome in several ways. First, an empirical fit to the calibration points can be hand drawn, or a polynomial equation generated using a computer can be used. Most integrators and computer integration programs are capable of fitting the points to a nonlinear least squares regression using a quadratic equation. Although a quadratic equation does not fit the points exactly, an excellent fit is obtained if the calibration range is limited to one order of magnitude or less. For computers lacking quadratic regression, point to point calibration can be used provided the points are not too far apart.

Conductivity is generally directly proportional to concentration in single column ion chromatography and in suppressed ion chromatography when the salts of acids weaker than carbonic acid are used as eluents. In single column ion chromatography this is because the counter-ion concentration does not change during the elution of analyte ions. In suppressed ion chromatography linear calibration curves are obtained when the eluent is either completely removed from the column effluent by the suppressor, or is totally neutralized by the suppressor. For example, the sodium from a sodium hydroxide eluent is exchanged in the suppressor with hydrogen ion from the regenerant and completely removed from the eluent. Strong-acid anions such as chloride or nitrate reach the detector with hydrogen counter-ions in a background of deionized water. When HCl is used as the eluent in cation exchange, the chloride is exchanged for hydroxide, neutralizing the hydrogen ions. Hydroxide eluents in anion exchange and hydrogen ion eluents in cation exchange produce linear calibration curves because they are completely removed by the suppressor. Anion-exchange eluents composed of salts of very weak acids are suppressed to very low background conductivities due to the low dissociation of the weak acid. For example, sodium borate is suppressed to boric acid. With a pKa of 9.1, the dissociation of boric acid is so slight as to contribute a negligible amount to the background conductivity, so linear calibration curves are the result.

F. GRADIENT ELUTION

Gradient elution in ion chromatography is performed by increasing the ionic strength of the eluent during the run. This allows ions of widely differing retention to be eluted in one run. Gradient elution, however, is only practical when steps are taken to minimize the change in background conductivity caused by the increasing eluent concentration. There are two methods for minimizing background shift. They are the use of a suppressor and the use of isoconductive eluents in nonsuppressed ion chromatography.

In suppressed ion chromatography gradients are performed by choosing eluents that can be suppressed to produce little or no background conductivity. For anion-exchange gradients, salts of weak bases with $pKa > 7$ can be used. These include the sodium salts of boric acid ($pKa = 9.1$), and p-cyanophenol ($pKa = 8.0$) Of course, the best weak acid to use is water itself; the salt being sodium hydroxide. As long as the capacity of the suppressor is not exceeded (about 100 m*M* NaOH at 1 ml/min), sodium hydroxide eluent is suppressed to water regardless of concentration. An example of the ability of gradient elution to separate and elute a large number of both organic and inorganic ions in a single run is shown in Figure 10. Here, the initial eluent is 0.75 m*M* NaOH, which is dilute enough to separate the weakly retained monoprotic acids. The eluent at the end of the gradient is 85 m*M* NaOH, concentrated enough to elute the much more strongly retained triprotic isomers of citric acid.

A comparison of isocratic to gradient elution is shown in Figure 11. The first peak in the isocratic separation is labeled fluoride. However, the gradient chromatogram separates three components early in the run: fluoride, acetate, and formate. These had all coeluted in the isocratic separation. For cation exchange, mixtures of HCl and 2,3-diaminopropionic acid are used. This eluent system is able to separate and elute a mix of alkali metals, alkaline earths, and amines (Figure 12).[12]

In addition to the obvious advantage of determining more ions in less time, gradient elution also provides more information than isocratic separation. Analytes which coelute in the column void volume can be separated, and strongly retained analytes which might otherwise never elute can be detected. Without gradient elution, the analytical chemist might not even know of the existence of these components.

When a suppressor is not used, the extent of increase in the eluent concentration from the beginning to the end of the gradient is much more limited. This greatly limits the range of ions which can be eluted in a single run. However, it is possible to decrease the run time of a long isocratic separation by about one third using isoconductive eluents.[18] In an anion-exchange gradient, the conductivity of the initial eluent is artificially raised by using a counter-ion with a high background conductivity, e.g., cesium or potassium. During the gradient, the conductivity increase caused by the increasing eluent concentration is balanced by gradually replacing the potassium counter-ion with the lower conductivity lithium ion. In this manner, it is possible to increase the eluent strength by about two time. However, this is far too small an increase to be useful for separating and eluting ions with widely differing retention.

G. ION EXCLUSION

1. Separation

The Donnan membrane equilibrium limits the concentration of co-ions within an ion-exchanger gel. This principle can be utilized to separate ionized solutes from nonionized

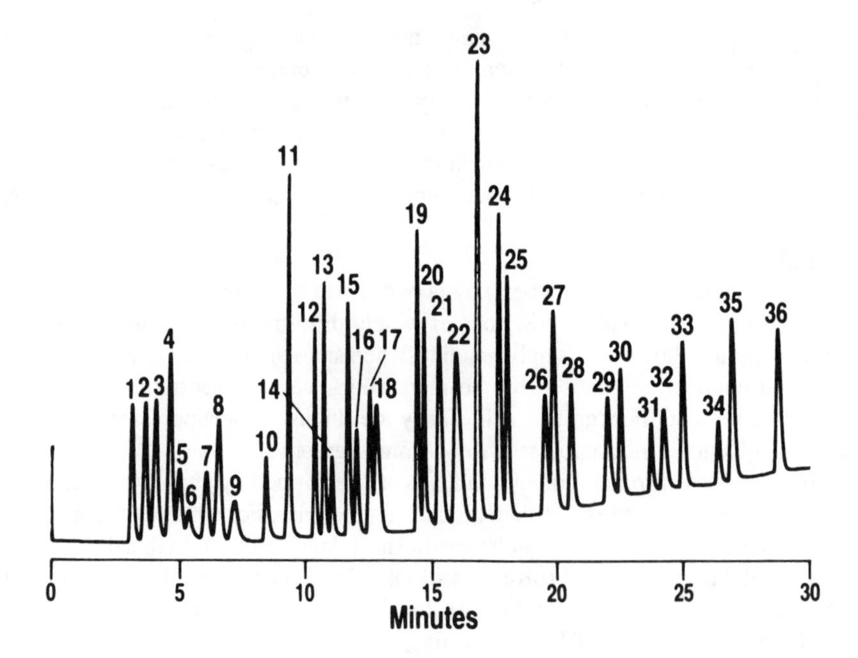

FIGURE 10. Gradient separation of anions using suppressed conductivity detection. Peaks (10 ppm unless otherwise noted): (1) Fluoride (1.5 ppm); (2) α-hydroxybutyrate; (3) acetate; (4) glycolate; (5) butyrate; (6) gluconate; (7) α-hydroxyvalerate; (8) formate (5 ppm); (9) valerate; (10) pyruvate; (11) monochloroacetate; (12) bromate; (13) chloride (3 ppm); (14) galacturonate; (15) nitrite (5 ppm); (16) glucoronate; (17) dichloroacetate; (18) trifluoroacetate; (19) phosphite; (20) selenite; (21) bromide; (22) nitrate; (23) sulfate; (24) oxalate; (25) selenate; (26) α-ketoglutarate; (27) fumarate; (28) phthalate; (29) oxalacetate; (30) phosphate; (31) arsenate; (32) chromate; (33) citrate; (34) isocitrate; (35) cis-aconitate; (36) trans-aconitate. Column: Dionex AS5A 4 × 150 mm, 5 μm latex-coated resin. Eluent: 0.75 m*M* NaOH held for 5 min, gradient to 85 m*M* NaOH in 30 min at 1 ml/min. (From Rocklin, R. D., Pohl, C. A., and Schibler, J. A., *J. Chromatogr.*, 411, 107, 1987. With permission.)

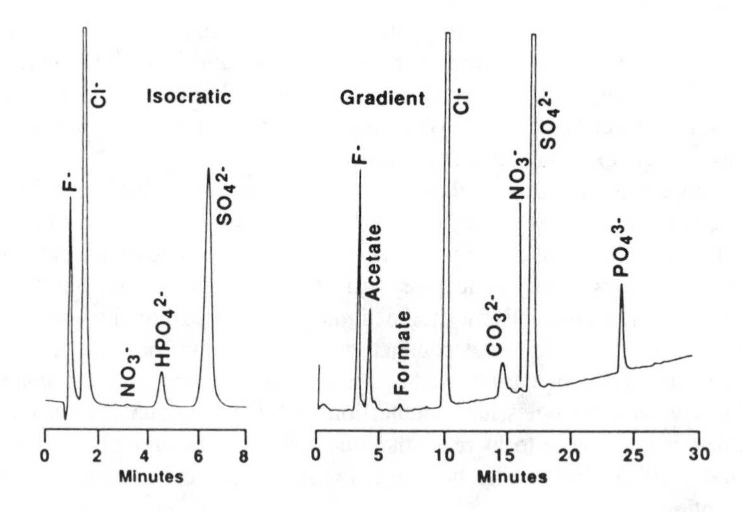

FIGURE 11. Separation of anions in waste water using isocratic and gradient elution. Gradient elution separates fluoride, acetate, and formate, which coelute in the void volume using isocratic separation. Conditions are similar to Figure 9 (isocratic) and Figure 10 (gradient). (From Rocklin, R. D., Pohl, C. A., and Schibler, J. A., *J Chromatogr.*, 411, 107, 1987. With permission.)

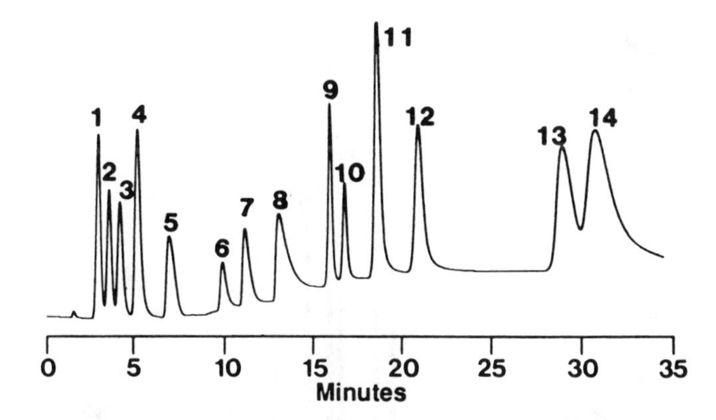

FIGURE 12. Elution of mono- and divalent cations using HCl/DAP gradient. Peaks: (1) Li^+; (2) Na^+; (3) NH_4^+; (4) K^+; (5) Triethylammonium$^+$; (6) Tripropylammonium$^+$; (7) Cyclohexylammonium$^+$; (8) Tetrabutylammonium$^+$; (9) Mg^{2+}; (10) Mn^{2+}; (11) Ca^{2+}; (12) Sr^{2+}; (13) Ethylenediammonium^{2+}; (14) Ba^{2+}. Column: Dionex Fast Cation I, 4 × 250 mm, 13 μm latex-coated resin. Flow rate: 1.0 ml/min.

solutes in a chromatographic column. Strong acids (such as mineral acids) are 100% ionized. Since they are anionic, they are excluded from the negatively charged pore structure of the sulfonated resin. Neutral molecules are not excluded. The volume inside the column that neutral molecules must traverse before they elute is much larger than the volume that anions must traverse. They elute near the totally permeated volume, which is the total volume inside the column that water molecules can permeate. Therefore, ionized strong acids will elute earlier than neutral molecules. This principle was used on a preparative scale in 1953 by Bauman and Wheaton[19] to separate hydrochloric and acetic acids. Fifteen ml of a solution containing 1.2 *M* hydrochloric and 0.7 *M* acetic acids were poured into a 100 ml bed of 8% crosslinked cation-exchange resin in its hydrogen form, then washed down with water. Fractions of 5 ml each were collected and analyzed. Hydrochloric acid came out of the column first, almost wholly separated from acetic acid, which came out afterwards. Hydrochloric acid is completely ionized and therefore excluded from the resin. It starts to elute in the effluent at the void volume of the column. At this pH of less than one, acetic acid is essentially totally protonated and is neutral. Acetic acid molecules enter the resin freely and must traverse not only the void volume, which is the volume between the resin beads, but also the volume of water internal to the resin beads.

Partially ionized acids elute between strong acids and neutral molecules. By controlling the pH of the eluent, the degree of dissociation of acids of intermediate strength, such as most carboxylic acids, can be controlled. Acids generally elute in order of their pKa with stronger acids eluting earlier than weaker acids. When a dilute strong acid is used as eluent, increasing the eluent strength will cause partially dissociated acids to elute later. The effect of eluent concentration on retention is the reverse of that observed in ion exchange.

Ion exclusion was used for analytical purposes in 1978 to separate a series of organic acids involved in the Krebs cycle.[20] A column of 4% crosslinked strong-acid cation-exchange resin, 1.3 × 30 cm, was used in the hydrogen form. The eluents were low-millimolar hydrochloric acid. The order of elution of the Krebs-cycle acids followed almost, but not exactly, the order of decreasing ionization constants. The same relation was found by Tanaka and colleagues[21] in the first of a series of papers on ion exclusion. Nitric, oxalic, phosphoric, hydrofluoric, formic, acetic, and carbonic acids were eluted by water in that order, and a straight-line graph was obtained by plotting the retention volume against pKa of the acid. In later work, these authors separated several of the condensed phosphoric acids by ion

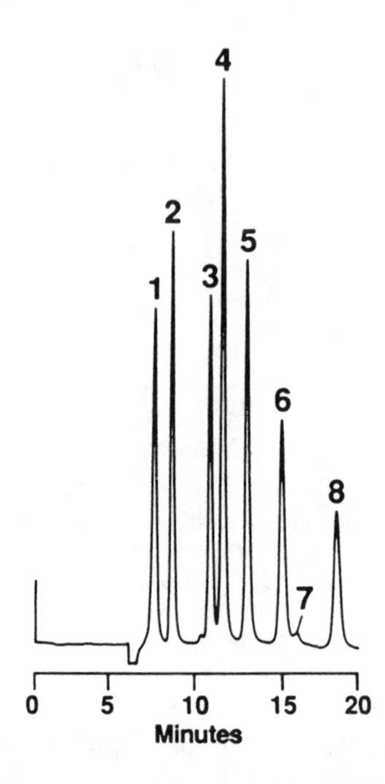

FIGURE 13. Ion-exclusion separation and suppressed conductivity detection of organic acids. Peaks (50 ppm each): (1) tartaric; (2) malic; (3) glycolic; (4) formic; (5) acetic; (6) propionic; (7) carbonate (contaminant, not added); (8) butyric. Column: Dionex ICE-AS1, 9 × 250 mm, 10 μm fully sulfonated PS/DVB resin. Eluent: 1 m*M* octanesulfonic acid, 2% 2-propanol at 1 ml/min.

exclusion.[22] They and others[23] found a correlation between ionic charge and retention volume. We noted in Chapter 3 that Donnan exclusion was more effective as the charge on the co-ion increased.

Acetic, propionic, and butyric acids are shown to be well resolved in the chromatogram in Figure 13. Since all three of these acids have very similar pKa and in fact are nearly completely protonated in the eluent used, the mechanism of their separation must be other than ion exclusion. Hydrophobic interaction between these molecules and the resin polymer is undoubtedly a factor. Mixed retention mechanisms are common when organic ions and molecules interact with organic polymers.

Eluent concentration can also affect peak shape. Figure 14 shows chromatograms of mono-, di-, and trichloroacetic acids on hydrogen-form cation exchanger.[20] The eluent for the first chromatogram is water, the second is 1 m*M* HCl, and the third, 10 m*M* HCl. Trichloroacetic acid elutes first, because it is the strongest of the three acids and is essentially fully ionized in each case. Dichloroacetic acid elutes second and monochloroacetic acid, the weakest acid, elutes last. Increasing HCl concentration represses the ionization and increases retention. The peak of monochloracetic acid in water and 1 m*M* HCl is wedge shaped with much "fronting". The fronting occurs because the proportion of acid which is undissociated rises with the acid concentration, and is therefore higher at the peak maximum than at the base. Since it is the undissociated acid that is retained, the peak maximum is retained more than the base. (The same peak shape is seen in reversed-phase chromatography of weak carboxylic acids on an alkyl bonded silica column with methanol-water eluents. Again the proportion of uncharged, undissociated acid increases as the concentration increases, and dimers may also form.) The cure for this situation is to add enough of a buffer

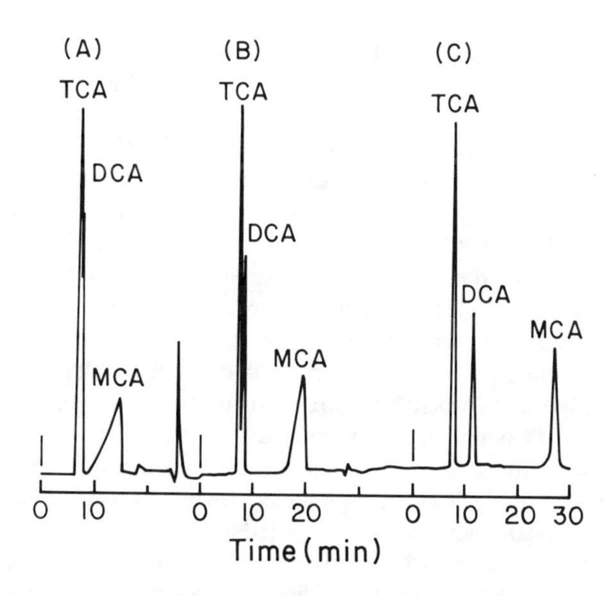

FIGURE 14. Peak shapes in ion-exclusion chromatography. MCA, DCA, and TCA are mono- di-, and trichloroacetic acids. Concentrations of HCl in eluents (m*M*): (A) O; (B) 1; (C) 10. Detection: UV absorbance at 210 nm. (From Turkelson, V. T. and Richards, M., *Anal. Chem.*, 50, 1420, 1978. Copyright, American Chemical Society, 1978. With permission.)

or an excess of a strong acid to stabilize the ratio of ionized to nonionized weak acid. The result is a symmetrical peak.

The exchangers used for ion-exclusion chromatography must be high-capacity gel-type resins with a uniform internal concentration of fixed ions. The low-capacity surface-functional exchangers used for ion chromatography will not work, because the functionalized surface of the resin is only a small fraction of the total internal volume. The column must also be fairly large. All the separation must take place within the internal water volume of the resin in the column. The degree of crosslinking is very important. Low-crosslinked resins have increased selectivity, but are less rigid and may collapse under pressure.

2. Ion Exclusion with Conductivity Detection

The form of detection used by Turkelson[20] was UV absorbance at 210 nm, where the carboxyl group absorbs light. Refractive index can also be used for detection, although its sensitivity is low and selectivity is poor. Electrical conductivity is generally impractical without suppression, except when the eluent is deionized water. When dilute strong-acid eluents are used, the conductivity rises as weak-acid analytes elute, but since they are weakly ionized, the conductivity rise is very small with respect to the eluent background. The solution to the problem is to use a membrane suppressor in which the regenerant is a strong base and the membranes are cation exchanging.[13] Tetrabutylammonium hydroxide regenerant is commonly used. Tetrabutylammonium cations diffuse from the regenerant across the membrane to the eluent side, and hydrogen ions diffuse from the eluent to the regenerant side. The driving force for the reaction is the neutralization of hydrogen ions and the concentration gradients across the membrane. Excess strong acid in the eluent is converted to its tetrabutylammonium salt, which has a much lower conductivity than the acid. The partially ionized organic-acid analytes are converted to their fully ionized tetrabutylammonium salts. This is essentially the reverse of the suppression process used in anion exchange. Eluents are chosen such that the anion of the strong acid has as low a conductivity as possible. Octanesulfonic acid is a good choice. Although the signal must still be measured against a large background, the ratio of signal to background is much greater than before suppression.

Another way to perform eluent suppression in ion-exclusion chromatography is to use hydrochloric acid as the eluent and a suppressor column packed with a high-capacity cation-exchange resin in the silver form.[24] Chloride from the eluent is precipitated by silver. Hydrogen ions from the eluent replace silver ions in the resin. Because hydrochloric acid is quantitatively removed from the eluent, background conductivity is near zero, making this a very sensitive method. However, the suppressor column will only last about 40 h, can not easily be regenerated, and must be replaced.

Simple one-component analyses can be made by conductivity detection without suppression. Ion exclusion is useful when a weak acid or its salt is sought in a matrix of strong acids and salts. A good example is the measurement of bicarbonate ions in blood plasma.[25] A column of high-capacity strong-acid cation-exchange resin is used in the hydrogen form; the eluent is deionized water. Salts and high molecular-weight material elute as a large peak at the void volume. Bicarbonate ions are converted to carbonic acid, which is not excluded from the resin and elutes later at the totally permeated volume. Because carbonic acid is only partly ionized the peak height is not directly proportional to the concentration, but rather to the square root of the concentration. Sensitivity could be increased and a linear calibration curve obtained by converting the carbonic acid to bicarbonate ions in a second cation-exchange column or suppressor device. Tanaka and Fritz[26] went further and followed the ion-exclusion separator column by two exchanger columns. The first converted carbonic acid to potassium bicarbonate and the second converted potassium bicarbonate by anion exchange to highly conducting potassium hydroxide.

Fluoride ions can be separated and measured by ion exclusion on a hydrogen form cation-exchanger column. Hydrofluoric acid is fairly weak; its pKa = 3.45. The advantage of ion-exclusion over anion-exchange chromatography is that in ion exchange, fluoride is held very weakly, and its peak is easily confused with the void peak and the peaks of other weakly retained materials. The ion exclusion method was used to determine fluoride in wastewater.[27]

Ion-exclusion chromatography with conductivity detection is a powerful method for determining carboxylic acids. Ion-exclusion separation is effective at separating weak-acid analytes from common strong-acid inorganic ions such as chloride and sulfate, and conductivity detection eliminates interferences from nonionic species that could interfere if UV detection were used. An application where both of these factors are important is the determination of organic acids in wine (Figure 15).[28] The separation eliminates interferences from inorganic salts and conductivity detection eliminates interferences from phenolic species.

H. CONDUCTIVITY DETECTORS

Detectors used in ion chromatography are the same as those commonly used in HPLC. These can be separated into two classes: optical and electrochemical. Optical detectors include the UV-visible absorbance detector, as well as fluorescence and refractive index (RI) detectors. All are commonly used in HPLC. Most electrochemical detectors measure current resulting from the application of potential across electrodes in a flow cell. Depending on how the potential is applied and the current measured, either the conductivity of the solution or the current caused by reduction or oxidation of analytes (amperometry) can be measured. Potentiometric detectors measure potential developed at an ion specific electrode.

The detector most commonly used in ion chromatography is the electrical conductivity detector. Because all of the species separated by ion exchange are partially or completely dissociated ions during their separation, they all share the capability of conducting electric charge in solution. Because of this, the electrical conductivity detector is the most generally useful detector following ion exchange separations.

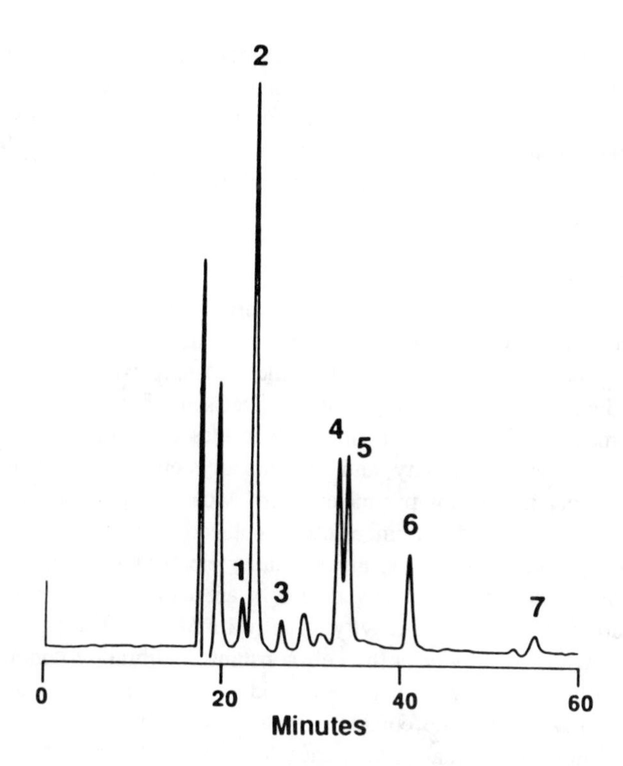

FIGURE 15. Separation of organic acids in wine using ion exclusion chromatography with suppressed conductivity detection. Peaks: (1) citric acid; (2) tartaric; (3) malic; (4) succinic; (5) lactic; (6) acetic; (7) carbonate. Conditions similar to Figure 13, except two ICE-AS1 columns in series were used, flow rate was 0.5 ml/min. (From Dionex Corp., Application Note 21, Sunnyvale, CA. With permission.)

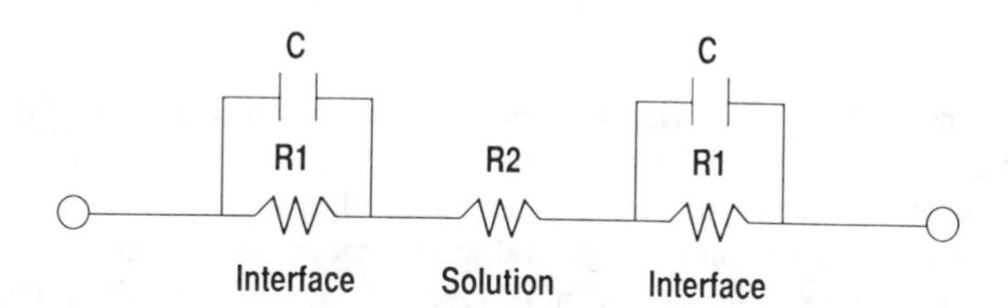

FIGURE 16. Equivalent circuit for conductivity detection. C represents the interfacial capacitance. R1 the resistance to oxidation or reduction at the electrode surface, and R2 the solution resistance.

1. Conductivity Measurements

A conductivity-detector cell can be simulated using resistors and capacitors in an equivalent circuit shown in Figure 16. The capacitance between the electrodes and the solution, called the interfacial capacitance, is represented by C. R1 is the resistance to electron transfer between the electrode and species in solution. R2 represents the resistance to ionic current in the solution between the electrodes, and it is this resistance we wish to measure. It is measured using an alternating electric field with a frequency of several kHz applied between the electrodes. The alternating potential usually has the waveform of either a sine wave or a square wave (also called a bipolar pulse).[29,30] At any instant in time, negatively charged anions migrate toward the positive electrode and positively charged cations migrate toward the negative electrode. At the frequency used, the resistance of the interfacial capacitance (more correctly the reactance) is low. Therefore, the potential drop across the electrode solution interface is too low to cause electron transfer reactions to occur between the electrode and species in solution (R1 is very high), and the large majority of the potential is applied across the solution (R2). The resistance is calculated using Ohm's law. The inverse of this resistance is the conductance of the solution.

2. Conductivity Cell Constants

The conductance of the solution is determined by many factors, such as the concentrations of the ions and the distance between the electrodes in the measuring cell. Since it is desirable to compare conductances of different detectors, the measured conductance is multiplied by a factor, the conductance cell constant, which converts the measured conductance to that which would be measured in a cell containing electrodes of 1 cm^2 surface area held 1 cm apart. This quantity is the conductivity, and the units are siemens per centimeter S/cm. (The SI unit for conductivity is siemens per meter, S/m. Most analytical instrument companies use S/cm.) Since a cell composed of 1 cm^2 electrodes placed 1 cm apart would be impractically large for use in liquid chromatography, much smaller electrodes placed closer together are used in flow-through cells. Conductivity detectors are calibrated by measuring the conductance of a solution of known conductivity, usually 1 m*M* KCl. The measured conductance divided by the known conductivity is the cell constant. Calibrating conductivity detectors usually involves filling the cell with 1 m*M* KCl and adjusting a trim pot until the displayed conductivity reads 147 uS/cm (the conductivity of 1 m*M* KCl). The user often does not know the value of either the actual conductance or of the cell constant.

3. Conductivity of Electrolytes

In dilute solutions, the conductivity of strong electrolytes is proportional to electrolyte concentration. Therefore, conductivity divided by concentration is constant. The quantity Λ is called the equivalent conductivity and is defined by the equation

$$\Lambda = \frac{1000\kappa}{c}$$

where κ is the measured conductivity in S/cm, and c is the concentration of the electrolyte in equiv/l. (Equiv/l equals mol/l times the charge on the ion.) The unit for Λ is S $\times$ cm^2/equiv. The analyte concentrations normally encountered in ion chromatography are below a few tenths of millimolars, and are generally low enough for Λ to be independent of concentration. For example, the equivalent conductivity at 25°C of KCl at infinite dilution is 149.9; and at 1m*M*, 146.9, a decrease of only 2%. (The conductivity of an eluting analyte can not be assumed to be directly proportional to concentration, because ionic components of the eluent may be contained in the eluting volume.) Measured values of Λ extrapolated to infinite dilution are called limiting equivalent conductivities, $\Lambda°$.

4. Conductivity of Ions

The conductivity of a dilute solution is the sum of the conductivities of all the ions in the solution multiplied by their concentration. This is called Kohlraush's law of independent migration. It states that each ion carries its portion of the total charge without being affected by any of the other ions in solution. Stated as an equation

$$\kappa = \frac{\Sigma_i \lambda_i^\circ c_i}{1000}$$

λ_i^o the ionic limiting equivalent conductivity, is similar to Λ^o for electrolytes, but is instead defined for individual ions. For example, the limiting equivalent conductivity at 25°C for NaCl is the sum of the ionic limiting equivalent conductivity for Na^+, 50.1, plus that of Cl^-, 76.4, or 126.5. A 0.1 m*M* solution of NaCl at 25°C will have a conductivity of 0.1 × 126.5, or 12.65 μS/cm. The conductivity of a solution containing 0.01 m*M* NaCl plus 0.01 m*M* Na_2SO_4 would be

$$0.03 \times 50.1 = 1.50\ (Na^+)$$

$$0.01 \times 76.4 = 0.76\ (Cl^-)$$

$$2 \times 0.01 \times 80.0 = \underline{1.60}\ (SO_4^{2-})$$

$$3.86\ \mu S/cm$$

Table 1 lists limiting equivalent conductivities for a number of organic and inorganic anions and cations.

If the electrolyte is a weak electrolyte such as an acid or base with only partial dissociation, then c_i must be replaced by the concentration of the dissociated ions only, since only they contribute to conductivity. For acids and bases, the pK values and the solution pH can be used to calculate the extent of dissociation.

5. Effect of Solvent and Temperature on Conductivity

The limiting equivalent conductivity of an ion, λ_i^o is essentially a measure of the mobility of the ion. Ionic mobility is greatly affected by the properties of the ion in the solvent, such as the extent of the ion's hydration sphere and the viscosity of the solvent. These properties are constant from run to run. It is not necessary to know these values, since quantitative analysis is generally performed by comparing the conductivity of the analyte in the sample to the conductivity of the same analyte in a standard (or standards). Even if a solvent gradient is used, the composition of the solvent during the elution of the analyte will be the same in both the sample and the standard. However, mobility, and therefore conductivity, is greatly affected by temperature. The conductivity of an aqueous solution is found experimentally to rise about 2% per degree C. (This dependence is described in an equation developed by Onsager.) Therefore, it is necessary to hold the temperature of the conductivity cell constant from run to run, or errors in concentration measurements will result. This is accomplished by thermostating the cell at a temperature above ambient. Alternatively, the measured conductivity can be corrected to a specific temperature (25°C) by measuring the cell temperature with a thermistor and multiplying the conductivity by a temperature-dependent constant. (More detailed discussions of the effect of solvent and temperature on conductivity can be found in References 31 and 32.)

6. Species Detected by Conductivity

Conductivity detection is used for species which are ionic when they enter the detector cell. This includes all anions and cations of strong acids and bases, such as chloride, nitrate, sulfate, sodium, and potassium. Since the eluent pH determines the extent of analyte ionization, the eluent pH must be chosen to ensure analyte dissociation. (When a suppressor is used, the eluent pH which determines whether or not an ion will be detected is the pH after suppression.) As analyte dissociation increases so does sensitivity. In anion exchange with chemical eluent suppression, sensitivity is good for anions with pKa below 5. Anions with pKa above 7 are not detected. This is generally the case with direct nonsuppressed detection (single column) using phthalate or benzoate eluents. However, the eluent pH can be adjusted to cause dissociation of weaker acids. Borate/gluconate eluents have a pH of about 8.5, allowing weaker acids to be detected. Fortunately, all organic acids with either carboxylate, sulfonate, or phosphonate functional groups have pKa below 4.75, so conductivity is often the preferred detection method for these species. Anions of weaker acids such as sulfide (pKa = 7.0), borate (9.1), cyanide (9.3), or silicate (9.7) can be detected using single column with a sodium hydroxide eluant.[4] These anions all have lower equivalent conductivities than hydroxide, so they cause dips instead of peaks. These as well as other weak acids such as carboxylic acids can be separated using ion exclusion and detected by suppressed conductivity after replacing hydrogen ions from the dilute strong-acid eluent with cations in a suppressor, as described earlier.

Nearly all organic cations are amines. Aliphatic amines have pKa around 10 and are easily detected. Aromatic and heterocyclic amines have pKa between 2 and 7. They can not be detected by suppressed conductivity detection following cation-exchange separation. With nonsuppressed detection, they cause dips. Because of their UV absorbance, photometric detection is the prefered method for these analytes.

I. APPLICATIONS

The range of analytical problems solved using ion chromatography with conductivity detection has grown enormously since introduction in 1975. Today, ion chromatography is used to analyzed many samples ranging from drinking water to electroplating baths. Ion chromatography has had the biggest impact on environmental analysis, where it is used most commonly to determine anions, particularly sulfate, nitrate, and chloride. Many of these applications were published in the late 1970s in two volumes edited by Mulik and Sawicki.[33] Today, most of these applications are still in use. The main difference is that newer, faster, columns and membrane suppressors have greatly increased the sample throughput rate. One of the most important applications is the tracking of acid rain by measuring the concentration of sulfate and nitrate from various sources near fossil fuel power plants.[34] In addition to inorganic ions, more recent investigations have shown that organic acids, particularly formic and acetic acids, are major contributors to acid rain.[35,36] Using preconcentration techniques, large increases in sensitivity can be achieved, as in the analysis of Antarctic snow and ice described earlier.[15] Other environmental applications include the analysis of soils,[37] and fluoride[38] and sulfur[39] from coal. Coal analysis is an environmental application, since these species enter the atmosphere and contribute to acid rain if they are not removed from the coal.

In addition to environmental applications, there are applications for ion chromatography in many other areas. The ability of ion chromatography to distinguish between species of the same element is applied to the analysis of metallurgical processing media, where both arsenic(III) (arsenite) and arsenic(V) (arsenate) are determined in one run.[40] Amperometric detection is used for arsenic(III) and suppressed conductivity detection for arsenic(V). Inorganic anions, especially species of sulfur, are determined in pulping liquors. Again, the ability of ion chromatography to determine independently species of the same element is used. UV is used to detect sulfide and conductivity to detect sulfate.[41] Many industries rely heavily on ion chromatography to determine impurities in deionized water. Most nuclear and many fossil fuel power plants ensure that impurities are kept below 0.1 ppb using ion chromatography with preconcentration.[42] In the same manner, semiconductor manufacturers increase product yield by using ion chromatography to measure the purity of deionized water used to rinse the devices during production.[43] Ion chromatography is now used in elemental analysis. The concentrations of heteroatoms in organic compounds can be determined by oxygen flask combustion followed by ion chromatography determination.[44,45] In the area of clinical analysis, ion chromatography is used to determine oxalate in urine, a major cause of kidney stones.[46] In addition to common ions, many less common ions can also be determined. For example, monofluorophosphate can be determined in toothpaste.[47] Another example is the determination of nitrogen-sulfur compounds such as hydroxysulfamate.[48]

It should be clear from the examples listed above the ion chromatography has been used for more applications than can be listed here. This is due to the major advantages ion chromatography has over competing analytical techniques: sensitivity, speed, and multi-species capability. In fact, ion chromatography with conductivity detection is usually the method of choice for the determination of most nonchromophoric ionic species, whether organic or inorganic.

Ion chromatography has made possible the analysis of mixtures of ions with similar properties that could hardly be analyzed in any other way. Phosphorus and sulfur form a great variety of anions in aqueous solutions. Polyphosphates are important as detergents and sequestering agents, and the oxy-anions of sulfur are formed in mine wastes. Generally these compounds are detected, not by conductivity, but by absorbance following post-column derivatization. They will be mentioned again below.

II. ION CHROMATOGRAPHY WITH AMPEROMETRIC DETECTION

Amperometric, pulsed amperometric, and coulometric detection are used to measure the current resulting from oxidation or reduction of analyte molecules at the surface of an electrode. For species which can be oxidized or reduced, detection is usually sensitive and highly selective, since many potentially interfering species cannot be oxidized or reduced, and are not detected. The major application is the detection of catecholamines.[49,50] These molecules are neurotransmitters, and are also used as drugs of abuse. Although they are separated most commonly by ion-pair reversed-phase chromatography, Kochi and Polzin[51] reported superior precision, specificity, and sample output using cation-exchange chromatography. A chromatogram of catecholamines from blood plasma determined using cation-exchange separation and amperometric detection is shown in Figure 17. There are many other applications of electrochemical detection based on oxidation or reduction. Table 2 lists the species most commonly determined and the working electrode materials used. The main categories are phenols (including catecholamines), aromatic amines, thiols; inorganic anions forming complexes with silver, such as cyanide, sulfide, and iodide; and carbohydrates, amines, and alcohols. Johnson and others[52] published a review of amperometric detection in liquid chromatography in 1986.

A. AMPEROMETRIC DETECTOR CELLS

Cells used for amperometric detection are miniature flow-through voltammetry cells. An equivalent circuit of the standard three electrode cell employing a working electrode, reference electrode, and counterelectrode is shown in Figure 18. Oxidation or reduction of analyte molecules is accomplished by applying either a single DC potential or a very low frequency (about 1 Hz) potential program between the working and reference electrodes. The reference electrode is chosen such that the potential difference between it and the solution is fixed by an electrochemical redox couple, such as the reaction for silver and silver chloride. Any changes in the potential applied between the working and reference electrode will be developed entirely between the working electrode (where analyte reduction or oxidation takes place) and the solution. To maintain a constant potential difference between the reference electrode and the solution, the cell current must be prevented from flowing through the reference electrode. An electronic circuit called a potentiostat causes the cell current to flow instead through the counterelectrode. The potentiostat automatically compensates for the solution resistance between the reference electrode and the counterelectrode (R3).

In contrast to conductivity detection where the solution resistance (R2) is measured, the quantity measured in an amperometric detector is the current resulting from analyte reduction or oxidation at the working electrode. R1 in the equivalent circuit represents the path of the

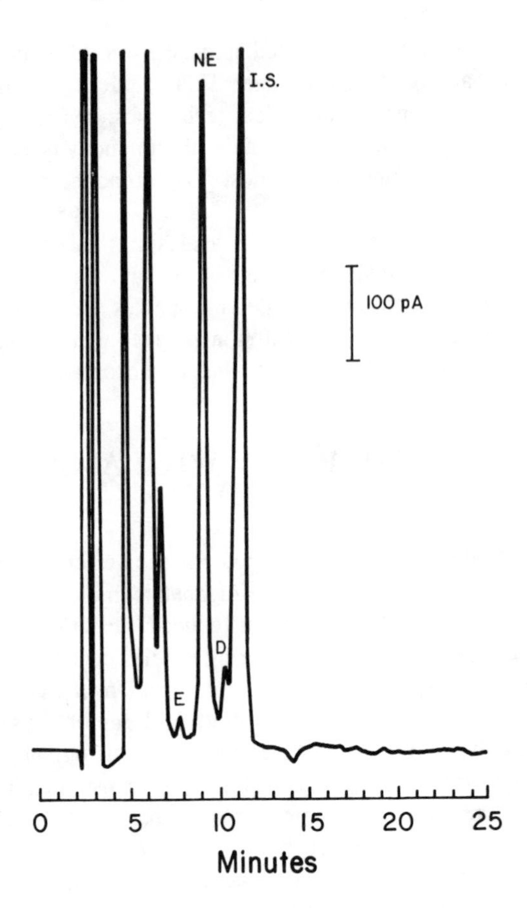

FIGURE 17. Plasma catecholamines separated on a silica-based weak cation-exchange column (Bio-Rad® clinical 4.6 × 150 mm). Peaks: E is epinephrine, NE is norepinephrine 423 pg/ml, D is dopamine 28 pg/ml. Eluent: 70 m*M* citrate adjusted to pH 6.4 with H_3PO_4, 15% CH_3CN at 0.55 ml/min. Detection: DC amperometric with glassy carbon electrode at 0.4 V vs. Ag/AgCl. (From Kochi, D. D. and Polzin, G. L., *J. Chromatogr.*, 386, 19, 1987. With permission.)

current. By using a DC or very low frequency potential program, the interfacial capacitance C has time to be fully charged. The result is that the majority of the potential is applied across the interface between the working electrode and the solution, and not across the solution. This can be ensured by proper cell design and by adding a conducting salt called a supporting electrolyte to the mobile phase, usually in the range of 10 to 100 m*M*. Most ion exchange eluents already contain an electrolyte solution such as an acid, base, buffer, or salt.

Amperometric detector cells are classified according to the manner in which the flowing stream impinges upon the working electrode. Two designs are in common usage. They are the thin-layer design and the wall-jet design. The thin-layer design is the most commonly used. The eluent flows in a thin channel parallel to the surface of a flat disk electrode. The resulting smooth flow produces a low noise level. In the wall-jet design, the eluent flows from a thin tube perpendicular to the electrode surface and impinges directly on the electrode. This results in a higher coulometric efficiency, that is a higher proportion of the analyte molecules reach the surface of the electrode to be oxidized or reduced. Researchers have

TABLE 2
Species Detected by Amperometry

Species	Electrode[a]
Oxidations	
Aromatics: phenols, catechols, catecholamines, aromatic amines	GC, Pt
Aliphatics: ascorbic, fumaric, oxalic acids	GC, Pt
Inorganics: arsenite, nitrite, sulfite, thiosulfate, iodide	Pt, GC
Species forming complexes with silver: halides, sulfide, cyanide, thiols	Ag
Pulsed amperometric detection of carbohydrates, amines, amino acids, organosulfur species	Au, Pt
Pulsed amperometric detection of alcohols, glycols, aldehydes, formic acid	Pt
Reversed pulse detection of amalgam forming metals: Pb, Cd, Cu, Zn, Tl	Hg
Reductions	
Aromatic nitros, quinones	GC, Hg, Au, Ag

[a] Listed in order of preference.

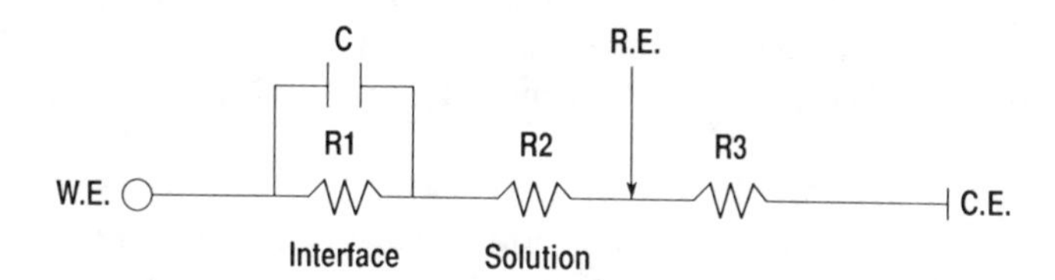

FIGURE 18. Equivalent circuit for amperometric detection. C represents the interfacial capacitance at the working electrode (W.E.), R1 the resistance to oxidation or reduction at the electrode surface, R2 the uncompensated solution resistance, and R3 the solution resistance compensated by the potentiostat. R. E. is the reference electrode; C. E. is the counterelectrode.

attempted to compare the performance of several different commercially available cells by measuring factors such as signal-to-noise ratio and linear dynamic range.[53,54] It appears from these reports that the performances of the cells are roughly equivalent.

The current resulting from oxidation or reduction of analyte molecules is dependent on many factors, the most important of which is the concentration of analyte. Other factors include temperature, the surface area of the working electrode, and the linear velocity of the flowing stream over the surface of the working electrode. Many of these factors vary from cell to cell, and there is currently no convention for normalizing cell current to a standard cell.

Cells are designed so that the electrical resistance between the working electrode and either the counterelectrode or reference electrode is as low as possible. This results in a wide linear dynamic range. In the thin-layer cell, low resistance is accomplished by locating the counterelectrode directly across the thin-layer channel from the working electrode.

B. COULOMETRIC DETECTOR CELLS

Coulometric detector cells have a much larger working electrode surface area than amperometric cells. This causes 100% of the analyte molecules to be oxidized or reduced. The main advantages to quantitative oxidation or reduction of analyte are a higher signal level (but not always a higher signal-to-noise level), and the ability to determine the concentration of analyte from Faraday's law simply by integrating the peak. Unlike amperometric detector cells, peak areas from coulometric detector cells should not be dependent on the cell.

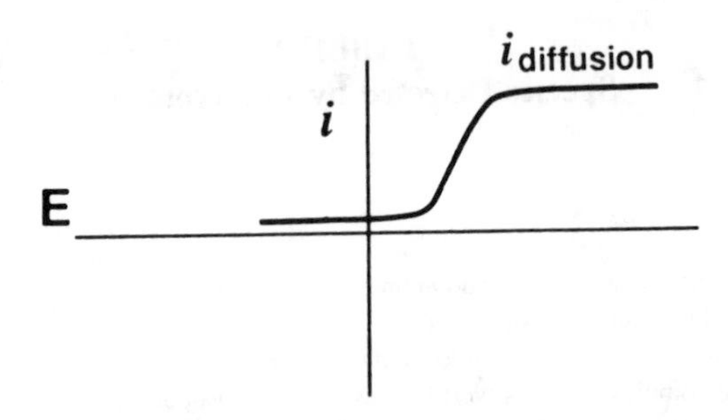

FIGURE 19. Peak current (i) as a function of applied potential (E) for a typical electroactive analyte. When the analyte molecules are oxidized (or reduced) instantaneously at the working electrode surface, the current is limited by their diffusion rate.

Because working electrodes in coulometric detector cells are often made out of graphite sponges, they can not be cleaned by mechanical polishing, as amperometric detector electrodes are. Cleaning must be done by chemical means. There are very few references to the use of coulometric detector cells in ion-exchange chromatography. This could be because amperometric detector cells are easier to manufacture and to clean.

C. DEPENDENCE OF CURRENT ON POTENTIAL

An example of the dependence of cell current on the potential for oxidation of a typical analyte in an amperometric cell is illustrated in Figure 19. The current vs. potential plot is generated by making multiple injections of analyte and increasing the potential after each injection. Peak height or area is plotted on the Y axis. (A similar plot is produced from other hydrodynamic voltammetric measurements such as rotated disk voltammetry.) At low potential, there is no oxidation. As the potential is increased, it approaches a level high enough to cause oxidation of a percentage of analyte molecules and the current increases. At a higher potential, 100% of the analyte molecules reaching the surface of the working electrode are oxidized, and the current is no longer dependent on potential. Since the current is now limited by the rate at which the analyte molecules are transported to the surface of the working electrode, and since this rate is largely dependent on diffusion, this maximum current is called the diffusion limited current. Amperometric detection using a single applied potential is called DC amperometry. For DC amperometric detection or for coulometric detection, the optimum potential is the lowest which will produce a diffusion limited current. Increasing the potential beyond this value will increase only the noise, and not the signal.

D. WORKING ELECTRODE MATERIALS

Of the many different materials that can be used as working electrodes, five are most commonly used. These are carbon, gold, silver, platinum, and mercury. The precious metal electrodes are very high purity solid metals. Mercury is much more difficult to use than the solid electrodes. Because of the high velocity of the flowing stream inside the cell, the dropping mercury electrode used in polarography is impractical. The high flow velocity also sweeps away the mercury drops formed on thin mercury film glassy carbon electrodes used in anodic stripping voltammetry. A commercially available flow-through cell using mercury is the static mercury drop electrode. One innovative use of a mercury electrode is the reverse-pulse detection of amalgam-forming transition metals.[55] There are numerous electrode materials based on carbon. Of these, glassy carbon, a hard graphitic substance, is the most frequently used. A carbon paste electrode is made by embedding powdered graphite in an inert matrix such a Nujol. It is reported to have a lower noise level than glassy carbon.[53]

TABLE 3
Potential Limits in Acidic and Basic Solutions vs. Ag/AgCl Reference Electrode[a]

Working electrode	Solution (0.1*N*)	Negative limit (volts)	Positive limit (volts)
Glassy carbon (GC)[b]	KOH	(−1.5)	(+0.6)
	$HClO_4$	(−0.8)	(−1.3)
Gold (Au)	KOH	−1.25	+0.75
	$HClO_4$	−0.35	+1.1
Silver (Ag)	KOH	−1.2	+0.1
	$HClO_4$	−0.55	+0.4
Platinum (Pt)	KOH	−0.90	+0.65
	$HClO_4$	−0.20	+1.30
Mercury (Hg)	KOH	−1.9	−0.05
	$HClO_4$	−1.1	+0.60

[a] These are approximate values. Potential limits in neutral solutions will be intermediate.
[b] Unlike metallic electrodes, the potential limits for the glassy-carbon electrode do not cut off sharply. The noise and background level will differ from application to application and must be determined experimentally.

From Rocklin, R. D., *LC Mag.*, 2, 588, 1984. With permission.

The main disadvantage of this electrode is that it disintegrates when organic solvent is added to the mobile phase. To avoid this problem Kel-F graphite composite[56] and carbon-polyethylene[57] electrodes were invented.

All solid electrodes must be periodically cleaned by mechanical polishing to remove products from oxidation reactions which coat the surface of the electrode, and to remove built-up surface oxide. Failure to clean the electrode produces a loss in coulometric efficiency, and therefore a low signal.

The choice of working electrode material for a given application is dependent on three factors:

1. The potential limits for the working electrode in the eluent
2. The involvement of the electrode itself in the electrochemical reaction
3. The kinetics of the electrode transfer reaction

1. Potential Limits

The negative potential limit is the potential at which the eluent or supporting electrolyte is reduced. At the positive potential limit, the eluent, the supporting electrolyte, or the electrode itself can be oxidized. Because these reactions will produce current far in excess of the analytical redox reaction, the potential used to detect the analyte must be within these limits. Table 3 lists the potential limits for the five common electrode materials in acidic and basic solutions. The potential limits are strongly affected by the pH of the eluent. Negative potential limits are more negative in base and more positive in acid. Conversely, positive limits are more positive in acid and more negative in base. In other words, the usable potential window shifts negative in basic solutions and positive in acidic solutions.

As the applied potential approaches the potential limit, the noise will increase as the background current increases. On metal electrodes, there is a sharp increase in background current as the potential limit is approached. On glassy carbon, the increase in background current is more gradual. Because the maximum applied potential that can be used is determined by the required signal-to-noise ratio, the values listed in Table 3 are only a rough guide. For some applications using glassy carbon, it may be necessary to exceed these limits.

The largest positive potential limits are obtained on glassy carbon and platinum. Accordingly, oxidations are generally performed using one of these two materials. The largest negative potential limits, listed in order, are obtained on mercury, glassy carbon, silver, and gold. Because of the ease of reducing hydrogen ion to hydrogen gas on a platinum electrode, platinum has a poor negative potential limit and is generally not used for reductions.

When potentials are used that are more negative than approximately 0.2 V vs. an Ag/AgCl reference electrode, a high background current is caused by the reduction of molecular oxygen dissolved in the eluent. This background current can be greatly reduced by degassing the eluent.

2. Involvement of the Electrode in the Redox Reaction

The reaction mechanism for the oxidation of many analytes is the transfer of electrons from the analyte molecules to the electrode. The electrode acts as an inert electron sink, and is otherwise not involved in the oxidation reaction. When this is the reaction mechanism, carbon is often the preferred electrode material. Examples include the detection of catecholamines, or the detection of nitrite.[58] In contrast, silver, gold, and mercury can be oxidized in the presence of complex or precipitate-forming ions. For the detection of these ions, the working-electrode material is directly involved in the reaction and is actually slowly consumed. For example, silver can be oxidized to silver cyanide in the presence of cyanide ion.[59] This reaction takes place at a much lower potential than the oxidation of cyanide to cyanate at a platinum electrode. The ability to use a lower applied potential increases the selectivity of the analysis, as fewer other species will be oxidized. Also, noise caused by the oxidation of trace contaminants in the eluent will be decreased at the lower applied potential. Silver is therefore the electrode of choice for the detection of cyanide. A silver electrode can also be used to detect other complex or precipitate-forming ions. These include bromide,[59,60] iodide,[59,61] sulfide,[59,62] sulfite, and thiosulfate. Figure 20 shows the detection of low levels of sulfide and cyanide separated by anion exchange and detected with a silver electrode. Minimum detection limits are approximately 1 ppb for sulfide and 2 ppb for cyanide. One disadvantage to the use of silver, gold, or mercury for oxidations is that the presence of halides in the eluent will greatly decrease the positive potential limit. Halides can usually be replaced by nonreacting anions such as acetate, perchlorate, nitrate, phosphate, or sulfate.

Another important case where the electrode acts as more than just an electron sink is the pulsed amperometric detection of carbohydrates, amines, organo-sulfur species and alcohols, described below. For this type of detection, adsorption of the analyte molecules on the surface of the electrode is necessary for detection.

3. Kinetics of the Electron-Transfer Reaction

For a kinetically fast electron-transfer reaction, the ratio of the oxidized to the reduced form of the analyte species will be described by the Nernst equation:

$$E = E^\circ + \frac{0.059}{n} \log \frac{[OX]}{[RED]}$$

where E is the applied potential and n is the number of electrons transferred. If the reaction rate is fast, then the reaction is said to be reversible. Note that if E is set 0.118 V positive of E°, then (for n = 1) the ratio of the oxidized to the reduced form of the analyte at the electrode surface will be 100:1. Nearly all of the analyte that reaches the electrode will be oxidized. The limiting current will have been attained, and no advantage will be gained

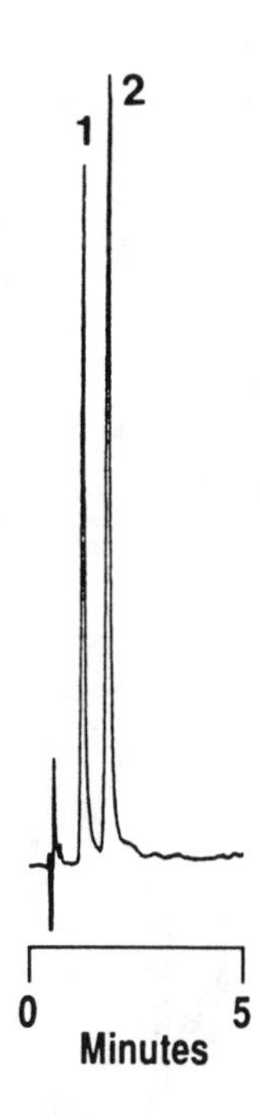

FIGURE 20. Anion exchange separation and DC amperometric detection of (1) sulfide, 50 ppb; and (2) cyanide, 100 ppb. A silver working electrode was used at O V vs. Ag/AgCl. (Reprinted with permission from Rocklin, R. D., *LC Mag.*, 2, 588, 1984. With permission.)

from a further increase in the applied potential. For a slow electron-transfer reaction, a potential considerably in excess of E° is required to drive the reaction at a fast rate. This excess voltage is called overpotential, and the reaction is said to be irreversible. The more irreversible the redox reaction, the greater the applied potential must be. This will result in more noise and less selectivity. The oxidation or reduction of many species is more facile on one electrode material than on another. This is particularly true for small inorganic species, many of which can be oxidized or reduced much more easily on platinum than on carbon. For example, platinum electrodes are used to detect arsenite,[40] iodide,[63] and sulfite.[64,65]

A major consideration when choosing an electrode material is its ability to maintain an active surface. Electrodes will develop a layer of surface oxide at positive applied potentials. This build-up will inhibit the oxidation of the analyte, often resulting in a decreasing response with repeated injections. The active surface can be renewed by polishing the electrode. Carbon electrodes are more resistant to poisoning by oxide formation than are platinum electrodes, and do not need to be polished as often. It is for this reason that carbon is used far more extensively than any other electrode material. Pulsed amperometric detection solves this problem for noble metal electrodes by pulsing to a reducing potential after each current measurement.

E. PULSED AMPEROMETRIC DETECTION

The development of pulsed amperometric detection grew from the need to detect carbohydrates. Since most carbohydrates contain no UV chromophore, UV absorbance detection can only be used at very low wavelength. The detection of carbohydrates at 210 nm is insensitive and nonselective. Refractive index detection, also insensitive and nonselective,

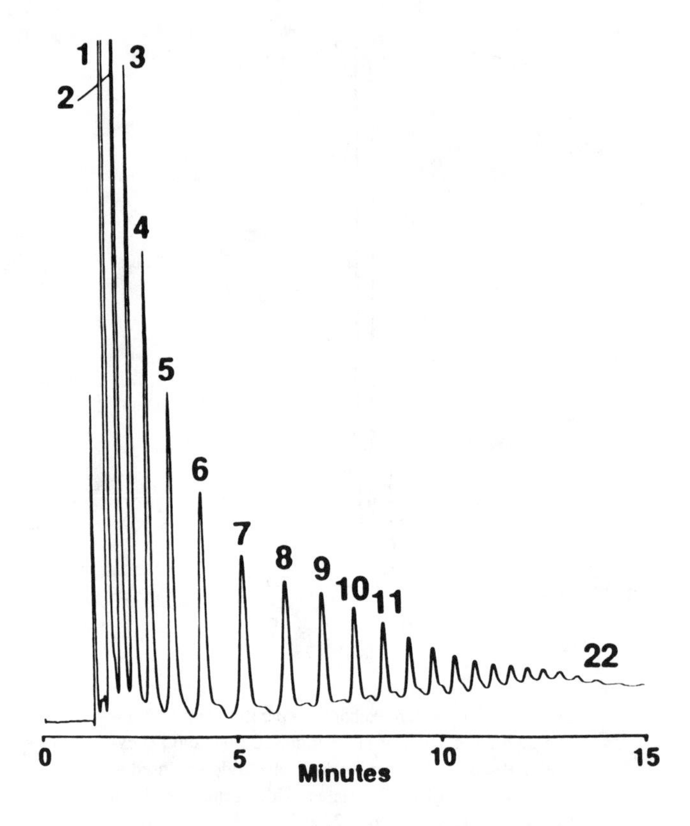

FIGURE 21. Oligo- and polysaccharides from hydrolyzed corn starch separated by anion exchange at pH 13.3 and detected by pulsed amperometric detection. Each number represents the degree of polymerization (DP, the number of glucose units in the saccharide chain). Column: Dionex CarboPac PA1, 4 × 250 mm, 10 μm latex-coated resin. Eluent: constant 0.15 *M* NaOH with gradient from 0.25 to 0.5 *M* $NaCH_3CO_2$ in 10 min at 1 ml/min. (Reprinted with permission, Dionex Corp., copyright 1987.)

had been the most commonly used detection method. Pulsed amperometric detection is replacing these optical methods for carbohydrates, and is now being used for other non-chromophoric molecules containing amine or sulfur functional groups.

Although carbohydrates can be oxidized at gold and platinum electrodes, the products of the oxidation reaction poison the surface of the electrode, inhibiting further analyte oxidation. By repeatedly pulsing between high positive and negative potentials, a stable and active electrode surface can be maintained. Pulsed amperometric detection has now emerged as the most sensitive and selective method for the detection of carbohydrates.[66-68] An example is the determination of oligo- and polysaccharides of hydrolyzed corn starch shown in Figure 21. Many new applications are being developed for the analysis of foods and beverages[69] and in biochemical research.[70-72] Other species which can be detected by pulsed amperometry include alcohols,[73] aldehydes,[74] amines[66,75] (primary, secondary and tertiary, including amino acids), and many organic sulfur species.[66,76,77] Thiols and mercaptans can be detected, but fully oxidized sulfur species such as sulfates and sulfonates cannot be.

1. Cyclic Voltammetry

To understand the mechanism of pulsed amperometric detection, it is first necessary to study the oxidation of an analyte using a conventional electrochemical technique such as cyclic voltammetry. For this experiment, the potential applied to the working electrode of

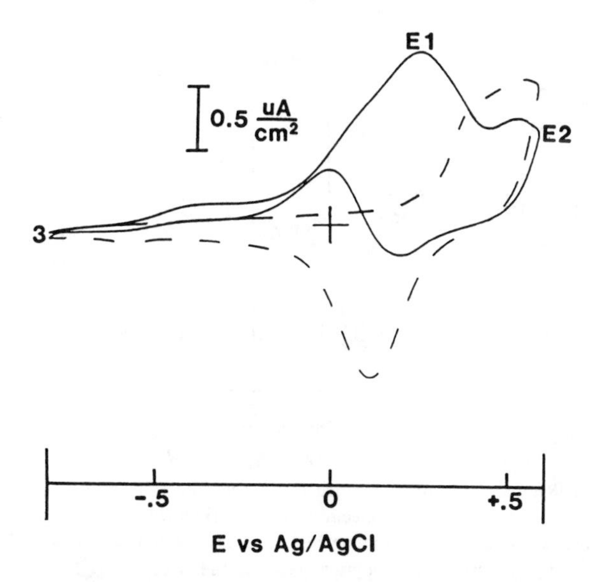

FIGURE 22. Cyclic voltammetry of 1 m*M* glucose on gold working electrode. Dashed line is 0.1 *M* NaOH supporting electrolyte, solid line is with addition of glucose. 0.2 V/sec sweep rate. (Reprinted with permission, Dionex Corp., copyright 1987.)

a standard three electrode cell is repeatedly cycled between positive and negative limits at a fixed sweep rate, and the current is plotted using an X-Y recorder. The cyclic voltammetry of glucose in 0.1 *M* sodium hydroxide using an Ag/AgCl reference electrode is shown as an example in Figure 22. The dashed line in the figure is the current resulting from the 0.1 *M* NaOH supporting electrolyte in the absence of glucose analyte; i.e., the background current. Beginning at −0.8 V and sweeping in a positive direction, the background current is flat until approximately 0.25 V, where oxidation of the surface of the gold electrode to gold oxide begins. Following reversal of the potential sweep direction at 0.6 V, the gold oxide is reduced back to gold, with the current peaking at 0.1 V.

With glucose added to the solution, the current rises slightly as the potential is swept in a positive direction from −0.8 V and remains unchanged until glucose oxidation begins. This causes the current to rise at −0.15 V towards a peak at 0.26 V. The current then decreases for two reasons. First, the concentration of glucose at the electrode surface has been depleted because much of it has been oxidized. Second, the formation of gold oxide inhibits further glucose oxidation. On the reverse scan, the current actually reverses from reducing to oxidizing at the onset of the gold oxide reduction. As soon as the reduction of gold oxide back to gold begins, oxidation of glucose also begins.

2. Choice of Potentials for Pulsed Amperometry

If single potential (DC) amperometric detection were used, the appropriate applied potential would be approximately 0.2 V. This is the potential at which the glucose oxidation current is the highest and the background current the lowest. However, the use of a single potential results in rapidly decreasing sensitivity as an oxide layer forms and products from the oxidation reaction coat and poison the electrode surface. This problem is solved by first measuring the oxidation current near 0.2 V, pulsing the potential to 0.6 V and then back to −0.8 V. The action of repeatedly forming and removing the metal oxide surface layer cleans the electrode surface and maintains an active and stable surface. A three-step program using these potentials is shown in Figure 23. The potentials are E1, E2, and E3. The current is measured at E1 (0.1 V) and the value held in a sample-and-hold amplifier until the next

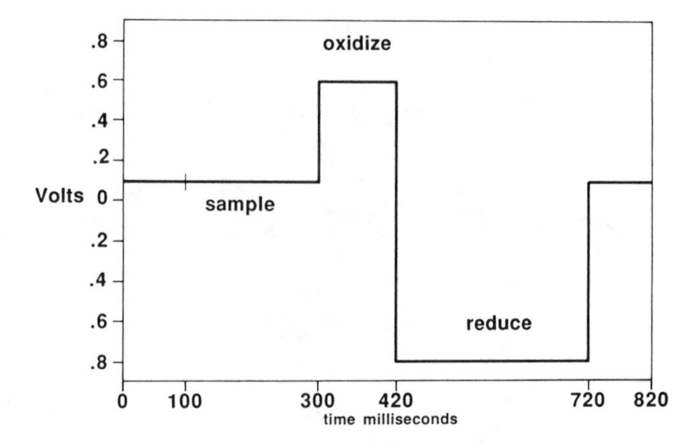

FIGURE 23. Potential sequence for pulsed amperometric detection of carbohydrates using a gold electrode at high pH showing the current sampling period and the alternating high and low cleaning potentials. (Reprinted with permission, copyright 1987, Dionex Corp.)

current measurement (0.1 V produces slightly better signal-to-noise ratio than 0.2 V). The step from E3 back to E1 is from −0.8 V to 0.1 V. This charges up the electrode/solution interfacial capacitance. The carbohydrate oxidation current is measured after a delay which allows the charging current to decay. Signal-to-noise ratios can be improved by integrating the current for several hundred milliseconds and reporting the average value to the detector. Pulsed amperometric detection at a gold electrode is a sensitive method for all carbohydrates of molecular weight less than several thousand.

Carbohydrates can only be detected by pulsed amperometry in high pH solutions, approximately above pH 11. Carbohydrates are also very weak acids with pKa around 12 and are easily separated on high efficiency anion-exchange columns in mobile phases of pH 11 to 13. The combination of anion-exchange separation with pulsed amperometric detection is a powerful method for the determination of carbohydrates. Although there can be some interference from amines and certain sulfur species, the technique is very sensitive and selective. Similar potential vs. time programs are used to detect amines and sulfur species. For molecules containing these functional groups, it is necessary to extend E1 to potentials where metal oxide formation occurs. Since carbohydrates are also detected under these conditions, the detection of amines and sulfur species is somewhat less selective than the detection of carbohydrates.

A new variant of pulsed amperometric detection is called Potential Sweep Pulsed Coulometric Detection (PS-PCD)[78] (also called Integrated Amperometric Detection). This method decreases the effect on the baseline of changes in pH, allowing moderate pH gradients to be performed without large shifts in the baseline. The potentials at which metal oxide is formed and reduced shift positive as pH is reduced. Using PS-PCD, the current is integrated while the potential is swept across the oxide formation wave and also during the reverse sweep across the oxide reduction wave. The net charge is approximately zero. The advantage is that the total current is less sensitive to the changes in oxide formation and reduction potentials caused by changes in pH. If the standard Ag/AgCl reference electrode is replaced with a pH electrode, the changes in oxide formation and reduction potentials caused by changes in eluent pH are also minimized. The combination of PS-PCD with a pH reference electrode can be used successfully to detect both primary and secondary amino acids separated by anion exchange, even though a succession of three buffers at different pH is used to elute the amino acids (Figure 24).[79]

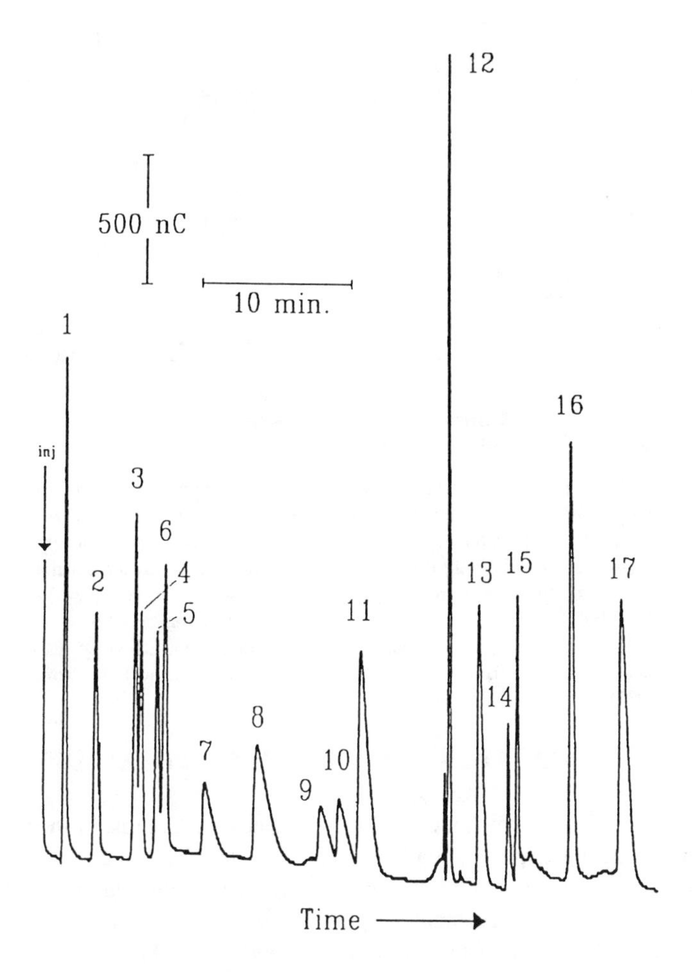

FIGURE 24. Anion-exchange separation and integrated amperometric detection of amino acids (25 nmol each). Peaks: (1) Arg; (2) Lys; (3) Gln; (4) Asn; (5) Thr; (6) Ala; (7) Gly; (8) Ser; (9) Val; (10) Pro; (11) Ile; (12) Leu; (13) Met; (14) His; (15) Phe; (16) Glu; (17) Asp. Column: Dionex AS8, 4 × 250 mm, 10 μm latex-coated resin. Eluent: hydroxide/borate/acetate gradient; 1 ml/min flow rate. (From Welch et al., *Anal. Chem.*, 61, 555, 1989. Copyright, American Chemical Society, 1989. With permission.)

F. POTENTIOMETRIC DETECTORS

Unlike amperometric detectors, potentiometric detectors measure the change in potential across an indicator electrode membrane. Concentration (actually activity) is related to potential through the Nernst equation. Indicator electrodes which can be used include the familiar pH electrode and other ion-specific electrodes.[80] Coated or bare metal electrodes such as silver[81] or copper[82] have been shown to detect numerous classes of ions. Although publications have been written on the use of potentiometric detectors, the publications focus on investigating potentiometric detection, rather than on finding solutions to analytical problems. Potentiometric detectors are slower to respond than other more commonly used detectors, are generally less sensitive, and are not available commercially for use in ion chromatography. More importantly, there are no important applications where potentiometric detection provides a significant advantage over the other forms of electrochemical detection, or over optical detection methods.

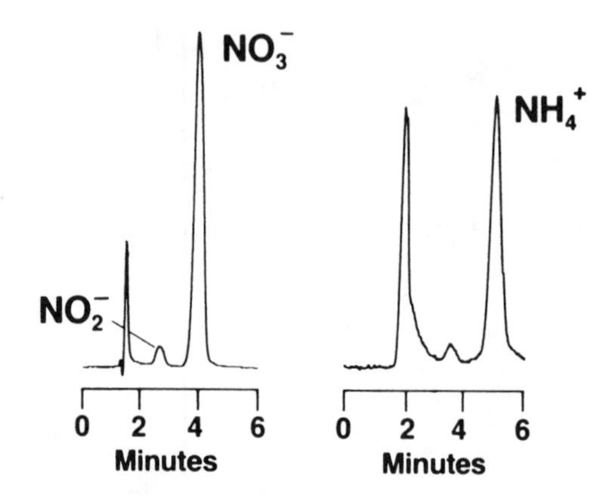

FIGURE 25. Determination of 2 ppm nitrite, 38 ppm nitrate, and 19 ppm ammonium in a 10% KCl soil extract using simultaneous UV absorbance (215 nm) for the anions and fluorescence for ammonium ion. Ammonium was derivitized with o-phthalaldehyde and 2-mercaptoethanol to form the fluorescent product; Ex = 230, Em = 400 nm. Sample was diluted 1/10, filtered, and injected. Column: Dionex IonPac CS5, 4 × 250 mm, 15 μm latex-coated anion/cation-exchange resin. Eluent: 35 m*M* KCl at 1 ml/min. (From Dionex Corp., Application Update 118, (Nov. 1987), Sunnyvale, CA. With permission.)

III. ION CHROMATOGRAPHY WITH OPTICAL DETECTION

There are many analytes which are more easily detected using one of the forms of optical detection. In fact, there are many situations when conductivity detection is used when UV absorbance or fluorescence would be preferable. This is probably due to the availability of conductivity detectors on ion chromatography instruments and to the familiarity of the analytical chemist with the use of this detection method. Conversely, many practitioners of HPLC attempt to detect everything with UV absorbance, even when amperometric or conductivity detection would be preferable.

Since the principles of optical detection are covered in many books on HPLC, it is not necessary to cover the subject here (see Reference 83 for a recent discussion of detectors). Instead, this section will present applications for which optical detectors are commonly used in ion chromatography. This includes the use of UV visible detectors for direct absorbance, indirect absorbance, and absorbance following postcolumn derivatization, and also fluorescence and refractive index detection.

A. DIRECT ABSORBANCE

Direct UV or visible absorbance detection is an alternative to conductivity detection for chromophoric molecules. For molecules which absorb UV or visible light, sensitivities of the two detection methods are generally comparable, with conductivity more sensitive for small molecules and UV absorbance more sensitive for large aromatic organic acids. However, UV absorbance detection is often much more selective than conductivity detection, because many of the samples commonly analyzed by ion chromatography contain high concentrations of inorganic salts. Typically, sodium, chloride, and sulfate are present in high concentrations in most environmental and biological samples. UV light-absorbing anions such as nitrite and nitrate are easily determined using absorbance detection at 215 nm, regardless of the concentrations of the non-UV light-absorbing anions chloride and sulfate.

An example of the selectivity of UV absorbance is the detection of nitrite and nitrate in soil extracts, shown in Figure 25.[84] Since soil particles are inorganic ion exchangers, the

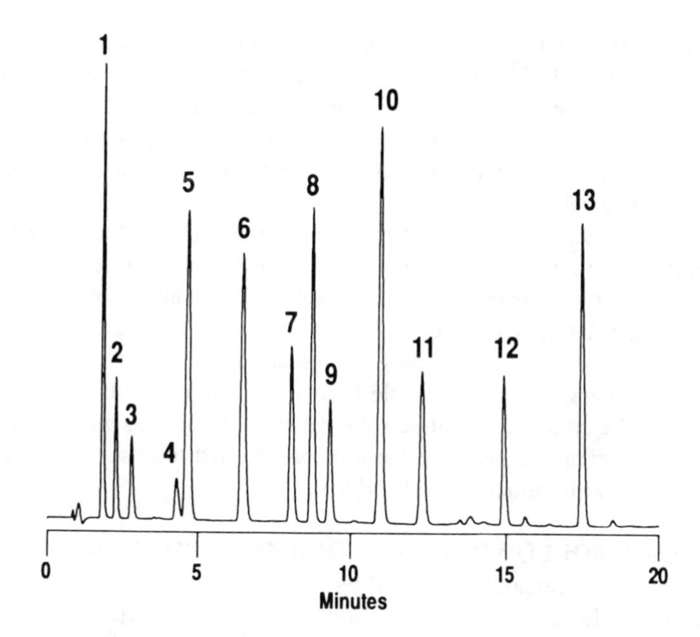

FIGURE 26. Anion-exchange separation of aromatic anions with UV detection at 254 nm. Peaks: (1) Benzoic acid 40 ppm; (2) benzenesulfonic 40; (3) p-toluenesulfonic 40; (4) p-chlorobenzenesulfonic 40; (5) p-bromobenzoic 20; (6) 3,4-dinitrobenzoic 10; (7) phthalic 20; (8) terephthalic 6; (9) p-hydroxybenzoic 2; (10) p-hydroxybenzenesulfonic 2; (11) gentisic 10; (12) trimesic 20; (13) pyromellitic 10 ppm. Column: Dionex OmniPac PAX-100, 4 × 250 mm, 8 μm latex-coated resin. Eluent: Constant 20% CH_3CN, 1 m*M* NaOH, 0.05 to 0.4 *M* NaCl gradient in 20 min; 1 ml/min flow rate.

analyte ions must be removed from the soil by ion exchange with a strong salt solution. A 10% KCl solution is commonly used. In this matrix, nitrite and nitrate can not be detected by conductivity, since the high concentration of chloride in the extract solution overwhelms the detector. UV absorbance provides a simple and highly selective detection method.

There are numerous aromatic anions and cations such as benzenesulfonate and nicotinamide which can easily be detected. These are commonly determined by standard HPLC: reversed phase (or ion pair) separation on silica-based C-18 columns with UV detection. The advantages of ion exchange have rarely been utilized. An example of the ability of high-performance ion-exchange resins to separate aromatic acids is shown in Figure 26. This separation required the use of acetonitrile organic solvent to minimize hydrophobic interaction between the aromatic acids and the PS-DVB resin. The column is packed with a surface functionalized macroporous resin which does not shrink or swell as the organic solvent content is changed.[85]

In addition to aromatic ions, many inorganic ions absorb UV light. Examples are nitrite, nitrate, bromide, iodide, sulfide, chromate, and numerous metal complexes such as ferrocyanide.

B. INDIRECT PHOTOMETRIC DETECTION

Aromatic compounds absorb UV light, and as eluents lend themselves to a detection method known as "vacancy chromatography" or "indirect photometric chromatography". The solution flowing through the column absorbs UV light in the region 250 to 270 nm. Common inorganic anions do not absorb at these wavelengths. Consequently, when an inorganic anion analyte elutes, there is a drop in the eluent ion concentration and hence a drop in the UV absorbance. (Recall that the ionic strength of the column effluent is constant. See Figure 4.) This method of detection was introduced by Small and Miller.[86] It is versatile and potentially sensitive and accurate, but it suffers from the same drawback the nonsuppressed conductivity detection does, namely, that one is reading small changes against a large background signal. To achieve a maximum signal to noise ratio, the pump and UV detector must be extremely stable.

With buffered solutions of benzoate, 4-hydroxybenzoate, phthalate, or any other aromatic acid eluent, one should remember that the undissociated acid is present and that it can be adsorbed by the polystyrene network of the ion exchanger. This adsorption has nothing to do with ion exchange, but it is a property of the polymer and its interaction with uncharged solute molecules. The partition of uncharged aromatic acid between the polymer and the flowing solution is changed when a sample is injected and a disturbance or concentration pulse travels along the column. If it affects the detector one sees as an extraneous ''system peak'' which may overlap the peaks of the ions one wished to observe. System peaks are especially noticeable with indirect UV absorbance.[87]

C. ABSORBANCE FOLLOWING POSTCOLUMN DERIVATIZATION

Many species which cannot be detected by any of the previously described techniques can be derivatized to form chromophoric complexes which are detected by UV or visible absorbance. Postcolumn derivatization is used to detect metals, silicate, phosphate, and polyphosphonate sequestering agents. The postcolumn reagent is pumped into a mixing T to mix with the column effluent. Adding the reagent through the walls of a membrane reduces the effects of pump pulsations, decreasing noise. For slow reactions, a heated reaction coil precedes the detector.

1. Metals

Metal ions can exist in several different forms. The factors which determine the form of the metal ion are the extent of complexation and the oxidation state. In many samples, metal ions are present as hydrated metal atoms. Hydrated metal ions are usually written without the water ligands included in the chemical formula. For example, chromic ion, Cr^{3+}, is actually the hexaquo complex $Cr(H_2O)_6^{3+}$. Hydrated metal ions can also be complexed by weak ligands such as organic acids or amino acids. These ligands are generally displaced by the complexing agents used in ion-chromatography eluents. Therefore, the methods described in this section will determine the total of both hydrated and weakly complexed metal ions. Since most weakly complexed metals will precipitate in a suppressor, the metals are detected instead by measuring the light absorbed by complexes formed in postcolumn reactions.

More powerful complexing agents present in samples will not be displaced during chromatography. Since most complexing agents are anionic, the resulting metal complex will have a net negative charge. An example is the trivalent anion $Fe(CN)_6{}^{3-}$. Metals can also exist as oxyanions. These are metals with oxygen atoms as ligands. An example is the molybdate ion $MoO_4{}^{2-}$. The chromatography of these ions will be different from the hydrated and weakly complexed metal ions, so they must be determined using different methods. Because of their inability to react with postcolumn reagents, metal oxyanions and other stable metal complexes are usually determined by anion-exchange chromatography with either direct UV or conductivity detection.

Ion exchangers generally have poor selectivity for transition metals and lanthanides. Selectivity is achieved by adding an anionic chelating agent to the eluent. The chelater is chosen to form weak complexes with the metal. Since the strength of the complex is different for each metal, the average number of ligands on each metal will be different, and therefore the net charge on each metal will be different. The differences in net charge provide the

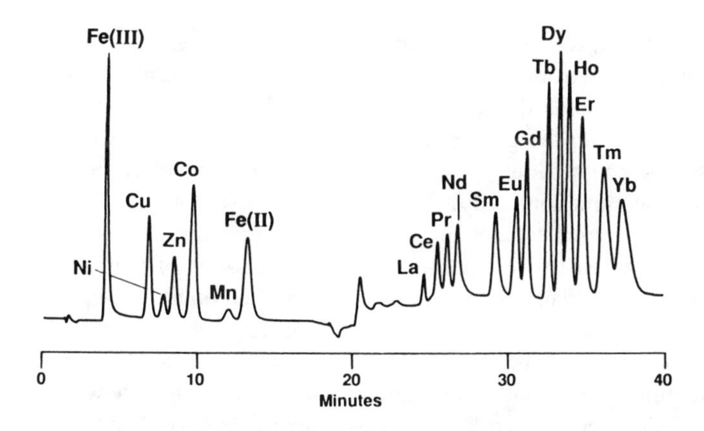

FIGURE 27. Separation of transition and lanthanide metals. Lanthanides are 7 ppm each. Column: Dionex IonPac® CS5, 4 × 250 mm, 15 μm latex-coated anion/cation-exchange resin. Eluent is a multistep gradient of PDCA (separates transition metals); and oxalic and diglycolic acids (separate lanthanides), 1 ml/min flow rate. Detection is visible absorbance at 520 nm following postcolumn addition of PAR. (From Heberling et al., *RD,* Sept., 74, 1987. With permission.)

majority of ion-exchange selectivity. Some commonly used organic acid chelators, listed in order of their complexing strength, are hydroxyisobutyric, tartaric, citric, oxalic, and pyridine-2,6-dicarboxylic acids. Low concentrations of the weaker chelators will diminish the net positive charges on the metals, allowing them to be separated on cation-exchange columns. If stronger chelating agents are used in higher concentrations, the net charges on the metals are negative, and anion-exchange columns are used.

The method of separating metals on a cation-exchange column with selectivity increased by a chelating agent was used in 1947 to determine lanthanides.[88] In this early demonstration of high-performance ion-exchange liquid chromatography, the authors packed a 0.6 × 120 cm column with 44 to 55 μm sulfonated PS-DVB resin. Since radioactive isotopes were being separated, a gamma-ray counter was used as a detector.

With few exceptions, most transition and lanthanide metals do not absorb UV light strongly enough for direct UV absorbance detection to be practical. So detection is accomplished by postcolumn addition of the metallochromic indicator 4-(2-pyridylazo)resorcinol (PAR).[89-91] The postcolumn reagent usually contains about 0.2 m*M* PAR dissolved in an ammonia/ammonium acetate buffer of molar concentration. The resulting complex has a λ_{max} at 520 nm, so a simple filter-based visible detector can be used. An example of the separation of seven transition metals and thirteen lanthanides is shown in Figure 27.[91] The determination of aluminum requires UV detection at 310 nm following postcolumn addition of 4,5-dihydroxy-*m*-benzenedisulfonic acid (Tiron).[92] Cr(VI) (chromate/dichromate) is detected at 520 nm following postcolumn addition of 1,5-diphenylcarbohydrazide.[93]

In recent years the most extensive study of lanthanide separations by ion exchange has been made by Cassidy, Elchuk, and colleagues. In an early paper[89] they compared, as stationary phases, two bonded silicas with a fully sulfonated, 8% crosslinked polystyrene resin of small particle size. The eluent was α-hydroxyisobutyric acid of pH 4.6, at concentrations that matched the exchanger capacities. For the fully sulfonated resin a gradient from 0.17 *M* to 1.0 *M* was used. For the bonded-silica cation exchanger Nucleosil SCX® the gradient was 0.018 *M* to 0.17 *M*. Nucleosil SCX® and the resin (Aminex A-5) gave equally good chromatograms, but Nucleosil®, particle size 5 μm, separated all the lanthanides in 18 min, while Aminex, swollen diameter 13 μm, took 28 min. Detection was by postcolumn reaction with Arsenazo I. In their later work, Cassidy's group favored dynamically coated ion exchangers (see Chapter 5, References 24 to 26).

As we have noted, the selectivity comes from the differences in stabilities of complex ions in solution, rather from ion-exchange selectivity. The exchanger is like an undiscriminating sponge that holds the ions while the complexing ligand makes them move. Lutecium, the element with the greatest atomic number and the smallest cation size (because of the "lanthanide contraction") forms the most stable complexes (because of Coulomb's law of electrostatic attraction) and therefore moves fastest, coming out of the column first. Lanthanum, with the smallest atomic number but the greatest ionic diameter, elutes last.

Since metals can easily be determined by several forms of atomic spectroscopy, it is useful to compare the advantages and disadvantages of chromatographic determination. For single metals, atomic absorption is much faster, usually more sensitive, and detects total metal, generally regardless of complexation or oxidation state. Chromatography has the ability to determine several metals in a single run. This is clearly shown in Figure 27. ICP-AES also has multi-element capability, but at much higher cost. For many samples, chromatographic separation minimizes problems from matrix interference, especially in brines and biological matrices. For example, the chromatographic method described above was used to determine copper and zinc in whole blood.[94] Matrix effects were minimal and sample pretreatment was simpler. Also, results from the chromatographic method were in close agreement with atomic absorption determination, detection limits and sensitivities were as good as electrothermal AA, and two metals could be determined simultaneously.

Unlike atomic spectroscopy, chromatographic determination of metals is species selective. Provided the complexed state of a metal is sufficiently inert, different complexes of the same metal may be separated. In some cases, different oxidation states of the same metal can be separated. However, oxidation or reduction of the metal ions on the column limits the accuracy of the measured oxidation state ratios.

2. Silicates

The detection of silicate is accomplished by postcolumn addition of an acidic solution of sodium molybdate. The absorbance of the silicomolybdate complex at 520 nm is monitored. This method is much more sensitive and selective than indirect conductivity, since the only other anion detected is phosphate. This technique is commonly used for trace determination of silicate in deionized water.[95]

3. Phosphates

Phosphorus forms a great variety of oxy-anions containing more than one phosphorus atom joined by P–P and P–O–P links. Phosphorus can be in the oxidation states +1, +3, +4, and +5. The +5 compounds, the polyphosphates, are separated by chromatography on a strong-base anion exchanger with sodium chloride eluents in the pH range 5 to 10. They are detected by postcolumn reactions. First the effluent is acidified and heated to hydrolyze polyphosphates to orthophosphoric acid, H_3PO_4; ammonium molybdate and a reducing agent are now added to produce molybdenum blue, whose light absorption is recorded. The longer the molecular chain, the more strongly the ion is retained. Cyclic polyphosphates (like hexametaphosphate, $Na_6P_6O_{18}$), are retained more strongly than their linear analogues. Linear phosphate oligomers, anions of $H_{n+2}P_nO_{3n+1}$, are beautifully separated with distinct peaks up to n = 30.[100,101] (Their separation by ion exclusion was mentioned above; see References 22 and 23.)

4. Sulfur Anions

The thionic acids have the general formula $HO_3S{\cdot}S_x{\cdot}SO_3H$, where x goes from zero to 4; that is, the number of sulfur atoms ranges from two to six. Anions of these acids, along with thiosulfate, $S_2O_3^{2-}$, are formed in mine tailings ponds by bacterial oxidation of sulfide minerals. They are separated chromatographically on a surface-functional, strong-base anion exchanger with 1.2 m*M* sodium citrate eluent (citrate, because the divalent sulfur anions are very strongly bound and must be displaced by a strongly bound anion, trivalent citrate) and they are detected by postcolumn reaction with 2.5×10^{-5} *M* cerium(IV) sulfate. Cerium(IV), which is not fluorescent, is reduced in acid solution to cerium(III), which is strongly fluorescent (excitation at 260 nm, emission at 350 nm). Thionate ions are first hydrolyzed in hot alkaline solution to sulfite and thiosulfate, then the solution is acidified and mixed with cerium(IV) sulfate. A continuous flow system is used with air segmentation and appropriate delay coils.[102,103] An interesting fact is that log k', the logarithm of the capacity factor (or the corrected retention volume), increases in a linear manner with *n*, the number of sulfur atoms. There is a linear relation between the free energy of retention and the ionic chain length.

5. Sequestering Agents

Polyphosphonate sequestering agents (these are organic derivatives of phosphorous acid) are detected following postcolumn addition of an acidic solution containing ferric ion.[96,97] The absorbance of the ferric ion at 330 nm is monitored. The mechanism of detection is not clear; however, the change in absorbance of ferric ion is probably caused more by the change in effluent pH in the volume of solution containing analyte as opposed to a complexation of the ferric ion by the analyte.

D. FLUORESCENCE DETECTION

Very few inorganic species fluoresce, so fluorescence detectors are rarely used in ion chromatography. The most important application of fluorescence detection in ion-exchange chromatography is the detection of primary amino acids following postcolumn derivitization with o-phthalaldehyde and 2-mercaptoethanol.[98] This method is extremely sensitive and responds to only primary amines and ammonia. Although there are few reports of the use of this detection method in ion chromatography, it should be the method of choice for all primary amines when high sensitivity and selectivity are required, whether they are separated by cation exchange, reversed phase, or ion pair. The ammonia from the KCl soil extract in Figure 25 was detected using this method. There is no interference from the sample matrix, which was 1% KCl.

One of the few inorganic ions that is fluorescent is cerium(III) and another is uranium(VI). Cerium(IV) is nonfluorescent and a strong oxidizing agent. Thus, post-column reaction with an acid solution of Ce(IV) could be a general-purpose method of detecting reducing species, organic or inorganic. It has been so used for phenols in polluted waters and for reducing substances in the ion-exchange chromatography of wastewater.[104] The application to sulfur anions was described above. However, the method is little used, perhaps for instrumental reasons (an acid Ce(IV) solution is very corrosive), perhaps because many substances, including trace transition-metal ions that could come from stainless-steel chromatographic fittings, quench fluorescence.

The fluorescence of Ce(III) is used for indirect detection in various ways. In suppressed chromatography of anions, the anions emerge from the suppressor as their acids. If the effluent is now passed through a short cation-exchanging column loaded with cerium(III), these ions are released into the solution and a peak of fluorescence results. High sensitivity is claimed by this procedure.[105] It is also possible to perform "indirect photometric detection" of cations (see above) by adding Ce(III) to the mobile phase and observing the drop in fluorescence that accompanies the elution of alkali-metal ions.[106]

E. REFRACTIVE INDEX DETECTION

For ions which can be detected by either conductivity, amperometry, UV absorbance, or fluorescence, these detection methods are generally more sensitive and much more selective than refractive index (RI) detection. Therefore, RI detection is rarely used in ion chromatography. However, it can be useful for species which are difficult to detect by other means. For example, an alternative to the postcolumn addition/UV absorbance method for polyphosphonate sequestering agents is their direct detection by refractive index.[99]

REFERENCES

1. **Beukenkamp, J. and Rieman, W.,** Determination of sodium and potassium, employing ion-exchange separation, *Anal. Chem.*, 22, 582, 1950.
2. **Foley, R. C. L. and Haddad, P. R.,** Conductivity and indirect ultraviolet adsorbance determination of inorganic cations in nonsuppressed ion chromatography using aromatic bases as eluents, *J. Chromatogr.*, 366, 13, 1986.
3. **Fritz, J. S., Gjerde, D. T., and Becker, R. M.,** Cation chromatography with a conductivity detector, *Anal. Chem.*, 52, 1519, 1980.
4. **Okada, T. and Kuwamoto, T.,** Ion chromatographic determination of silicic acid in natural water, *Anal. Chem.*, 57, 258, 1985, and Potassium hydroxide eluent for nonsuppressed anion chromatography of cyanide, sulfide, arsenite, and other weak acids, *Anal. Chem.*, 57, 829, 1985.
5. **Gjerde, J. T. and Fritz, J. S.,** *Ion Chromatography,* 2nd ed., Heuthig, New York, 1987.
6. **Gjerde, D. T., Fritz, J. S., and Schmuckler, G.,** Anion chromatography with low-conductivity eluents, *J. Chromatogr.*, 186, 509, 1979.
7. **Gjerde, J. T., Schmuckler, G., and Fritz, J. S.,** Anion exchange with low conductivity eluents. II., *J. Chromatogr.*, 187, 35, 1980.
8. **Schmuckler, G., Jagoe, A. L., Girard, J. E., and Buell, P. E.,** Gluconate-borate eluent for anion chromatography. Nature of the complex and comparison with other eluents, *J. Chromatogr.*, 356, 413, 1986.
9. **Okada, T. and Kuwamoto, T.,** Sensitivity of nonsuppressed ion chromatography using divalent organic acids as eluents, *J. Chromatogr.*, 284, 149, 1984.
10. **Golombek, R. and Schwedt, G.,** 2,4-Dihydroxybenzoic acid as a novel eluent in single-column ion chromatography, *J. Chromatogr.*, 367, 69, 1986.
11. **Stillian, J.,** An improved suppressor for ion chromatography, *LC Mag.*, 3, 802, 1985.
12. **Rocklin, R. D., Rey, M. A., Stillman, J. R., and Campbell, D. L.,** Ion chromatography of monovalent and divalent cations, *J. Chromatogr. Sci.*, 27, 474, 1989.
13. **Rocklin, R. D., Slingsby, R. W., and Pohl, C. A.,** Separation and detection of carboxylic acids by ion chromatography, *J. Liq. Chromatogr.*, 9, 757, 1986.
14. **Wetzel, R. A., Anderson, C. L., Schleicher, H., and Cook, G. D.,** Determination of trace level ions by ion chromatography with concentrator columns, *Anal. Chem.*, 51, 1532, 1979.
15. **Legrand, M., DeAngelis, M., and Delmas, R. J.,** Ion chromatographic determination of common ions at ultractrace levels in Antarctic snow and ice, *Anal. Chim. Acta,* 156, 181, 1984.
16. **Hill, R. and Lieser, K. H.,** Bestimmung von Alkali-und Erdalkaliionen in Regenwasserproben mit der Ionen Chromatographie, *Fresenius Z. Anal. Chem.*, 327, 165, 1987.
17. **Doury-Berthod, M., Giampaoli, P., Pitsch, H., Sella, C., and Poitrenaud, C.,** Theoretical approach of dual-column ion chromatography, *Anal. Chem.*, 57, 2257, 1985.
18. **Jones, W. R., Jandik, P., and Heckenberg, A. L.,** Gradient elution of anions in single column ion chromatography, *Anal. Chem.*, 60, 1977, 1988.
19. **Wheaton, R. M. and Bauman, W. C.,** Ion exclusion, *Ind. Eng. Chem.*, 45, 228, 1953.
20. **Turkelson, V. T. and Richards, M.,** Separation of the citric acid cycle acids by liquid chromatography, *Anal. Chem.*, 50, 1420, 1978.
21. **Tanaka, K., Ishizuka, T., and Sunahara, H.,** Elution behaviour of acids in ion-exclusion chromatography using a cation-exchange resin, *J. Chromatogr.*, 174, 153, 1979.
22. **Tanaka, K. and Ishizuka, T.,** Ion-exclusion chromatography of condensed phosphates on a cation-exchange resin, *J. Chromatogr.*, 190, 77, 1980.
23. **Waki, H. and Tokunaga, Y.,** Donnan exclusion chromatography: theory and application to the separation of phosphorus oxoanions or metal cations, *J. Chromatogr.*, 201, 259, 1980.
24. **Rich, W., Johnson, E., and Sidebottom T.,** U.S. Patent #4,242,097.
25. **Kreling, J. R. and DeZwaan, J.,** Ion chromatographic procedure for bicarbonate determination in biological fluids, *Anal. Chem.*, 58, 3028, 1986.

26. **Tanaka, K and Fritz, J. S.,** Determination of bicarbonate by ion-exclusion chromatography with ion-exchange enhancement of conductivity detection, *Anal. Chem.*, 59, 708, 1987.
27. **Hannah, R. E.,** Fluoride by anion exclusion in wastewater and effluents, *J. Chromatog. Sci.*, 24, 336, 1986.
28. **Dionex Corp.,** Determination of organic acids in wine, Application Note 21, Sunnyvale, CA.
29. **Johnson, D. E and Enke, C. G.,** Bipolar pulse technique for fast conductance measurements, *Anal. Chem.*, 42, 329, 1970.
30. **Keller, J. M.** Bipolar-pulse conductivity detector for ion chromatography, *Anal. Chem.*, 53, 344, 1981.
31. **Plambeck, J. A.,** *Electroanalytical Chemistry, Basic Principles and Applications,* John Wiley & Sons, New York, 1982, chap. 4.
32. **Shedlovsky, T. and Shedlovsky, L.,** Conductimetry, in *Techniques of Chemistry,* Vol. 1, *Physical Methods of Chemistry,* Part II A, *Electrochemical Methods,* Weissberger, A. and Rossiter, B. W., Eds., John Wiley & Sons, New York, 1971.
33. **Mulik, J. D. and Sawicki, E., Eds.,** Ion Chromatographic Analysis of Environmental Pollutants, Vols. I & II Ann Arbor Science, Ann Arbor, MI, 1978.
34. **Fitchett, A. W.,** Analysis of rain by ion chromatography, ASTM Spec. Tech. Publ. 823, Sampling Anal. Rain, 29, 1983.
35. **Bachman, S. R. and Peden, M. E.,** Determination of organic acid anions in precipitation by ion chromatography exclusion, *Water, Air, Soil Pollut.*, 33, 191, 1987.
36. **Brocco, D. and Tappa, R.,** Determination of organic and inorganic acid species in the atmosphere and in rain-water by ion chromatography, *J. Chromatogr.*, 367, 240, 1986.
37. **Karlson, U. and Frankenberger, W. T.,** Single-column ion chromatography of selenite in soil extracts, *Anal. Chem.*, 58, 2704, 1986.
38. **Conrad, V. B. and Brownlee, W. D.,** Hydropyrolytic-ion chromatographic determination of fluoride in coal and geological materials, *Anal. Chem.*, 60, 365, 1988.
39. **Chriswell, C. D., Mroch, D. R., and Markuszewski, R.,** Determination of total sulfur by ion chromatography following peroxide oxidation in spent caustic from the peroxide oxidation of coal, *Anal. Chem.*, 58, 319, 1986.
40. **Tan, L. K. and Dutrizac, J. E.,** Simultaneous determination of arsenic(III) and arsenic(V) in metallurgical processing media by ion chromatography with electrochemical and conductivity detectors, *Anal. Chem.*, 58, 1383, 1986.
41. **Easty, D. G. and Johnson, J. E.,** Recent progress in ion chromatographic analysis of pulping liquors: determination of sulfide and sulfate, *Tappi J.*, March, 109, 1979.
42. **Strauss, S. D.,** Polishing cuts condensate impurities below 0.1-ppb level, *Power,* Oct., 18, 1984.
43. **Plechaty, M. M.,** Measurements of anions in high-purity water by ion chromatography, *LC Mag.*, 2, 684, 1984.
44. **Kreling, J. R., Block, F., Louthan, G. T., and DeZwann, J.,** Sulfur and chlorine organic microanalysis using ion chromatography, *Microchem. J.*, 34, 158, 1986.
45. **Quinn, A. M., Siu, K. W. M., Gardner, G. J., and Berman, S. S.,** Determination of heteroatoms in organic compounds by ion chromatography after Schoniger flask decomposition, *J. Chromatogr.*, 370, 203, 1986.
46. **Classen, A. and Hesse, A.,** Measurement of urinary oxalate: an enzymatic and an ion chromatographic method compared, *J. Clin. Chem. Clin. Biochem.*, 25, 95, 1987.
47. **Talmage, J. M. and Biemer, T. A.,** Determination of potassium nitrate and sodium monofluorophosphate in the presence of phosphate and sulfate by high-resolution ion chromatography, *J. Chromatogr.*, 410, 494, 1987.
48. **Littlejohn, D. and Chang, S.-G.,** Determination of nitrogen-sulfur compounds by ion chromatography, *Anal. Chem.*, 58, 158, 1986.
49. **Parvez, H. et al., Eds.,** *Progress in HPLC,* Vol. 2, *Electrochemical Detection in Medicine and Chemistry,* VNU Science Press, Utrecht, Netherlands, 1987.
50. **Selavka, C. M. and Krull, I. S.,** The forensic determination of drugs of abuse using liquid chromatography with electrochemical detection: a review, *J. Liq. Chromatogr.*, 10, 345, 1987.
51. **Kochi, D. D. and Polzin, G. L.,** Effect of sample preparation and liquid chromatography column choice on selectivity and precision of plasma catecholamine determination, *J. Chromatogr.*, 386, 19, 1987.
52. **Johnson, D. C., Weber, S. G., Bond, A. M., Wightman, R. M., Shoup, R. E., and Krull, I. S.,** Electroanalytical voltammetry in flowing solutions, *Anal. Chim. Acta,* 180, 187, 1986.
53. **Patthy, M., Gyenge, R., and Salat, J.,** Comparison of the design and performance characteristics of the wall-jet type and thin-layer type electrochemical detectors. Separation of catecholamines and phenothiazines, *J. Chromatogr.*, 241, 131, 1982.
54. **Elbicki, J. M., Morgan, D. M., and Weber, S. G.,** Theoretical and practical limitations on the optimization of amperometric detectors, *Anal. Chem.*, 56, 978, 1984.

55. **Maitoza, P. and Johnson, D. C.,** Detection of metal ions without interference from dissolved oxygen by reverse pulse amperometry in flow injection systems and liquid chromatography, *Anal. Chim. Acta,* 118, 233, 1980.
56. **Weisshaar, D. E., Tallman, D. E., and Anderson, J. L.,** Kel-F-graphite composite electrode as an electrochemical detector for liquid chromatography and application to phenolic compounds, *Anal. Chem.,* 53, 1809, 1981.
57. **Armentrout, D. N., McLean, J. D., and Long, M. W.,** Trace determination of phenolic compounds in water by reversed phase liquid chromatography with electrochemical detection using a carbon-polyethylene tubular anode, *Anal. Chem.,* 51, 1039, 1979.
58. **Newberry, J. E. and Lopez de Haddad, M. P.,** Amperometric determination of nitrite by oxidation at a glassy carbon electrode, *Analyst,* 110, 81, 1985.
59. **Rocklin, R. D. and Johnson, E. L.,** Determination of cyanide, sulfide, bromide and iodide by ion chromatography with electrochemical detection, *Anal. Chem.,* 55, 4, 1983.
60. **Pyen, G. S. and Erdmann, D. E.,** Automated determination of bromide in waters by ion chromatography with an amperometric detector, *Anal. Chim. Acta,* 149, 355, 1983.
61. **Yang, X. H. and Zhang, H.,** Determination of trace amounts of iodine in sedimentary rocks by ion chromatography with amperometric detection, *J. Chromatogr.,* 436, 107, 1988.
62. **Han, K. and Koch, W. F.,** Determination of sulfide at the parts-per-billion level by ion chromatography with electrochemical detection, *Anal. Chem.,* 59, 1016, 1987.
63. **Han, K., Koch, W. F., and Pratt, K. W.,** Improved procedure for the determination of iodide by ion chromatography with electrochemical detection, *Anal. Chem.,* 59, 731, 1987.
64. **Kim, H. and Kim, Y.,** Analysis of free and total sulfites in foods by ion chromatography and with electrochemical detection, *J. Food Sci.,* 51, 1360, 1986.
65. **Kim, H., Park, G., and Kim, Y.,** Analysis of sulfites in foods by ion exclusion chromatography with electrochemical detection, *Food Technol.,* 41, 85, 1987.
66. **Johnson, D. C. and Polta, T. Z.,** Amperometric detection in liquid chromatography with pulsed cleaning and reaction of noble metal electrodes, *Chromatogr. Forum,* 1, 37, 1986.
67. **Rocklin, R. D. and Pohl, C. A.,** Determination of carbohydrates by anion exchange chromatography with pulsed amperometric detection, *J. Liquid Chromatogr.,* 6, 1577, 1983.
68. **Olechno, J. D., Carter, S. R., Edwards, W. T., and Gillen, D. G.,** Developments in the chromatographic determination of carbohydrates, *Anal. Biotechnol. Lab.,* 5, 38, 1987.
69. **Rocklin, R. D.,** Ion chromatography: a versatile technique for the analysis of beer, *LC Mag.,* 1, 504, 1983.
70. **Chen, L. -M., Yet, M -G., and Shao, M. -C.,** New methods for rapid separation and detection of oligosaccharides from glycoproteins, *FASEB,* 2, 2819, 1988.
71. **Hardy, M. R. and Townsend, R. R.,** Separation of positional isomers of oligosaccharides and glycopeptides by high-performance anion-exchange chromatography with pulsed amperometric detection, *Proc. Natl. Acad. Sci.,* 85, 3289, 1988.
72. **Hardy, M. R., Townsend, R. R., and Lee, Y. C.,** Monosaccharide analysis of glycoconjugates by anion exchange chromatography with pulsed amperometric detection, *Anal. Biochem.,* 170, 54, 1988.
73. **Hughes, S. Meschi, P. L., and Johnson, D. C.,** Amperometric detection of simple alcohols in aqueous solutions by application of a triple-pulse potential waveform at a platinum electrode, *Anal. Chim. Acta,* 132, 1, 1981.
74. **Rocklin, R. D.,** Ion chromatography with pulsed amperometric detection. Simultaneous determination of formic acid, formaldehyde, acetaldehyde, propionaldehyde, and butyraldehyde, in Adv. in Chem. Series No. 210, (ACS) *Formaldehyde: Analytical Chemistry and Toxicology,* Turoski, V. Ed., 1985, 13.
75. **Polta, J. A. and Johnson, D. C.,** The direct electrochemical detection of amino acids at a platinum electrode in an alkaline chromatographic effluent, *J. Liquid Chromatogr.,* 6, 1727, 1983.
76. **Polta, T. Z. and Johnson, D. C.,** Pulsed amperometric detection of sulfur compounds. Part I. Initial studies at platinum electrodes in alkaline solutions, *J. Electroanal. Chem.,* 209, 159, 1986.
77. **Polta, T. Z., Johnson, D. C., and Luecke, G. R.,** Pulsed amperometric detection of sulfur compounds. Part II. Dependence of response on adsorption time, *J. Electroanal. Chem.,* 209, 171, 1986.
78. **Neuburger, G. G. and Johnson, D. C.,** Pulsed coulometric detection with automatic rejection of background signal in surface-oxide-catalyzed anodic detections at gold electrodes in flow-through cells, *Anal. Chem.,* 60, 2288, 1988.
79. **Welch, L. E., LaCourse, W. R., Mead, D. A., Hu, T., and Johnson, D. C.,** Comparison of pulsed coulometric detection and potential-sweep pulsed coulometric detection for underivatized amino acids in liquid chromatography, *Anal. Chem.,* 61, 555, 1989.
80. **Suzuki, K., Aruga, H., and Sharai, T.,** Determination of monovalent cations by ion chromatography with ion-selective electrode detection, *Anal. Chem.,* 55, 2011, 1983.
81. **Lockridge, J., Fortier, N., Schmuckler, G., and Fritz, J. S.,** Potentiometric detection of halides and pseudohalides in anion chromatography, *Anal. Chim. Acta,* 192, 41, 1987.

82. **Haddad, P. R., Alexander, P. W., and Trojanowicz, M.,** Ion chromatography of inorganic anions with potentiometric detection using a metallic copper electrode, *J. Chromatogr.*, 321, 363, 1985.
83. **Scott, R. P. W.,** *Journal of Chromatography Vol. 33: Liquid Chromatography Detectors,* 2nd ed., Elsevier, Amsterdam, 1986.
84. **Dionex Corp.,** Determination of nitrite, nitrate, and ammonia in KCl soil extracts; Application Update #118, (Nov., 1987), Sunnyvale, CA.
85. **Stillman, J. R. and Pohl, C. A.,** New latex-bonded pellicular anion-exchangers with multi-phase selectivity for high-performance chromatographic separations, *J. Chromatogr.*, 499, 249, 1990.
86. **Small, H. and Miller, T. E.** Indirect photometric chromatography, *Anal. Chem.*, 54, 462, 1982.
87. **Jackson, P. E. and Haddad, P. R.,** The origin of system peaks in nonsuppressed ion chromatography of inorganic anions with indirect UV absorbance detection, *J. Chromatogr.*, 346, 125, 1985.
88. **Ketelle, B. H. and Boyd, G. E.,** The exchange adsorption of ions from aqueous solutions by organic zeolites. IV. The separation of the yttrium group rare earths, *J. Am. Chem. Soc.*, 69, 2800, 1947.
89. **Elchuk, S. and Cassidy, R. M.,** Separation of the lanthanides on high-efficiency bonded phases and conventional ion-exchange resins, *Anal. Chem.*, 51, 1434, 1979.
90. **Rubin, R. B. and Heberling, S. S.,** Metal determination by ion chromatography, *Am. Lab.*, May, 46, 1987.
91. **Heberling, S. S., Riviello, J. M., Mou, S., and Ip, A. W.,** Separate lanthanides by ion chromatography, *R&D,* Sept., 1987, p. 74.
92. **Bertsch, P. M. and Anderson, M. A.,** Speciation of aluminum in aqueous solutions using ion chromatography, *Anal. Chem.*, 61, 535, 1989.
93. **Dionex Corp.,** Determination of Chromium, Technical Note 24 (May, 1987), Sunnyvale, CA.
94. **Ong, C. N. Ong, H. Y., and Chua, L. A.,** Determination of copper and zinc in serum and whole blood by ion chromatography, *Anal. Biochem.*, 173, 64, 1988.
95. **Potts, M. E., Gavin, E. J., Angers, L. O., and Johnson, E. L.,** Monitoring anions and silica in high-purity water using on-line ion chromatography, *LCGC,* 4, 912, 1986.
96. **Fitchett, A. W. and Woodruff, A.,** Determination of polyvalent anions by ion chromatography, *LC Mag.*, 1, 48, 1983.
97. **Pacholec, F., Rossi, D. T., Ray, L. D., and Vazopolos, S.,,** Characterization of a phosphonate-based sequestering agent by ion chromatography, *LC Mag.*, 3, 1068, 1985.
98. **Roth, M.,** Fluorescence reaction for amino acids, *Anal. Chem.*, 43, 880, 1971.
99. **Wong, D., Jandik, P., Jones, W. R., and Haganaars, A.,** Ion chromatography of polyphosphonates with direct refractive index detection, *J. Chromatogr.*, 389, 279, 1987.
100. **Yamaguchi, H., Nakamura, T., Hirai, Y., and Ohashi, S.,** HPLC separation of linear and cyclic condensed phosphates, *J. Chromatogr.*, 172, 131, 1979.
101. **Ramsey, R. S.,** Liquid chromatographic analysis of oxo acids of phosphorus, in *Advances in Chromatography,* Giddings, J. C. and Keller, R. A., Eds., Marcel Dekker, New York, 1986, 219.
102. **Wolkoff, A. W. and Larose, R. H.,** Separation and detection of low concentrations of polythionates by high-speed anion-exchange liquid chromatography, *Anal. Chem.*, 47, 1003, 1975.
103. **Wolkoff, A. W. and Larose, R. H.,** Effect of chain length on the retention times of polythionates in high-pressure liquid chromatography, *J. Chromatogr. Sci.*, 14, 353, 1976.
104. **Pitt, W. W., Jr., Jolley, R. L., and Scott, C. D.,** Determination of trace organic contaminants in natural waters by high-resolution liquid chromatography, *Environ. Sci. Technol.*, 9, 1068, 1975.
105. **Shintani, H. and Dasgupta, P. K.,** High-sensitivity optical detection methods in hydroxide eluent suppressed anion chromatography with postsuppression ion exchange, *Anal. Chem.*, 59, 1963, 1987.
106. **Sherman, J. H. and Danielson, N. D.,** Comparison of inorganic mobile phase counterions for cationic indirect photometric chromatography, *Anal. Chem.*, 59, 490, 1987.

Chapter 5

ION-PAIR CHROMATOGRAPHY

I. INTRODUCTION

This chapter is about a group of methods that are called by various names, including paired-ion chromatography, ion-interaction chromatography, mobile-phase ion chromatography, soap chromatography, chromatography on dynamically coated ion exchangers. They are not ion exchange in the strict sense of the term, because the ion population in the stationary phase is not constant except in extreme cases, but they complement ion exchange, especially for large organic anions and cations. They have in common the use of reagents that have large hydrophobic ions that generally are singly charged, and charged oppositely to the ions that one wishes to determine. Examples of large hydrophobic ions are long-chain alkyl sulfonates (to determine cations) and long-chain quaternary ammonium salts (to determine anions). In the usual form of ion-pair chromatography these ions are added to the mobile phase. They accompany the analyte ions and are called "pairing ions" or "hetaerons", meaning companions. This term was introduced by Horvath et al.[1] The hetaerons go with the analyte ions into the stationary phase or cause them to be adsorbed on the surface. The stationary phase itself is a non-ionic, hydrophobic material like alkyl-bonded silica or a porous styrene-DVB polymer. In another form of ion-pair chromatography the stationary phase is porous silica coated with an aqueous solution of the hetaeron, and the mobile phase is nonaqueous.

Ion-pair chromatography was first used by Haney and collaborators[2] to analyze a food coloring, tartrazine, which contained unreacted precursors. These were complex aromatic compounds that were sulfonate salts. The authors wanted to take advantage of the good chromatographic characteristics of alkyl-bonded, totally porous silica, but they knew that highly polar, ionized substances were not retained on alkyl-bonded silica. They found that the negative sulfonate ions were retained if large positive ions were added to the eluent. The mobile phase that they used was methanol-water, 1:1 made 0.001 *M* with tetrabutylammonium hydroxide or tridecylamine and acidified with formic acid. They got good chromatograms in run times less than 10 min. They used the same system to determine ascorbic acid in fruit and vegetable juices.[3]

Almost simultaneously in Sweden, Schill and Wahlund[4] developed a variation on this technique in which they used bare porous silica as the stationary phase and coated it with an aqueous solution of the pairing ion or hetaeron. To retain negative analyte ions they used 0.1 *M* tetrapropylammonium salt; to retain positive analyte ions they used 0.1 *M* sodium naphthalene sulfonate. In each case the mobile phase was 9:1 chloroform:1-pentanol. The analyte, the anion of an acid or the cation of a base such as phenylalanine, entered the stationary phase in company of the pairing ion, left it in the company of the pairing ion, and if the pairing ion absorbed ultraviolet light, a UV detector could be used even if the analyte ion itself did not absorb.

The drawback to the bare-silica technique is that the water coating and the hetaeron salt are gradually stripped off and must be renewed from time to time. However, the recoating is easy to do. One pumps through the column an aqueous solution of the hetaeron salt, then displaces this solution by pumping the nonaqueous mobile phase until the stationary and mobile phases are again in equilibrium.

Important studies of ion-pair chromatography have been made by Knox and co-workers. In 1976 Knox and Laird[5] separated a number of dyestuff intermediates, including 4-aminobenzenesulfonic acid (sulfanilic acid) and 2-hydroxynaphthalene-6-sulfonic acid

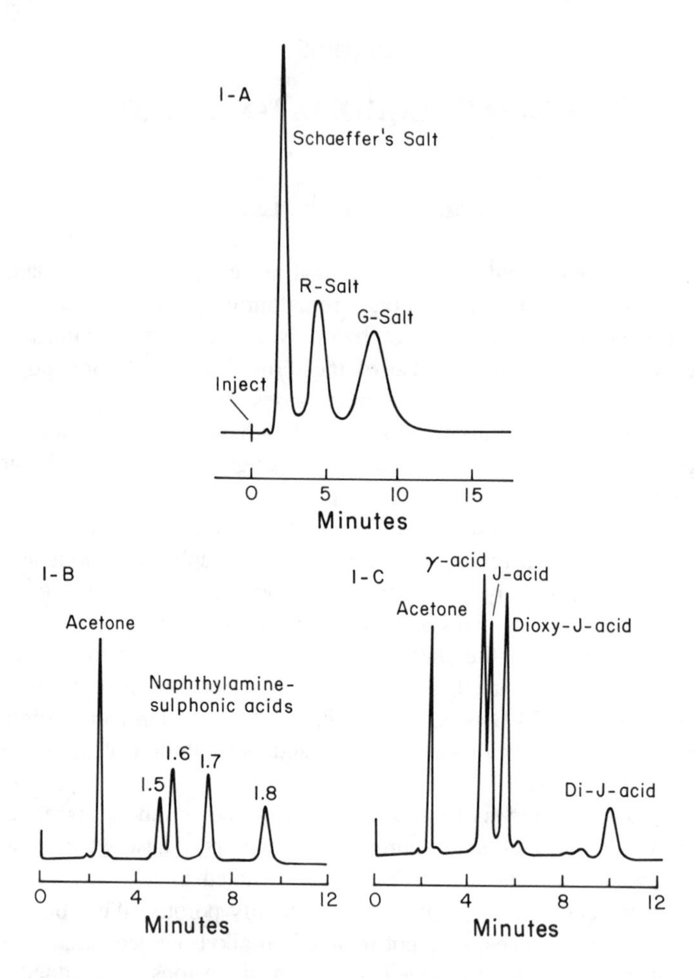

FIGURE 1. Comparison of paired-ion chromatography with chromatography on an anion-exchange resin. Figure 1-A, on ion-exchange resin. (From Schmidt, J. A. and Henry, R. A., *Chromatographia,* 3, 497, 1976. With permission.) Figures 1-B and 1-C, by ion pairs (From Knox, J. H. and Laird, G. R., *J. Chromatogr.*, 122, 17, 1976. With permission.)

(Schaeffer's acid), on a statinary phase of bonded alkyl silica (a reversed-phase packing), using as the mobile phase a 5:2 water-1-propanol mixture containing 0.1 to 1% of hexadecyl trimethylammonium bromide (cetyl trimethylammonium bromide or cetrimide). The large cetyl trimethylammonium cations formed ion pairs with the large anions of the various sulfonic acid dye intermediates and caused them to be attached to the nonpolar stationary phase. They also used porous silica as the stationary phase, with a mobile phase containing much more propanol: the water to propanol ratio was 1:3. Cetrimide was added to the mobile phase as before. The effective stationary phase was now a layer of aqueous cetrimide solution held on the silica surface. This system which is like that developed by Schill is not used much today; almost always the stationary phase is alkyl-bonded silica or a porous polymer, and the pairing ions are added to the mobile phase.

Knox's[6] new technique gave much better chromatography of the organic acids than had conventional ion exchange. Figure 1 compares Knox's "soap chromatography" with anion-exchange chromatography on a Zipax® surface-modified silica-based exchanger. Soap chromatography, as Knox called it, gave much narrower and more symmetrical peaks with 10 to 20 times as many theoretical plates. The reason for this big difference is undoubtedly the slow diffusion of large organic ions in and out of the ion exchanger. Another factor that makes conventional ion exchange inefficient for large organic ions that have aromatic character is interaction between the aromatic rings of the solutes and those of a polystyrene-based ion exchanger.

In a later paper Knox and Jurand[7] described the chromatography of catecholamines on a reversed-phase bonded silica packing, using as eluent 0.02% sodium dodecyl sulfate in 27% methanol-73% water, made 1 m*M* in sulfuric acid to ensure that the catecholamines were cationic. Again, pairs of ions formed between the large analyte cations and the large hetaeron dodecyl sulfate anions. In the "normal-phase" version of this separation, Knox and Jurand used porous silica coated with an aqueous solution 0.9 *M* in sodium perchlorate and 0.1 *M* in perchloric acid. The eluent was a mixture of *n*-butanol and methylene chloride. The perchlorate ion was the hetaeron. It held the catecholamine cations into the stationary phase and caused them to be retained. Raising the proportion of butanol in the mobile phase reduced the retention, apparently by solubilizing the ion pairs.

II. ION PAIRS

Before going into details of the effects of various parameters on retention, it will be well to explain what is meant by the term *ion pair*.

The first mathematical description of pairs of oppositely charged ions in an electrolyte solution was due to Bjerrum[8] in 1926. Consider the ions to be points in a medium of dielectric constant D, and for simplicity let their charges be equal, $\pm q$. Let the averate concentration of each kind of ion be C. Consider the probability of finding a negatively charged ion at a given distance, r, from a central positive ion. In a spherical shell of radius r and thickness dr the average number of negative ions is:

$$dn = C4\pi r^2 dr \exp \frac{q^2}{rDkT} \tag{1}$$

The exponential factor contains the electrical potential energy, q^2/Dr, and it increases rapidly as r becomes small. The graph of dn/dr against r looks like Figure 2. To find the minimum in the curve we take the second derivative, d^2n/dr^2, set it equal to zero, and find:

$$r_{min} = \frac{q^2}{DkT} \tag{2}$$

Ions that approach closer than this critical distance, r_{min} are considered to form ion pairs.

For singly charged ions in water (D = 80) at 25°C, the critical distance is 3.7 A or 0.37 nm. For ions of higher charge this number is multiplied by the product of the valences. Now the sum of the radii of hydrated univalent cations and anions in aqueous solution is always considerably greater than this critical distance. (The hydrated radius of Na^+, for instance, is 3.3 A, estimated from its equivalent conductance). Thus, according to Bjerrum's definition, ion pairs can only form in aqueous solution with ions of high charge, for example in yttrium ferricyanide, $YFe(CN)_6$. They can, however, form in solvents or solvent mixtures whose dielectric constant is lower than water's. In mixtures of water with dioxane (D = 2.2) ion pairs do form; evidence comes from the abnormally low electrical conductivity of dissolved salts.

Ion pairs of another kind may exist in water. Water is an unusual liquid. There is much evidence that hydrogen bonding, directed tetrahedrally around the oxygen atoms, causes liquid water to have a structure, a residue of the structure of the ice crystal. Ions and molecules dissolved in water alter that structure. Small cations draw water dipoles around

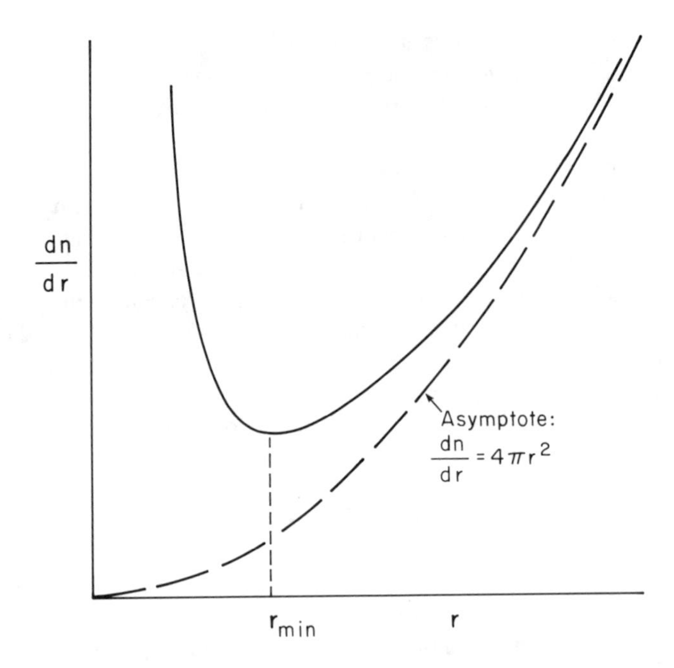

FIGURE 2. Ion pairs. Abscissae are distances from the central ion. (Adapted from Robinson, R. A. and Stokes, R. H., *Electrolyte Solutions*, 2nd ed., 1959, chap. 14.)

them, in an oriented fashion, as water of hydration. Then, it seems, there is a transition zone of randomly oriented water molecules outside wich there is "normal", or ice-like, hydrogen-bonded water.[9] Large inert molecules like aliphatic hydrocarbons, krypton, and xenon cause an entropy decrease with Diamond[10] has called a "tightening" of the water structure. The water molecules immediately next to the large dissolved molecule are driven to form their full complement of four hydrogen bonds, rather than a lesser number. Large ions of single charge have the same effect. (The electric charge favors dipole orientation, breaking hydrogen bonds; in ions of higher charge than 1 ion-dipole interaction could counteract the tightening effect.)

A large ion or molecule, dissolving in water, must create a cavity within the water structure, and this process requires energy. For two ions of opposite charge it takes less energy for them to share the same cavity than to create two separate cavities. Two large ions of low charge, brought together in this way, constitute an ion pair. To quote Diamond[10], "This water structure-enforced ion pairing has the opposite requirements of ordinary electrostatic ion pairing, the larger the ions and the smaller their charges, and can occur only in water or some equally structured solvent." As evidence for the existence of such ion pairs he cites lowered osmotic and activity coefficients, (but not lowered conductivity), the "salting in" of uncharged molecules by solutes having large ions, solvent extraction phenomena, and above all, anion-exchange selectivities. Anion exchangers with quaternary ammonium functional groups bind large singly charged anions very strongly, including perchlorate ions and singly charged metal chlorocomplexes like $AuCl_4^-$ and $FeCl_4^-$, and bind them strongly even in resins that are very weakly crosslinked; (see Chapter 8). In cation exchange of simple hydrated ions the selectivity drops to near zero as the crosslinking is decreased (see Chapter 3). In anion exchange of large anions this is not so. It seems that in anion exchange, ion pairing or ion interaction is responsible for selectivity, not the work of stretching the swollen polymer.

III. ION-PAIR EXTRACTION

A phenomenon well known in analytical chemistry is the extraction of pairs of large ions of single charge out of water and into an immiscible solvent that has a somewhat polar character. An efficient way to extract gold(III) from water is to make the solution 0.1 to 1 *M* in hydrochloric acid, add methyl violet, which is an aminotrimethyl-methane that forms positive ions in acid solution, and then to shake with chloroform in a separating funnel. The aqueous solution is a pale greenish-blue; the chloroform extract is a deep blue-purple and contains the chloraurate ion, $AuCl_4^-$ combined with the cation of methyl violet. Iron(III) chloride, as $FeCl_4^-$, is also extracted, but to a much lesser extent. The extraction of Fe(III), Ga(III), In(III), and Au(III) from aqueous hydrochloric acid by ethyl acetate or methyl isobutyl ketone is well known and much used in chemical separations.

It appears that ion pairs are being formed in the aqueous phase in accordance with the water-induced structural tightening invoked by Diamond[10] and then extracted as ion pairs into the organic solvents, which have no structure to cause ion pairing, though they do have low dielectric constants. Rules for finding ion pairs in water are being extrapolated to predict ion pairs in organic solvents and they appear to be successful.

IV. THEORIES OF ION-PAIR CHROMATOGRAPHY

A. EXPERIMENTAL TESTS

Thus, it is natural to reason from ion-pair extraction to ion-pair chromatography. The first mechanism proposed for the retention of sample ions, S^+, by pairing ions, P^-, was that ions S^+ and P^- (the signs of the charges may of course be reversed) combine in the mobile phase and immediately plunge into the stationary phase to form an ion pair in the stationary phase:

$$P_m^- + S_m^+ = PS_s$$

It was considered, for a start, that ion pairs existed only in the stationary phase, not in the mobile or aqueous phase, and that the pairing ions of hetaerons did not enter the stationary phase except as ion pairs. The next step was to modify the model to take account of ion pairs in the aqueous phase as well. Then the role of the small counter-ions in the aqueous phase, which we shall call B^- and C^+, that accompanied the sample and pairing ions, had to be considered; if the sample salt is SB and the pairing reagent CP, then these salts could be absorbed into the stationary phase. Knox and Hartwick[11] set up equations expressing the formation of the ion pair SP and the combination of ions with the ligand, L, of the stationary phase. L would be the octadecyl chain in the common C18-bonded silica. The equations are as follows.

$$\text{Adsorption of solute ions: } S_m^+ + B_m^- + L_s = SBL_s; \quad K_1 \tag{3a}$$

$$\text{Ion-pair formation in solution: } S_m^+ + P_m^- = SP_m; \quad K_2 \tag{3b}$$

$$\text{Adsorption of hetaeron: } P_m^- + C_m^+ + L_s = PCL_s; \quad K_3 \tag{3c}$$

$$\text{Adsorption of solute-hetaeron ionpair: } S_m^+ + P_m^- + L_s = SPL_s; \quad K_4 \tag{3d}$$

Bearing in mind that the concentrations of S and B are small compared with those of P and C, we derive this equation for the capacity factor, k′

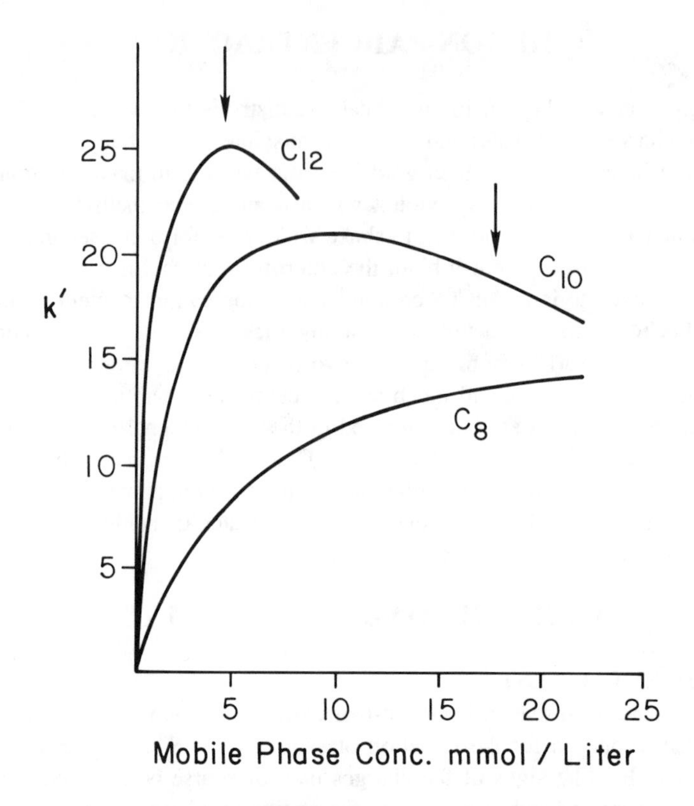

FIGURE 3. Retention of tyrosine amide on C18-bonded silica in solutions of alkyl sulfates. Arrows indicate critical micelle concentrations. (From Knox, J. H. and Hartwick, R. A., *J. Chromatogr.*, 204, 3, 1981. With permission.)

$$k' = \phi[L]\frac{K_1[B_m^-] + K_4[P_m^-]}{(1 + K_2[P_m^-])(1 + K_3[P_m^-][C_m^+])} \tag{4}$$

ϕ is the phase ratio, the ratio of the volumes of the stationary and mobile phases. Note the distinction between the total concentration of the ligand, L, and the concentration of unbound ligand, L_s

$$[L] = [L_s] + [PCL_s] + [PSL_s] \tag{5}$$

the last term being negligibly small.

Equation 4 predicts that k′ will rise with increasing hetaeron concentration and will be directly proportional to this concentration as long as it is small. At higher hetaeron concentrations k′ will rise less rapidly and will eventually fall. This relation is found in practice (see Figure 3). Similar graphs appear in Reference 1. If the capacity factor k′ is plotted against the concentration of hetaeron in the stationary phase, the graph is linear over a considerable range (see Figure 4). The concentration of hetaeron in the stationary phase is measured by pumping a standard solution of the hetaeron salt of known concentration through the column and noting the breakthrough volume. Repeating this experiment at different hetaeron concentrations, Knox and Hartwick[11] obtained the adsorption isotherm, and found that it followed the Freundlich equation:

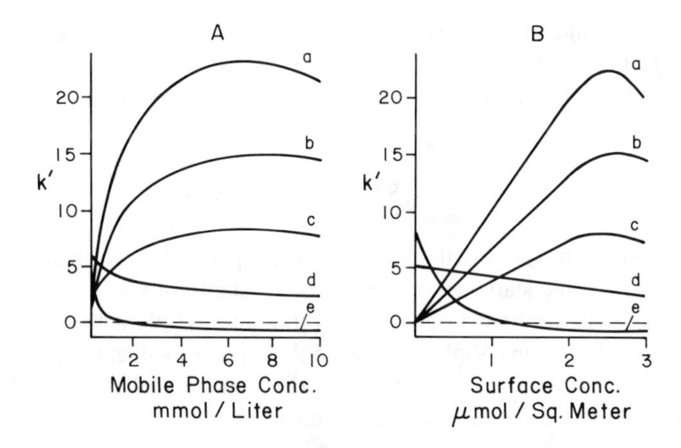

FIGURE 4. Retention of solutes on C18-bonded silica in octyl sulfate solutions. The capacity factor k′, is related to the mobile phase concentration in graph A, to the stationary phase concentration in graph B. Solutes: (a) tyrosine amide; (b) normetadrenaline; (c) adrenaline; (d) benzyl alcohol; and (e) naphthalene-2-sulfonate. (From Knox, J. H. and Hartwick, R. A., *J. Chromatogr.*, 204, 3, 1981. With permission.)

$$\log(\text{conc. in stationary phase}) = A + B \log(\text{conc. in mobile phase}) \tag{6}$$

The same relation was found by other workers.[12,13]

In Figure 3, arrows point to the critical micelle concentrations. As the concentration of a solution of a surface-active electrolyte like sodium decyl sulfate is increased, and the ions come closer together, the hydrocarbon "tails" of the surfactant ions attract one another, or perhaps it is better to say that the water structure pushes them together. Each "tail" must create a cavity in the water, and this process requires energy, as we have noted above. When it takes less energy to form one cavity for 10 to 20 surfactant ions than to form 10 to 20 separate cavities, aggregation takes place; the aggregate is called a "micelle", and the concentration at which micelles start to form is called the "critical micelle concentration". The onset of micelle formation is quite sharp; association takes place in a narrow concentration range. In a cooperative association where several molecules act together, the equilibrium depends on a high power of the concentration. This phenomenon is seen in the attachment of proteins to ion exchangers (see Chapter 7).

The formation of micelles causes a drop in k′ because the micelles act as little drops of nonpolar solvent; they can extract the solute ions from the surrounding water. This phenomenon is the basis of micellar mobile phases in liquid chromatography.[14]

We note in Figure 4 that there are graphs for retention of benzyl alcohol, a neutral uncharged species, and naphthalene sulfonate, which is negatively charged. Retention of the uncharged species is almost unaffected by the hetaeron ions. It falls slightly as the hetaeron ions take up more and more of the adsorbent surface. The neagative sulfonate ions are repelled by the hetaeron ions (octyl sulfate) and are quickly driven out of the stationary phase.

The first ideas about ion-pair chromatography implied that the hetaeron ion, octyl sulfonate in Figure 4, only entered the stationary phase in company with the hydrophobic sample ion. We have seen that this idea is false, or only represents an extreme case. Early in the development of this kind of chromatography, Kissinger[15] pointed out that the pairing agent or hetaeron was adsorbed by the nonpolar stationary phase, because several column volumes of hetaeron solution had to be passed before the retention of a solute became constant, and he commented that the alkyl sulfonate in his experiments, by becoming adsorbed on the C18-silica, had turned the stationary phase into an ion exchanger. Reversed-phase ion-pair partition chromatography could therefore be regarded as ion exchange. Referring back to Equations 3c and 3d, one may combine them to give an equation of ion exchange:

$$C^+_m + SPL_s = S^+_m + CPL_s; \quad K_5 = K_3/K_4 \tag{7}$$

Certainly this is ion exchange, but the ion-exchange capacity, which is the concentration of bound hetaerons, P, in the stationary phase is not constant but varies with the composition of the mobile phase. Nevertheless one may call the process "dynamic ion exchange".

Experiments of Bartha and Vigh[41,42] showed the effect of solvent modifiers and stressed the importance of the hetaeron concentration in the stationary phase, and also the concentrations of counter-ions. Equation 7 was confirmed.

V. ION-INTERACTION MODEL

The ion-pair concept has shortcomings that led Bidlingmeyer et al.[16,17] to develop an ion-interaction theory that did not require the formation of ion pairs. In the first place, careful conductivity measurements showed no indication of ion-pair formation between the octanesulfonate anion and the octylammonium cation at the concentrations used in chromatography (0.01 to 0.05 *M* in 35% methanol). In the second place, mutual interactions between the reagent ions (pairing ions or hetaerons) in the mobile phase and surface-active (or hydrophobic) solute ions went beyond what could be expected from the ion-pair model. Not only were hydrophobic solute ions having the same charge as the reagent ions repelled and driven off the stationary phase, as seen in Figure 4, but they were made to elute before the void volume; k′ actually became negative. Hydrophobic solute ions of charge opposite to the reagent ions (hetaerons) caused increased adsorption of the hetaerons; hydrophobic solute ions of the same charge as the reagent ions caused decreased adsorption of these reagent ions. To take account of less observations Bidlingmeyer and his co-workers postulated an electrical double layer at the surface of the adsorbent. The primary layer consisted of ions of the hetaeron, which Bidlingmeyer called the *ion-interaction reagent*. The secondary ion layer contained ions of opposite charge, including the hydrophobic (or "adsorbophilic") ions of the sample to be analyzed. The classical theory of the electrical double layer pictures this secondary ion layer as a diffuse zone containing ions of both charge signs, distributed according to the Poisson-Boltzmann relation, and extending some distance out into the solution; the effective thickness is smaller, the greater the ionic strength of the solution. Immediately we have an explanation for the remarkable fact shown in Figure 4b, that analyte ions having the same charge sign as the hetaeron ions (ion-interaction reagent, in Bidlingmeyer's nomenclature) can elute before the void volume. If the double-layer thickness is commensurate with the pore diameters of the bonded-silica adsorbent, these ions will be excluded from the pores.

The ion-interaction model requires that the ionic strength shall affect the retention of both the reagent and the solutes. Studies of Bidlingmeyer and Warren[18] showed that the ionic strength affects the adsorption of the reagent; cetyl pyridinium ions were more strongly adsorbed by an octadecyl silica stationary phase, the higher the ionic strength; increasing the ionic strength from 0.0001 to 0.01 raised the adsorption of cetyl pyridinium ions by a factor of three. Raising the ionic strength reduces the repulsion between neighboring adsorbed ions and allows a greater surface density of these ions on the stationary phase. Raising the ionic strength of the mobile phase, while keeping the concentration of cetyl pyridinium ions in the mobile phase constant, reduced the retention of long-chain alkyl sulfonate ions, in spite of the increased concentration of cetyl pyridinium ions in the stationary phase. (One

TABLE 1
Retention of Anions on PRP-1 Porous Polymer in Presence of 10^{-4} *M* $Fe(phen)_3^{2+}$

A. Effect of counter-anion

Mobile phase: $10^{-4}M$ sodium succinate, pH 6.1, $2 \times 10^{-4}M$ in anions X

		k'	
Analyte ion	**X = ClO_4^-**	**X = Cl^-**	**X = F^-**
IO_3^-	5.4	6.05	8.2
NO_2^-	9.5	10.9	11.3
NO_3^-	14.5	18.1	18.3

B. Effect of other mobile-phase anions

		k'	
Analyte ion	**A**	**B**	**C**
Cl^-	7.4	1.18	1.24
NO_2^-	9.5	2.7	2.1
NO_3^-	14.5	4.9	4.6

Mobile phase anions, each $1 \times 10^{-4}M$ except sulfate:

A, succinate, perchlorate, pH 6.1; B, Citrate, perchlorate, pH 7.2; C, sulfate, $1 \times 10^{-3}M$ pH 5.5.

Adapted from Rigas, P. G. and Pietrzyk, D. J., *Anal. Chem.*, 58, 2226, 1986.

could, of course, consider this to be anion-exchange effect caused by the increasing concentration of anions in the mobile phase). One strange effect of ionic strength is reported in this paper. The ion-interaction reagent (which we have been calling the hetaeron) comes out of the column along with the sample or solute ion, and if the reagent ions absorb light, the emergence of the sample ions is marked by an increase of absorbance even if the sample ions themselves do not absorb. This effect occurred as expected if the ionic strength was low (10^{-5} *M* phosphate); the absorbance rose as pentyl, hexyl, heptyl, and octyl sulfonates were eluted in this order. If the ionic strength was high, however, 10^{-3} *M*, the absorbance fell as the samples eluted; the peaks were negative. At the same time the retentions were significantly reduced.

A study by Iskandarani and Pietrzyk[19] confirmed the ion-interaction model. They used quaternary ammonium salts as interaction agents in solutions of pH 11 containing acetonitrile and various kinds and amounts of added salts to control the ionic strength. The stationary phase was PRP-1, a macroporous styrene-divinylbenzene polymer. Test solutes were aromatic sulfonic and carboxylic acids. These workers confirmed the Freundlich isotherm for the adsorption of quaternary ammonium salts, and made the very important point that the nature of the anion in the eluent affected the strength of binding of the quaternary ammonium cation and also the retention of test solutes. Of the three halide ions, F, Cl, and Br, the fluoride ion gave the weakest retention of the quaternary ammonium ions (in the absence of other salts in the mobile phase) and it gave the strongest retention of the test solutes. This point was made even more clearly in a subsequent publication[27] from which the data of Table 1 are taken. A similar observation was made by Dreux et al.,[20] who included the perchlorate anion in their study and found that perchlorate caused the least retention of test solutes, which were inorganic anions including iodide. The anions of the mobile phase must compete with solute anions for adsorption sites. In this sense their effect can be anticipated from the treatment of Knox and Hartwick[11] (see Equation 7). Perchlorate ions are held very strongly and fluoride ions very weakly; thus, perchlorate ions will compete strongly with the sample ions and cause them to be less retained.

The effect can also be interpreted by the ion-interaction, electrical double layer theory. Iskandarani and Pietrzyk[19] showed that the adsorption of quaternary ammonium ions, and also the capacity factor of aromatic acid anions in eluents containing these quaternary ammonium ions, depended on ionic strength in the manner predicted by the electrical double-layer theory: $1/k'$ rose linearly with the reciprocal of the square root of ionic strength. It is clear that ionic strength is an important factor in ion pair chromatography.

VI. SOLVENT EFFECTS

A. SUMMARY OF EXPERIMENTAL DATA

The theoretical interpretation of ion-pair chromatography is difficult. The phenomena are complex, and no one theory explains all of them, even experimental data do not agree. Melander et al.[21] concluded from conductivity data that octylamine and octyl sulfonate do indeed form ion pairs in 35% methanol, under the conditions used by Bidlingmeyer and Warren[16] and that their formation constant was several liters per mole. In this section we shall return to the experimental relationships and the information that the practicioner needs in order to use this kind of chromatography effectively.

Ion-pair chromatography is generally performed in mixed solvents, water plus an organic modifier, which is usually methanol or acetonitrile. One reason for this choice is to raise the solubility of the analyte. Another reason is that C18-silica and even more so, porous styrene-DVB polymers is not wet by water, though surface-active ion-pairing reagents promote wetting. C18-silica is used to concentrate trace organic copounds from water, but it must first be conditioned by wetting with methanol. A third reason is that retention can be brought within the desired range for chromatography, not too weak nor too strong, by adjusting the solvent composition. Adding organic modifier reduces the retention.

Several authors have studied the effect of the solvent in paired-ion chromatography. Bidlingmeyer et al.[16] injected small amounts of alkyl sulfonates as samples into mobile phases of methanol-water mixtures each 0.001 *M* in hydrochloric acid and measured their retention of C18-silica. The capacity factor, k', gives the partition ratio between stationary and mobile phases. Increasing the methanol concentration reduced the retention, and the graph of $\log k'$ against the volume fraction of methanol was a descending straight line. At constant methanol concentration $\log k'$ rose linearly with the carbon number, the number of carbon atoms in the alkyl sulfonate sample, which ranged from 5 to 8 (see Figure 5). Both these relations are familiar relations in reversed-phase chromatography, and they mean that the free energy of adsorption of the alkyl sulfonates rises linearly with the volume fraction of water and with the number of carbon atoms in the sulfonate ions.

Vera-Avila et al.[13] measured the adsorption of several alkyl sulfates and sulfonates on C8-silica (LichrosorbRP8) in mixtures of acetonitrile and water made 0.04 *M* in perchloric acid. They used the breakthrough method, i.e., they pumped solutions of known concentration and noted the volumes that were passed before the sulfates or sulfonates appeared in the (refractive index) detector; thus, they found the quantities adsorbed and could plot the adsorption isotherms. Some are shown in Figure 6. The Freundlich isotherm, Equation 6 above, was obeyed. The coefficients A and B were related in a linear manner to the volume fraction of acetonitrile. The coefficient A fell with increasing acetonitrile concentration, i.e., the adsorption decreased, while the coefficient B increased, i.e., the isotherms became straighter as the acetonitrile concentration increased. These effects are seen in Figure

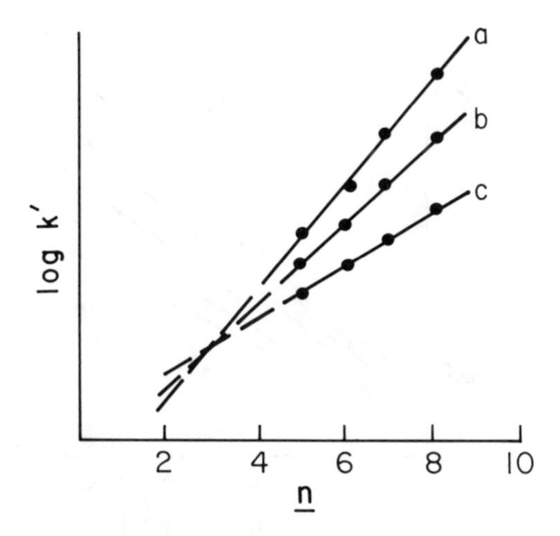

FIGURE 5. Dependence of capacity factor, k', on carbon number, n, for sodium alkyl sulfonates. Alkyl sulfonate samples were injected into methanol-water mobile phases containing no other solutes. Line (a) 20% methanol; (b) 35% methanol; and (c) 50% methanol. (From Bidlingmeyer et al., *J. Chromatogr.*, 186, 419, 1979. With permission.)

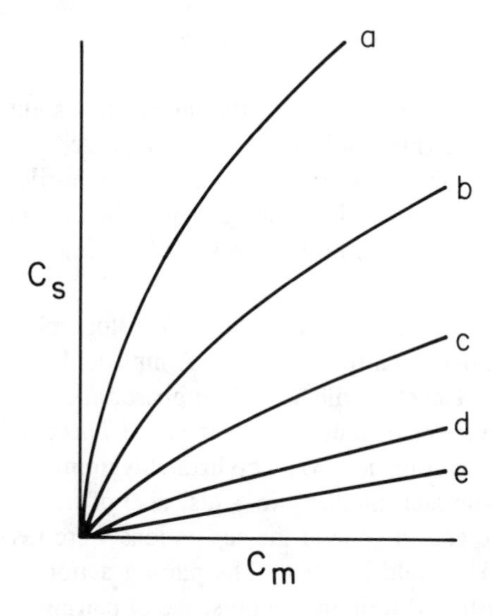

FIGURE 6. Adsorption isotherms of sodium dodecane sulfonate on C18-silica. The C_s and C_m are concentrations in the stationary and mobile phases; solvents are acetonitrile-water mixtures, 0.04 *M* in perchloric acid, temp. 40°C. Curve a, 20% acetonitrile; curves b, c, d, and e, 25, 30, 35, and 40% acetonitrile. Scales are linear. (From Vera-Avila, L. E., Caude, M., and Rosset, R., *Analusis,* 10, 36, 43, 1982. With permission.)

6. Having measured the adsorption isotherms, the authors next measured the retention of pyridine and some substituted pyridines in water-acetonitrile mobile phases that were all 0.04 *M* in perchloric acid, but had different acetonitrile concentrations and different alkyl sulfate or sulfonate concentrations. Plotting the capacity factor, k', against the concentration of sulfate or sulfonate in the stationary phase they got straight lines (see Figure 7). To summarize their conclusions:

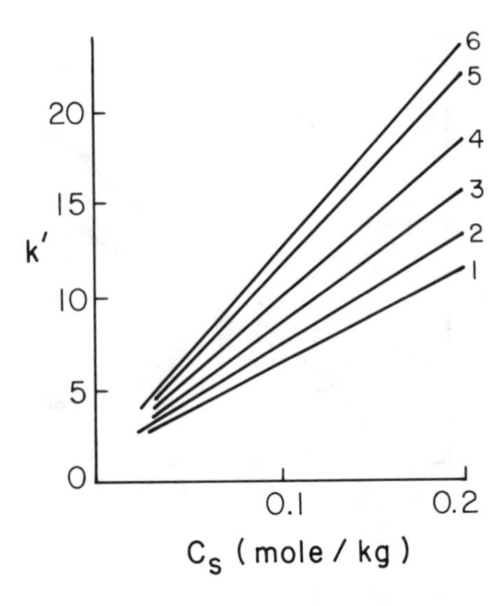

FIGURE 7. Capacity factors and pairing-ion concentrations in the stationary phase. Stationary phase is C8-silica, mobile phase 10% acetonitrile, 0.04 *M* in $HClO_4$, containing sodium octane sulfonate. Abscissae are concentrations of sodium octane sulfonate. Solutes are: (1) pyridine; (2) 2-methylpyridine; (3) 3-methylpyridine; (4) 2,4-dimethylpyridine; (5) 3,4-dimethylpyridine; (6) 2,4,6-trimethylpyridine. (From Vera-Avila, L. E., Caude, M., and Rosset, R., *Analusis,* 10, 36, 43, 1982. With permission.)

1. The retention, expressed by k', of the different pyridine solutes increased linearly with the concentration of pairing ion in the stationary phase
2. Log k' fell linearly with the volume fraction of acetonitrile in the mobile phase
3. Log k' - log k'_o rose linearly with the number of carbon atoms in the pairing ion. Here, k'_o is the capacity factor found in the absence of pairing agent.

These authors used the relationships to design chromatographic separations of the cations of amino acids and the anions of carboxylic acids. Amino acids were made cationic by using 0.05 *M* sulfuric acid in 20% acetonitrile as mobile phase, and the pairing ion was dodecane sulfate; carboxylic acids were made anionic by using a pH 7 acetate buffer in 35% acetonitrile as mobile phase, and the pairing ion was cetyltrimethylammonium (cetrimide). They got better separations of acrylic and methacrylic acids, also of maleic and fumaric acids, than they had by ion exchange and in general the separations were faster.

The importance of pH should be clear. The pairing action requires ions. Retention of uncharged molecules is affected little by the presence of pairing ions;[11,16] thus, the pH must be chosen and stabilized when necessary by buffering the mobile phase. If the analyte is a weak acid, HA, then if pH = pKa half of it is in the ionic form, A^-. The fraction in the ionic form increases as the pH increases. If the analyte is a weak base, B, then half it is in the ionic form if pH = pKa where pKa refers to the conjugate acid, BH^+, and the fraction in the ionic form *decreases* as the pH increases. The pKa values in the literature refer to aqueous solutions. In the mixed aqueous-nonaqueous solutions used in ion-pair chromatography, pKA for protonated bases BH^+ should be about the same as in water because there is no separation of charges on ionization; however, an uncharged acid HA will ionize less in the mixed solvent than in water because the dielectric constant is smaller and the work of separating the ionic charges is greater than in water. Also the pH scale in another solvent is not the same as in water. Hence, the ionization constants of acids and bases given in the literature are only rough guides to the ionization in the mixed solvents used in ion-pair chromatography.

Most of the pairing ions used in ion-pair chromatography are the ions of strong acids or strong bases and they remain in ionic form over the whole pH range. Sometimes this is not so; e.g., the octylammonium ion, $C_8H_{17}NH_3^+$, may be used. If this ion is to join with the anion of a weak acid there will be an optimum pH, somewhere in the neutral region, where both the acid and the base are largely in their ionic forms. Ion-pair formation and chromatographic retention will fall as the pH is raised above this value or lowered below it.

VII. DYNAMICALLY COATED ION EXCHANGERS

Some of the easiest ion-pair chromatography to understand and perhaps some of the most useful involves pairing agents or hetaerons that are sufficiently strongly adsorbed that they make the stationary phase a surface-functional ion exchanger. Cassidy and co-workers[22-26] have produced a large body of research directed primarily to the analysis of mixtures of the lanthanides, uranium, and thorium, and also to the analysis of inorganic anions and simple anions of organic acids. In their earlier work[22] they used cation exchangers of various types, including gel-type sulfonated polystyrene resins of small particle size, to separate lanthanide ions and had great success, getting separate peaks for all 14 lanthanide elements (excluding promethium) in less than 30 min, using gradient elution with *alpha*-hydroxy-isobutyric acid. Detection was by post-column derivatization with PAR or Arsenazo and was very sensitive. The high-capacity gel-type resin, Aminex A-5, 8% crosslinked, 13 μm diameter when swollen in water, performed as well as, or better than, bonded-silica ion exchangers. Later these authors looked into ion-pair chromatography. With a stationary phase of C18-silica and a mobile phase of 0.005 *M* in sodium *n*-octane sulfonate and containing *alpha*-hydroxyisobutyric acid (HIBA) whose concentration rose in a gradient from 0.05 to 0.4 *M*, they separated the lanthanides in 2 min.[23] To coat the stationary phase with octanesulfonate ions from a 0.005 M solution took a long time and a large volume of this dilute solution; a faster way to prepare the column was to pass a more concentrated octanesulfonate solution for a while and then bring it to equilibrium with the dilute eluent.

Longer-chain sulfonates and sulfates stuck to the stationary phase so strongly that they gave "permanent" coatings that did not wash off and needed no addition of pairing agent to the mobile phase. Cassidy and Elchuk[26] described the use of these coating agents, and compared them with the shorter-chain pairing agents that do wash off and do need the presence of the pairing agent in the mobile phase. In this work they used not only alkyl-bonded silica as the stationary phase, but also the macroporous styrene-DVB polymer, PRP-1, which was more stable and reproducible than bonded silica and could be used over the complete pH range. To coat the stationary phase with C_{20} sulfate they pumped 500 ml of 2.5×10^{-4} *M* sulfate in 25% acetonitrile through the column, followed by water. This treatment, repeated as needed, gave a permanently coated ion exchanger whose capacity was 0.1 to 1 meq for a 30-cm column of bulk volume 3.8 ml. To make a positively charged, anion-exchanging coating they deposited quaternary amine salts by pumping solutions of these salts, 0.001 *M* or so, in 80 to 90% methanol or acetonitrile, followed by a gradient of decreasing methanol concentration to "lock" the coating in place.[26] Mention was made of this procedure in Chapter 2, section III.C.

Aqueous eluents could be passed through the coated columns almost indefinitely without any change in ion-exchange capacity. Up to 5% methanol or acetonitrile could be used in the eluent. This small proportion of organic modifier gave better contact between the stationary and mobile phases and more symmetrical peaks.

These dynamically coated or permanently coated columns gave faster ion exchange than fixed-site ion exchangers, which meant faster chromatography with more symmetrical peaks and less cross-contamination. Because the column capacity could be made quite high, the

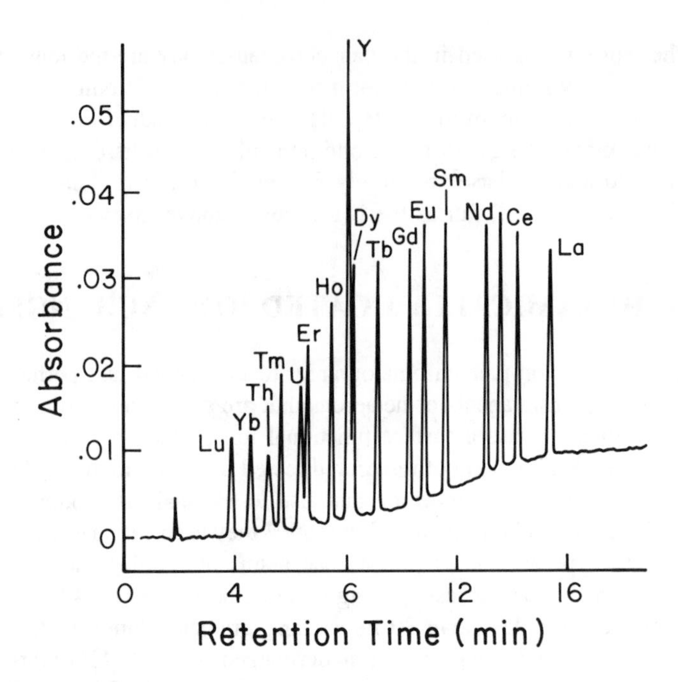

FIGURE 8. Separation of lanthanides, yttrium, thorium, and uranium by "dynamic ion exchange". Stationary phase, 5-μm C18-bonded silica; mobile phase, α-hydroxyisobutyric acid, pH 4.2, gradient from 0.05 to 0.4 *M*, with 7.5% methanol and 0.03 *M* sodium octyl sulfonate throughout. Detection by post-column reaction with Arsenazo III. (From Barkley et al., *Anal. Chem.*, 58, 2222, 1986. With permission.)

technique is very good for separating and measuring very small amounts of certain constituents such as lanthanide elements in the low nanogram range in the presence of much larger amounts of others, like uranium and thorium. It is well suited to the demands of geochemical analysis. Another great advantage of permanently coated ion exchangers over the conventional fixed-site exchangers is that the capacity of the column can be varied at will. The coating of long-chain alkyl sulfate can be stripped off by passing 70% acetonitrile, and the column can be recoated at another level. Changing the ion-exchanger capacity changes elution volumes as was noted in Chapter 3. The extent of the change depends on the charges on the ions being exchanged, and it is possible to move chromatographic peaks around and place them where one wishes. Thorium and uranium in HIBA solutions form uncharged complexes that are retained by PRP-1 in the absence of ion-pair reagents, and almost independently of these reagents; the lanthanides form charged or ionic complexes. With the proper proportion of long-chain sulfate or sulfonate in the stationary phase, the thorium peak was placed between ytterbium and terbium, the uranium peak between terbium and erbium.[25] The point of doing this was that terbium is very rare and of minor importance in uranium leach liquors. A chromatogram of the lanthanides, yttrium, uranium, and thorium is shown in Figure 8. Here, the pairing agent was octyl sulfonate, the stationary phase C18-silica. The eluent was 0.03 *M* in sodium octyl sulfonate and 7.5% in methanol, and a gradient with hydroxyisobutyric acid was used.

What can be done with anionic coatings can of course be done with cationic coatings, and long-chain amines with up to 37 carbon atoms in the molecule $(C_{12}H_{25})_3\ CH_3N^+$ were attached to the stationary phase by passing solutions in 80% methanol, followed by water. As was found with alkyl sulfonates and sulfates, carbon chains as short as eight atoms did not give permanently bound coatings, and where these reagents were used it was necessary to add some of the reagent to the mobile phase. These reagents were used, however, when they gave sharper peaks than the longer-chain, permanently bound pairing reagents. The reference cited shows chromatographic separations of strongly bound inorganic anions (iodide, nitrate, thiosulfate) on quaternary ammonium ion coatings as well as transition metals on C_{20} sulfate. Cetyl pyridinium ions attached to C18-silica or PRP-1 (porous polymer) made a good stationary phase for the chromatography of inorganic anions; they were applied to the adsorbents by passing a solution of cetyl pyridinium salt in 80 to 90% acetonitrile.[26] The eluent was tetramethylammonium salicylate, 0.0005 *M* in 7% acetonitrile; detection was done by conductivity. The elution order of common anions was F^-, Cl^-, NO_2^-, Br^-, NO_3^-, SO_4^{2-}, just as in conventional anion chromatography.

VIII. DETECTION IN PAIRED-ION CHROMATOGRAPHY

A. OPTICAL

An all-purpose method of detection in liquid chromatography is by the refractive index. Refractive index is a bulk property to which all the components of the solution contribute. It is insensitive, and not the method of choice unless there is no other way to detect the substances of interest; however, it is used in ion-pair chromatography to follow breakthrough tests that tell how much reagent ion is adsorbed on a column. Also, aromatic compounds, as a class, have higher refractive indices than aliphatic compounds, and refractive index detection may sometimes be used for special problems.

Much more often, the absorption of UV or visible light is used for detection. There are three possibilities. The first is direct detection, where the compound of interest absorbs and the pairing ion (hetaeron) does not. The second is indirect detection, where the hetaeron absorbs and the compound of interest does not. The third is an indirect method too; the counter-ion to the hetaeron, the ion C in Equation 3c above, absorbs.

In the first case, detection is simple. It works especially well in normal-phase ion-pair chromatography, where the stationary phase is an aqueous solution of sodium perchlorate and perchloric acid attached to porous silica, and the mobile phase is methylene chloride mixed with *n*-butanol (Knox and Jurand[7]). The mobile phase is transparent to UV light and the detector registers absorbance only when the analyte, a catecholamine cation, emerges as an ion pair with perchlorate. The perchlorate ion is likewise transparent at the wavelengths usually used in UV detection. In the commonly used "reversed phase" mode, the hetaeron could be dodecyl sulfate, and this, too, is transparent in the UV.

The second case is that in which the hetaeron absorbs and the analyte ions do not. An interesting example of such a system is provided by Rigas and Pietrzyk.[27] The hetaeron (or ion-interaction agent) was the cationic complex of iron(II) with 1,10-phenanthroline, $Fe(phen)_3^{+2}$, which has an intense red color. (It is used for the photometric determination of iron). It was adsorbed into the porous styrene-DVP polymer, PRP-1. It was very strongly adsorbed, so that the concentration of the complex in the eluent could be low; it was 10^{-4} molar. The column effluent was monitored at 505 nm, the wavelength of maximum absorbance of the red complex. There was a background absorbance and an increase in absorbance when the anions F^-, Cl^-, NO_2^-, NO_2^-, ClO_3^-, etc. came out. As the sample anions left the column they brought hetaeron ions, $Fe(phen)_3^{+2}$, with them.

An interesting feature of this system is that the hetaeron is doubly charged. Usually the pairing ions have a single charge, because one wants the creation of a cavity in the water structure to require energy; higher charges would cause orientation of water dipoles with an accompanying decrease in free energy. The complex of iron with 1,10-phenanthroline, however, is very large and symmetrical, and the electrostatic charge density at its surface is everywhere very low. Another interesting feature of this work is that the retention of sample anions was shown to depend greatly on the nature and concentration of other competing mobile-phase anions (see Table 1). Fluoride ions gave stronger retention than perchlorate, for perchlorate is a more strongly adsorbed competitor. Adding citrate, a highly charged, strongly adsorbed anion, to the mobile phase reduced the retention of all the test anions. Adding sodium sulfate to increase the ionic strength caused large drops in retention. These effects can be understood by the double-layer ion-interaction theory as well as by the approach of Knox and Hartwick[11] cited above.

A more orthodox example of ultraviolet detection with an absorbing hetaeron is the work of Barber and Carr[28] on the separation of inorganic anions on C18-silica with a mobile phase 4 m*M* in the light-absorbing cation, benzyltrimethyl-ammonium, plus 0.25 m*M* hexanesulfonate as a competing anion, plus acetate buffer, pH 4.75. As the anions appeared the light-absorbing cation appeared also and the absorbance rose; the peaks were positive.

Another way to use UV absorbance indirectly is to use an absorbing counter-ion to the hetaeron. Thus Dreux et al.[20] used the *p*-toluenesulfonate of a long-chain amine cation, N-methyloctylamine, in acid solution as the eluent with C8-silica as the stationary phase. They found that C8 gave sharper peaks than C18. The test anions, chloride, sulfate, nitrite, nitrate, and others, were transparent at the detection wavelength (254 nm). When they appeared, they took the place of the light-absorbing toluenesulfonate anion, and the peaks were marked by a *decrease* in absorbance. Of course, the toluenesulfonate ion, being strongly adsorbed by the stationary phase, lowered the capacity factors of the test anions by comparison with the perchlorate ion, which was less strongly adsorbed.

Another example of the same method is the work of Bidlingmeyer et al.[29] on the chromatography of simple inorganic anions on C18-silica with 0.4 m*M* tetrabutylammonium salicylate, pH 4.6, as the mobile phase. The detection wavelength was 288 nm, where salicylate absorbs but the tetramethylammonium ion does not. As the inorganic ions appear in the effluent the concentration of salicylate drops, and the absorbance decreases, giving negative peaks (see Figure 9). The sensitivity is very good; the chloride peak shown was given by 100 ng. The resolution is about the same as that found in ion chromatography.

Some inorganic anions, including nitrate, nitrite, bromide, iodide, bromate, iodate, and thiocyanate, absorb UV light at 205 nm and so can be determined directly. The eluent must then be transparent to UV. Skelly[39] used 0.01 *M* *n*-octylamine adjusted to pH 6 to 6.5 with hydrochloric, sulfuric, or phosphoric acids (all transparent in the UV), and columns of bonded silica, C_8 and C_{18}. The columns took a long time to come to equilibrium with the eluent; eluents were circulated overnight. During that time the alkyl-bonded silica became coated with octylammonium ions and became a surface-functional anion exchanger. Inorganic anions could be determined chromatographically with good sensitivity; 40 ng of bromide gave a good peak, and bromide, iodide, nitrite, nitrate, and iodate were determined in a 10 min run. Several organic anions were determined: formate, acetate, propionate, and butyrate were eluted in this order, as were acrylate (first) and methacrylate ions; these are the elution orders expected from the theory of Chu and Diamond that relates anion-exchange selectivity to ionic size and the breaking of hydrogen bonds (Chapter 3).

An annoying feature of chromatograms obtained with indirect UV absorbance is the appearance of system peaks, usually dips in the absorbance, that do not coincide with the elution of a solute. They come from disturbances caused when the sample is injected. Every component of the sample, including the solvent, perturbs the distributions of the various substances between stationary and mobile phases and these perturbations travel along the column at rates that depend, not on the distribution ratio of the substance between the two phases, but on the dependence of this ratio on the concentration, dc_s/dc_m. System peaks appear when any bulk property, such as nonsuppressed conductivity, is used for detection. Chromatographic conditions must be chosen so that they do not interfere with the peaks of interest.

A recent review by Haddad and Heckenburg[40] treats ion-interaction methods, indirect UV detection, and system peaks.

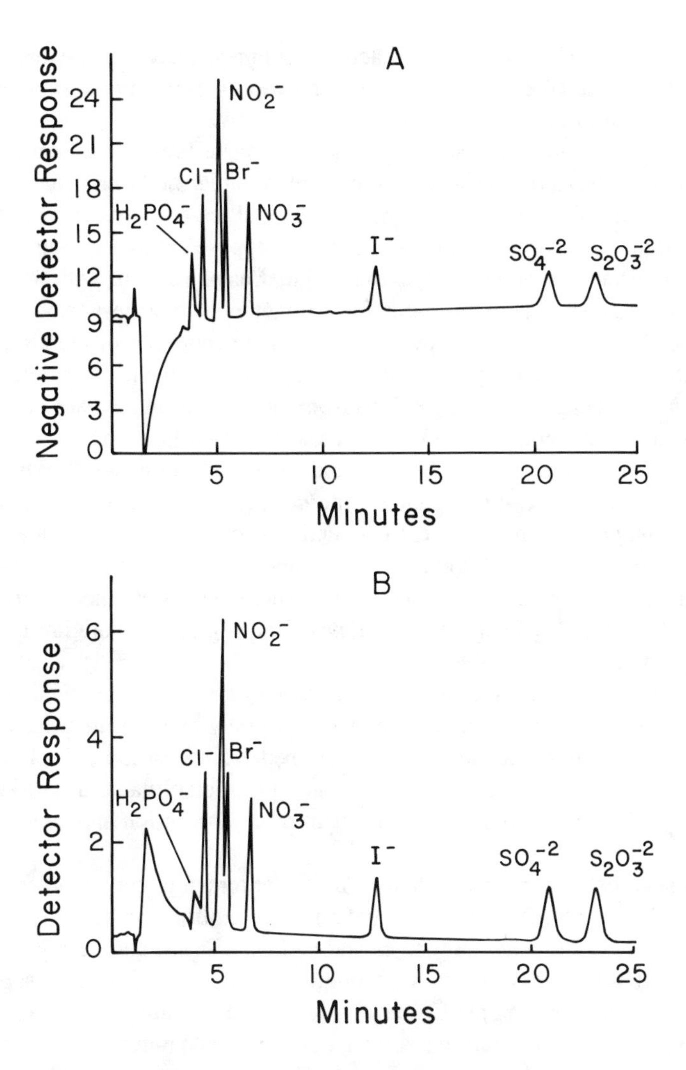

FIGURE 9. Chromatograms of inorganic ions by (A) UV visualization and (B) conductivity detection. Stationary phase 5-μm C18-bonded silica; mobile phase, 0.4 m*M* tetrabutylammonium salicylate in water, pH 4.2. Amounts of ions; 100 ng Cl^-, 300 ng SO_4^{2-}, and $S_2O_3^{2-}$, 200 ng for others. (From Bidlingmeyer, B. A., Santasania, C. T., and Warren, F. V., *Anal. Chem.*, 59, 1843, 1987. With permission.)

B. CONDUCTIMETRIC

Figure 9 shows besides the absorbance, a recording of electrical conductivity for the same chromatogram. Faster-moving chloride, nitrite, bromide, nitrate ions, and others were replacing the more slowly moving salicylate ions. Cassidy and Elchuk[26] used the same eluent, tetrabutylammonium salicylate, for the chromatography of anions and used conductivity detection. Molnar et al.[30] used the tetrabutylammonium ion as pairing ion with a smaller concentration of heptanesulfonate to serve as the competing or displacing ion; this ion is large and slow-moving, and when it was replaced by simple inorganic anions the conductivity rose. These authors described a new design of conductivity cell that was very sensitive.

The conditions for sensitive conductivity detection in ion-pair chromatography are just like those in ion chromatography described in Chapter 4. If no chemical suppression is to be used, the eluent must be so chosen that the ions to be determined shall replace other ions of very different equivalent conductance. Chemical suppression can, however, be used. Here it is necessary to choose an eluent that can be removed by a membrane suppressor or other device. A description of this technique is given by Weiss.[31]

To separate anions, inorganic anions or alkyl sulfonate ions, the eluent is 2 m*M* tetrapropylammonium hydroxide with a few percent of organic modifier if desired; this gives faster elution. Because this solution is highly alkaline, silica-based stationary phases cannot be used; the stationary phase is a porous polymer. Suppose sodium nitrate is injected into the column. The nitrate ion is retained by the polymer as an ion pair with the tetraalkylammonium ion, and leaves the column in this form, together with excess tetraalkylammonium hydroxide. The eluent now passes through a membrane suppressor of the kind described in Chapter 4, with cation-exchanging membranes, that have a regenerant solution of sulfuric acid flowing on the other side. Hydrogen ions from the regenerant pass across the membrane into the eluent stream and combine with the hydroxide ions, while tetraalkylammonium ions cross the membrane in the opposite direction and are swept out to waste. The solution that leaves the membrane suppressor and enters the conductivity detector contains nitric acid and, ideally, nothing else. The electrical conductivity gives a direct measure of the nitrate ions, and of all other strong-acid anions as they elute.

The chromatography of cations is analogous. The eluent is an alkyl sulfonic acid, and the suppressor has anion-exchanging membranes. The regenerant solution is a quaternary ammonium hydroxide.

The membranes that are used in these suppressors must tolerate small proportions of orgnic solvents. Acetonitrile, in particular, attacks most polymers that are used in membranes, and special kinds of membranes had to be developed for use in ion-pair chromatography. The term ''mobile-phase ion chromatography'' is a proprietary name used to describe ion-pair or ion-interaction chromatography with membrane suppression and conductivity detection.

In the course of developing this technique for the chromatography of anions it was found that ammonium hydroxide. 0.01 *M* or so, gave good retention and separation of the larger anions, such as hexyl and heptyl sulfonates and sulfates. The ammonium ion is the pairing ion or hetaeron. It was more effective than quaternary ammonium ions in separating pairs that were very similar, for example C_6 and C_7, or C_7 and C_8 sulfonates. The supposition is that the small ammonium ion, sharing a solvent cavity with the much larger sulfonate anions, allowed small differences in the sulfonate ions to affect the surrounding solvent or stationary phase to an extent that would not have happened with a large quaternary ammonium pairing ion. Separation factors were greatly diminished by adding organic solvents, as were the retention times.

C. ELECTROCHEMICAL

Any method of detection may be used that depends on a specific property of the analyte. One such method is amperometric. Kissinger et al.[32] detected catecholamines and vanillin by electrochemical oxidation at a carbon anode, following chromatographic separation on C18-silica in a citrate-phosphate buffer of pH 3.25, with 0.001 *M* octyl sulfate as the pairing ion. This method has become very popular for the determination of catecholamines in biological fluids, and there are many references in the literature to ion-pair chromatography and amperometric detection.

IX. SUMMARY AND CONCLUSION

In this chapter the emphasis has been on theory and methodology, but illustrative examples of actual applications have been given. Since 1970 there have been so many publications on ion-pair and ion-interaction chromatography that it would take several pages to list them all, and then the list would be incomplete. A comprehensive review of the principles and practice of ion-pair chromatography up to 1978 was written by Tomlinson et al.[33] A table of published applications lists over 100 references. Bidlingmeyer[17] in 1980 listed almost 100 more and another excellent review appeared by Jira et al.[34] in 1984, with a table of 150 publications that had appeared since Tomlinson's review. Most of the applications are of pharmaceutical or biochemical interest. Many are to the detergent and dyestuffs industries. In Jira's review nearly all of them use the ''reversed-phase'' technique, where the pairing ion or ions are supplied in the effluent and the stationary phase is an alkyl-bonded silica. Some workers[13] have noted that shorter alkyl chains, for example C8 in place of C18, give sharper peaks, though they also give less retention per carbon chain. (The use of porous polymers, styrene-DVB or other, has only appeared since 1982).

Cationic pairing ions in use are largely quaternary ammonium ions; tetrabutylammonium ions are the smallest as a rule, although we have mentioned the use of ammonium ions for chromatography of long-chain sulfonic acid anions. Cations or anions with long carbon ''tails'', cetyl (hexadecyl) trimethylammonium, for example, have different properties from their isomers with shorter chains. Primary amines can also be used (in acid solution) as pairing ions; octylammonium ions are used in some applications, such as the chromatography of large inorganic anions (see below). We have noted the special use of the iron(II)-1,10-phenanthroline complex.[27]

Anionic pairing species are more varied. Long-chain alkyl sulfates and sulfonates with C_6 to C_{12} are widely used; sulfonates have terminal groups $-SO_3^-$, sulfates $-O-SO_3^-$; sulfates are more strongly absorbed. Aryl sulfonates and sulfates are used where their UV absorbance helps in detection. Naphthalene-2-sulfonate is often used. Alkyl naphthalene sulfonates like dinonylnaphthalene sulfonate are extremely hydrophobic and soluble only in organic solvents, but they may be used to prepare permanently coated ion exchangers. Camphorsulfonic acid is sometimes used in pharmaceutical applications. Picric acid was used in the normal-phase mode, that is as an aqueous solution adhering to porous silica.

The ion-pairing ability of inorganic anions should not be overlooked. Perchlorate has been used in the normal-phase mode, and is added to mobile phases for anion chromatography that contain a quaternary ammonium salt as the pairing agent, in order to moderate or reduce the retention of the analyte anions.[13] Perchlorate forms strong ion pairs with quaternary ammonium ions.

Comparing ion-pair chromatography with orthodox ion chromatography on fixed-site surface-functional ion-exchange resins, we have seen that ion-pair chromatography is much better for large hydrophobic organic anions and cations. We have also seen that it can be used for inorganic anions and cations. It is most useful for larger ions of high charge that are strongly retained on fixed-site ion exchangers. Examples are thiosulfate and the anions of thionic acids, and metal cyanide complexes. Ferrocyanide and ferricyanide, and gold complexes Au $(CN)_4^-$ and Au $(CN)_2^-$, are efficiently separated by mobile-phase ion chromatography[31] using tetrabutylammonium hydroxide as the pairing reagent plus 0.0002 *M* sodium carbonate in 35% acetonitrile.[31] Not only are these ions very strongly bound by ion-exchange resins, their mass transfer is very slow and the bands are broad. This behavior is characteristic of ions of high charge. We have noted (Chapter 3) the very slow diffusion rates of highly charged cations in gel-type ion-exchange resins and it can be attributed to, first, the high activation energy required to make the cation jump from one negative fixed ion to another, and second, the shrinkage of the resin that is caused by high electrostatic fields around the ions.

To analyze complicated mixtures of coordination complexes of chromium and cobalt, Searle et al.[35] turned to ion exchange. Ion exchange has been effective but very slow and tedious, because of the slowness of exchange of these ions, which often carry high charges and are very large. Ion-pair chromatography works much better. Buckingham et al.[36,37]

separated the complex cations $[Co(NH_3)_5X]^{2+}$, where X was a halogen atom, nitrate or nitro-, thiocyanate or isothiocyanate, on C18-silica with 0.025 *M* toluene sulfonate as the pairing ion, with a water-to-methanol gradient. Separations were accomplished in 20 min that formerly took hours, or were impossible. Cobalt(III) complexes with charges ranging from +4 to −3 were separated in 15 min; pairing ions were butanesulfonate for cations, octylammonium for anions; competing ions, triethylammonium for cations and citrate for anions, were added to moderate the attachment.[38]

It is interesting that the ion-pair method works so well with ions of charge greater than 1. In Diamond's water-structure theory ion pairs are favored by single charges, and in nearly all the applications to organic ions, the ions are large and singly charged.

These last examples show the versatility of ion-pair chromatography. No technique is without its disadvantages, and one difficulty that will have become apparent is the complexity of this method, compared with ordinary ion exchange. Eluents have many components and all must be controlled. Large volumes of dilute mobile phases may have to be pumped before the column packing comes to equilibrium with the mobile phase. Surface-acting pairing ions, like anionic and cationic soaps, become permanently attached to silica-based adsorbents, so that a column once used for a particular type of ion-pair analysis cannot be used for anything else. In this respect the porous polymers seem to be better.

REFERENCES

1. **Horvath, C., Melander, W., Molnar, I., and Molnar, P.,** Enhancement of retention by ion-pair formation in liquid chromatography with nonpolar stationary phases, *Anal. Chem.*, 49, 2295, 1977.
2. **Wittmer, D. P., Nuessle, N. O., and Haney, W. G.,** Simultaneous analysis of tartrazine and its intermediates by reversed-phase liquid chromatography, *Anal. Chem.*, 47, 1422, 1975.
3. **Sood, S. P., Sartori, L. E., Wittmer, D. P., and Haney, W. G.,** High-pressure liquid chromatography determination of ascorbic acid, *Anal. Chem.*, 48, 796, 1976.
4. **Schill, G. and Wahlund, K. G.,** Ion-pair HPLC of drugs and related organic compounds in Trace Organic Analysis, Special Publ. No. 519, National Bureau of Standards, U.S. Government Printing Office, 1978, 509.
5. **Knox, J. H. and Laird, G. R.,** Soap chromatography; A new high-performance liquid chromatographic technique for separation of ionizable materials, *J. Chromatogr.*, 122, 17, 1976.
6. **Schmidt, J. A. and Henry, R. A.,** Applications of a new anion-exchange column for liquid chromatography, *Chromatographia*, 3, 497, 1976.
7. **Knox, J. H. and Jurand, J.,** Ion-pair or soap chromatography: separation of catecholamines, *J. Chromatogr.*, 125, 89, 1976.
8. **Robinson, R. A. and Stokes, R. H.,** *Electrolyte Solutions*, 2nd ed., Butterworths, London, 1959, chap. 14.
9. **Frank, H. S. and Wen, W.-Y.,** Structural aspects of ion-solvent interaction in aqueous solution; a suggested picture of water structure, *Faraday Discussions Chem.*, Soc., 24, 133, 1957.
10. **Diamond, R. M.,** The aqueous solution behaviour of large univalent ions. A new type of ion-pairing, *J. Phys. Chem.*, 67, 2513, 1963.
11. **Knox, J. H. and Hartwick, R. A.,** Mechanisms of ion-pair liquid chromatography of amines, neutrals, zwitterions and acids using anionic hetaerons, *J. Chromatogr.*, 204, 3, 1981.
12. **Van de Venne, J. L. M., Hendrikx, J. L. H. M., and Deelder, R. S.,** Retention behaviour of carboxylic acids in reversed-phase liquid chromatography, *J. Chromatogr.*, 167, 1, 1978.
13. **Vera-Avila, L. E., Caude, M., and Rosset, R.,** Chromatographie de paires d'ions, *Analusis*, 10, 36, 43, 1982.
14. **Armstrong, D. W.,** Micelles in separations: a practical and theoretical review, *Sep. Purif. Methods*, 14, 213, 1985.
15. **Kissinger, P. T.,** Comments on reversed-phase ion-pair partition chromatography, *Anal. Chem.*, 49, 883, 1977.
16. **Bidlingmeyer, B. A., Deming, S. N., Price, W. P., Sachok, B., and Petrusek, M.,** Retention mechanism for reversed-phase ion-pair liquid chromatography, *J. Chromatogr.*, 186, 419, 1979.

17. **Bidlingmeyer, B. A.,** Separation of ionic compounds by reversed-phase liquid chromatograhy: an update of ion-pairing techniques, *J. Chromatog. Sci.,* 18, 525, 1980.
18. **Bidlingmeyer, B. A. and Warren, F. V.,** Effect of ionic strength on retention and detector response in reversed-phase ion-pair liquid chromatography with ultraviolet-absorbing reagents, *Anal. Chem.,* 54, 2351, 1982.
19. **Iskandarani, A. and Pietrzyk, D. J.,** Ion interaction chromatography of organic ions on a poly(styrene-divinylbenzene) adsorbent in the presence of tetraalkylammonium salts, *Anal. Chem.,* 54, 1065, 1982.
20. **Dreux, M., Lafosse, M., and Pequignot, M.,** Separation of inorganic anions by ion-pair reverse phase liquid chromatography, *Chromatographia,* 15, 653, 1982.
21. **Melander, W. R., Kalgheti, K., and Horvath, C.,** Formation of ion pairs under conditions employed in reversed-phase chromatography, *J. Chromatogr.,* 201, 201, 1980.
22. **Elchuk, S. and Cassidy, R. M.,** Separation of the lanthanides on high-efficiency bonded phases and conventional cation-exchange resins, *Anal. Chem.,* 51, 1434, 1979.
23. **Cassidy, R. M. and Fraser, M.,** Equilibria effects in the dynamic ion-exchange separation of metal ions, *Chromatographia,* 18, 369, 1984.
24. **Cassidy, R. M. and Elchuk, S.,** Dynamically coated columns for the separation of metal ions and anions by ion chromatography, *Anal. Chem.,* 54, 1558, 1982.
25. **Barkley, D. J., Blanchette, M., Cassidy, R. M., and Elchuk, S.,** Dynamic chromatographic systems for the determination of rate earths and thorium in samples from uranium ore refining processes, *Anal. Chem.,* 58, 2222, 1986.
26. **Cassidy, R. M. and Elchuk, S.,** Dynamic and fixed-site ion-exchange columns with conductometric detection for the separation of inorganic anions, *J. Chromatog. Sci.,* 21, 454, 1983.
27. **Rigas, P. G. and Pietrzyk, D. J.,** Liquid chromatographic separation and indirect detection of inorganic anions using iron(II) 1,10-phenanthroline as a mobile phase additive, *Anal. Chem.,* 58, 2226, 1986.
28. **Barber, W. E. and Carr, P. W.,** Ultraviolet visualization of inorganic ions by reversed-phase ion-interaction chromatography, *J. Chromatogr.,* 260, 89, 1983.
29. **Bidlingmeyer, B. A., Santanasia, C. T., and Warren, F. V.,** Ion-pair chromatographic determination of anions using an ultraviolet-absorbing co-ion in the mobile phase, *Anal. Chem.,* 59, 1843, 1987.
30. **Molnar, I., Knauer, H., and Wilk, D.,** High-performance liquid chromatography of ions, *J. Chromatogr.,* 201, 225, 1980.
31. **Weiss, J.,** Handbook of Ion Chromatography, Dionex Corp., Sunnyvale, CA, 1986.
32. **Kissinger, P. T., Bruntlett, C. S., Davis, G. C., Riggin, R. M., and Shoup, R. E.,** Recent developments in the clinical assessment of the metabolism of aromatics by high-pressure, reversed-phase chromatography, *Clin. Chem.,* 23, 1449, 1977.
33. **Tomlinson, E., Jefferies, T. M., and Riley, C. M.,** Ion-pair high-performance liquid chromatography, *J. Chromatogr. Chromatogr. Rev.,* 159, 315, 1978.
34. **Jira, Th., Beyrich, Th., and Lemke, E.,** Ionenpaar-HPLC, *Pharmazie,* 39, 141, 1984.
35. **Searle, G. H.,** The role of ion-association in the chromatographic separation of isomeric cationic Co(III)-amine complexes on cation-exchange resins, *Aust. J. Chem.,* 30, 2625, 1977.
36. **Buckingham, D. A., Clark, C. R., Deva, M. M., and Tasker, R. F.,** Reversed-phase HPLC of pentammine cobalt(III) complexes, *J. Chromatogr.,* 262, 219, 1983.
37. **Buckingham, D. A.,** Reversed-phase high- performance ion-pair chromatography of cobalt(III) coordination compounds, *J. Chromatogr. Chromatogr. Rev.,* 313, 93, 1984.
38. **Kirk, A. D. and Hewavitharana, A. K.,** Reversed-phase chromatographic separation of highly charged inorganic cations and anions using ion interaction reagents and competing ions, *Anal. Chem.,* 60, 797, 1988.
39. **Skelly, N. E.,** Separation of inorganic and organic anions on reversed-phase liquid chromatography columns, *Anal. Chem.,* 54, 712, 1982.
40. **Haddad, P. R. and Heckenburg, A. L.,** Determination of inorganic anions by high-performance liquid chromatography, *J. Chromatogr.,* 300, 357, 1984.
41. **Bartha, A. and Vigh, Gy.,** Studies in ion-pair chromatography. II. Retention of positive and negative ions and neutral solutes, *J. Chromatogr.,* 265, 171, 1983.
42. **Bartha, A., Billiet, H. A. H., de Galan, L., and Vigh, Gy.,** Studies in reversed-phase ion-pair chromatography. III. The effect of counter-ion concentration, *J. Chromatogr.,* 291, 91, 1984.

Chapter 6

LIGAND-EXCHANGE CHROMATOGRAPHY

I. INTRODUCTION

The interior of a bead of strong-acid sulfonated polystyrene ion-exchange resin, swollen in water, is like a drop of concentrated electrolyte solution. The cations are hydrated but are not attached covalently to the sulfonate ions. Water molecules are coordinated as ligands to the metal ions. The word ligand simply means something that is "tied on", or attached to the central atom or ion by coordinate valences. If water molecules are ligands, then other molecules or ions can be ligands too.

A bead of crosslinked sulfonated polystyrene resin carrying copper(II) ions and swollen with water has the characteristic light blue color of the copper(II) ion in solution. If it is now placed in a solution of ammonia, the color becomes dark blue, the color of copper(II)-ammonia complexes in aqueous solution. One ligand, the water molecule, has been exchanged for another, the ammonia molecule. The exchange is rapid and reversible. This is not true of all metal-ligand complexes. The complexes of cobalt(III) and chromium(III) are well known to be kinetically stable or inert. Their ligands can be exchanged for others but only slowly. Exchange of ligands on the copper ion can be the basis for chromatography but exchange of ligands on the ions Co(III) and Cr(III) cannot.

The first use of ligand exchange for chemical analysis was made in 1960 by Tsuji and Sekiguchi.[1] They used cation-exchange resins loaded with ions of transition metals, including Cu, Ni, Cd, Zn, and Pb to adsorb isonicotinic acid hydrazide from water through the formation of coordination complexes within the resins. Copper gave the strongest adsorption. Though this attachment was not chromatography, it did show that metal-ligand coordination could occur within an ion-exchange resin just as it did in water. The first use of ligand exchange for chromatography was announced by Helfferich in 1962.[2] Helfferich took a polyacrylate resin, Bio-Rex 70, with functional carboxylate groups, and loaded it in a column with copper-ammonia complex ions by pouring through it a dilute solution of copper sulfate to which excess ammonia was added. The color of the resin became a rather light blue, not the deep blue of the $Cu(NH_3)_4^{2+}$ ions, suggesting that the copper ions in the resin were not fully coordinated with ammonia, but partly coordinated to the fixed carboxylate ions of the resin. Nevertheless, the coordinated ammonia molecules, two at least for every copper ion, could be replaced by other ligands. A dilute solution of a diamine, 1,3-diamino-2-propanol, which also contained ammonia was passed through the column. The diamine stuck tightly to the copper-loaded resin, displacing ammonia, and turning the resin deep blue. The front was very sharp. The diamine was displaced and recovered in concentrated form by passing a small volume of concentrated ammonia solution. We have here an equilibrium in which two ligands are exchanged for one:

$$Res_2Cu(NH_3)_2 + da = Res_2Cu(da) + 2NH_3$$

where da = the diamine. This is an exchange like the exchange of one calcium ion for two sodium ions that is used in water softening, where a high concentration favors the univalent species in the exchanger, and a dilute solution favors the divalent species in the exchanger. Thus, Helfferich had devised a simple and efficient process for recovering the divalent ligand, the diamine, in concentrated form from a dilute solution that also contained ammonia. He also saw the potential of ligand exchange in analytical chromatography. Two amines, XNH_2 and YNH_2, could be separated by injecting a small sample of the mixture on a column

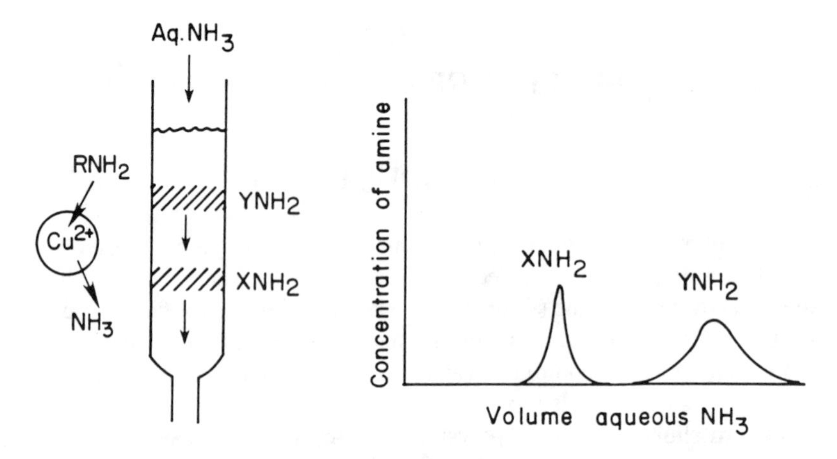

FIGURE 1. Ligand-exchange chromatography, schematic. (Reprinted with permission from Walton, H. F., *Ion-Exchange Chromatography* ACS Audio Courses C-93; Society: Washington, DC, 1987. Copyright 1987 American Chemical Society.)

of copper-loaded resin and passing a solution of ammonia. If XNH_2 forms a stronger copper complex than YNH_2, it will pass through the column faster and be eluted first (see Figure 1).

Ligand-exchange chromatography has the potential for being very versatile. The metal ion can be varied and so can the kind of ion exchanger, both the polymer type and the functional group. A variety of selectivity orders can be expected. Not only amines, but other types of ligands with other electron-donor atoms, like oxygen and sulfur, can be exchanged, provided only that they do not strip the metal ion from the exchanger. This condition requires that the ligands be uncharged, or at least that they do not completely neutralize the positive charge of the metal ion. Helfferich wrote, ''the method combines two fields of chemistry, namely, ion exchange and coordination chemistry, in order to accomplish a task that neither could do alone.''

In ligand-exchange chromatography as it was first visualized, the metal ions would not move. They would be held firmly in place in the exchanger, while only the labile ligands changed places and moved with the mobile phase. Of course this would be an ideal situation. The dilute ammonia solution used as the eluent in Figure 1 is partly ionized. It contains cations NH_4^+ which can replace copper ions in the exchanger by orthodox ion exchange. The metal ions are bound to move to some extent. In general one must add a low concentration of metal ions to the mobile phase to compensate for this displacement, but chromatography is simpler if the metal ions are attached firmly to the exchanger. It was for this reason that Helfferich chose an exchanger with functional carboxylic groups, rather than the usual sulfonic groups. He discussed the various equilibria in this system, and in the replacement of ammonia by a diamine, in two papers.[3] The choice of copper(II), rather than another transition-metal ion, was made primarily because, in the series of transition-metal ions from Ti to Zn in the periodic table, Cu(II) forms by far the most stable complexes with nitrogen and oxygen ligands. In the Irving-Williams series, the stability of the complexes with divalent ions rises slowly in the order $Mn<Fe<Co<Ni$, rises considerably from Ni to Cu, then falls to Zn. The square planar configuration of copper(II) complexes may be a factor in the stability. Another consideration is the color of copper ions, which helps the experimenter to see what is going on.

II. STABILITY OF METAL-LIGAND COMPLEXES

In 1954 Stokes and Walton[4] measured the stability of copper-ammonia complexes in sulfonated polystyrene cation-exchange resins by bringing to equilibrium known quantities of resin, copper salt, ammonia, and ammonium salt in aqueous solution, and then analyzing the solution and resin. A measurement of pH, combined with the known concentration of ammonium ions, gave pNH_3, the negative logarithm of the ammonia activity in solution. They plotted the formation curve, (ratio of bound ammonia to total copper in the resin) against pNH_3, and found that the curve was identical with the formation curve in aqueous 5 *M* ammonium nitrate solution, measured earlier by J. Bjerrum. The same relation was found for silver ions in the resin. The implication was that the interior of the resin was indeed like a drop of concentrated salt solution; the ammonia complexes were as stable inside the resin as they were in water containing an equivalent concentration of salt.

In the carboxylic (acrylic) resin this was not so. The formation curves were displaced by a factor of 30 in the ammonia concentration (or 1.5 logarithmic units) to less stable values. The presumption would be that the carboxylate ion combines with the copper ion and blocks two of its coordinate valences, leaving only two to bind with ammonia; however, the copper ion could bind more than two ammonia molecules, but with less stability than in the sulfonic acid resin.

In an exchanger with functional iminodiacetate groups (Chelex®-100) the metal ion is held tightly by three of its coordinate valences, leaving only one bond (in Cu(II)) to combine with ammonia and exchange ligands. The low ligand-binding capacity makes this resin unsuitable for chromatography. Another difficulty with using a polystyrene-based iminodiacetate resin for ligand exchange is that diffusion of the ligands in and out of the polymer is very slow. Other iminodiacetate polymers with the functional group on the end of a long spacer chain do find use in the chromatography of proteins; see below.

Turning from ammonia as a ligand to other nitrogen ligands, we find that the complex-ion stabilities are generally not the same inside the resin as out. In some cases the complexes are less stable in the resin. This is generally true of secondary and tertiary aliphatic amines, an example being piperidine. The reason seems to be steric hindrance; a nitrogen donor atom that has more than one carbon atom attached to it has difficulty in approaching the metal ion within the confines of the resin polymer. The presence of hydroxyl groups in the amine makes the complex more stable in the surrounding water and less stable in the resin. A simple example of this effect is the comparison of the nickel(II) complexes of ethanolamine and methylamine. The stability constants of the 1:1 complexes in water are respectively 950 and 50 l/mol, yet in a sulfonated polystyrene resin methylamine is held more strongly than ethanolamine,[5] and binding is weaker. In some cases the complexes are more stable within the resin. This is true of aromatic amines, where there is pi-electron interaction between the aromatic ring of the amine and the aromatic rings of the resin polymer. Spectacular increases in stability within the resin are found with 1,2-diamines. The 1:1 silver-ethylenediamine complex is a thousand times more stable in a sulfonated polystyrene resin than in water,[6] yet the 1:2 complex, which forms freely in water, does not form at all in the resin. The cause of this anomaly is not known. In montmorillonite clay, which is a cation exchanger, the 1:2 copper-ethylenediamine complex is 1000 times as stable as in aqueous solution; in a macroporous sulfonated polystyrene ion exchanger it is 20 times as stable.[7] Again, the reason is unknown, but it has been suggested that a delocalization of the ionic charges in montmorillonite may be responsible.[8,9] Whatever the cause, the chromatographer should know that a copper-loaded polystyrene-based ion exchanger holds 1,2-diamines so strongly that the only way to get them off is to strip the copper off the resin with acid. The 1,3-diamines are held less strongly.

III. ELUTION ORDERS OF AMINES

Elution orders will naturally follow the order of stability of the complexes within the exchanger. Many experiments have been made by injecting mixtures of amines into columns

TABLE 1
Selectivity Values, k'_D/k'_L, for Amino Acids on Copper-Loaded Chiral Polystyrene-Based Ion Exchangers Having Different Fuctional Groups

	Group				
Amino Acid	Proline	Hydroxyproline	Allo-hydroxyproline	Azetidine carboxylic acid	Benzyl propane diamine
Alanine	1.08	1.04	1.04	1.06	0.70
Aminobutyric acid	1.17	1.22	1.18	1.29	0.49
Norvaline	1.34	1.65	1.42	1.24	0.48
Norleucine	1.54	2.20	1.46	1.40	0.49
Valine	1.29	1.61	1.58	1.76	0.31
Leucine	1.27	1.70	1.54	1.24	0.49
Isoleucine	1.50	1.89	1.74	1.68	0.62
Serine	1.09	1.29	1.24	2.15	0.62
Threonine	1.38	1.52	1.48	0.78	0.44
Methionine	1.04	1.22	1.52	1.29	0.60
Phenylalanine	1.61	2.89	3.10	1.86	0.52
Tyrosine	2.46	2.23	2.36	1.78	—
Proline	4.05	3.95	1.84	2.48	0.47
Hydroxyproline	3.85	3.17	1.63	2.25	—
Allo-hydroxyproline	0.43	0.61	1.48	1.46	—
Azetidine carboxylic acid	—	2.25	—	—	—
Ornithine	1.0	1.0	1.20	1.0	1.0
Lysine	1.10	1.22	1.33	1.06	1.0
Histidine	0.37	0.36	1.32	0.56	0.85
Tryptophan	1.40	1.77	1.10	1.13	—
Aspartic acid	0.91	1.0	0.81	0.88	1.0
Glutamic acid	0.62	0.82	0.69	0.77	—

From Davankov, V. A., Navratil, J. D., and Walton, H. F., *Ligand Exchange Chromatography*, CRC Press, Boca Raton, Fl, 1988. With permission.

of metal-loaded ion exchangers, with aqueous ammonia solutions as eluents. The column effluent was monitored by refractive index, an insensitive method but adequate to establish elution orders. Shimomura et al.[10] took four amines, dimethylamine, *n*-butylamine, ethanolamine, and diethanolamine, and observed the elution orders on five exchangers with Ni(II) and Cu(II) as their counter-ions. The elution orders are shown in Table 1. Of the twenty-four orders possible, six were observed.[10] A similar study with diamines and polyamines as solutes, Cu and Zn as metal ions, and carboxylate and sulfonate as fixed ions, showed several different elution orders.[11] Evidently, different orders may be obtained by varying the type of exchanger, its functional group, and the metal ion.

In this early work metal salts were generally not added to the mobile phase. Copper ions bled rapidly from sulfonated polystyrene, but nickel ions did not, so many early studies were made with nickel-loaded sulfonated polystyrene and with copper-loaded polyacrylate resin.

Comparing different amino compounds, two structural factors were found to be important for their retention on metal-loaded resins. The first was steric hindrance around the terminal amino group. Of the four isomeric primary butylamines, the elution order on nickel- or copper-loaded polystyrene-based resins was as follows.

Isopropylamine was held more weakly than *n*-propylamine; methyl hydrazines were held much more weakly than unsubstituted hydrazine; phenethylamine was held much more strongly than its methyl-substituted derivatives, amphetamine, and metamphetamine.[12] There are two ways of looking at these effects. It may be that crowding of carbon atoms around

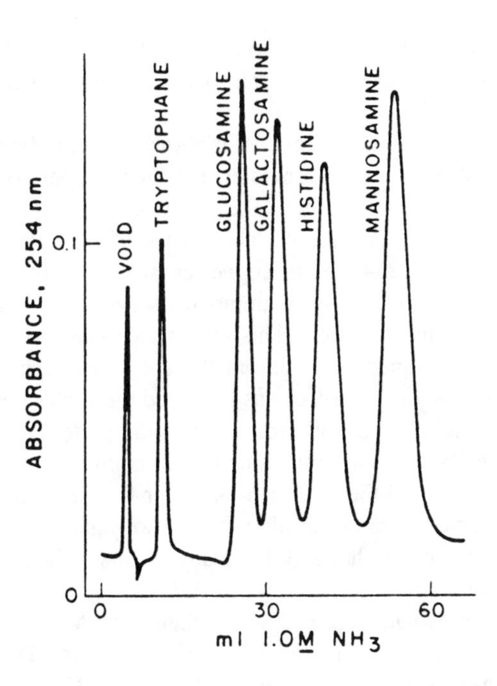

FIGURE 2. Ligand-exchange chromatography of amino sugars. Stationary phase, polyacrylate resin, Bio-Rex 70, loaded with Cu(II). (From Navratil, J. D., Murgia, E., and Walton, H. F., *Anal. Chem.*, 47, 890, 1972. With permission.)

the amine nitrogen makes it harder for the nitrogen to approach the metal ion in the resin, and it may also be that in the aqueous phase, outside the resin polymer, larger and more extended molecules require more energy to make cavities in the hydrogen-bonded water structure (see Chapter 3). Certainly the steric effect is very evident in ligand-exchange selectivity.

The second effect of the structure of the amine is the presence of hydroxyl groups. Hydroxyl groups interact with water though hydrogen bonds, and as we have seen, hydrogen bonding tends to bring the solute out of the resin and into the aqueous phase. Studies with amphetamines and derivatives of phenethylamine with mobile phases rich in ethanol and propanol showed that the group $-CHOH-CHNH_2-$ bound more tightly to the metal ions in the resin than the group $-CH_2-CHNH_2-$ owing, presumably, to the coordination of the metal ion (Cu or Ni) in a five-membered ring.[12]

A good example of steric selectivity is the ligand-exchange chromatography of amino sugars on copper-loaded Bio-Rex 70, an acrylate ion exchanger, with 1 *M* ammonia, 0.001 *M* in copper sulfate, as the eluent (see Figure 2). Three isomeric amino hexoses, glucosamine, galactosamine, and mannosamine, whose structures difer only in the orientation of hydroxyl and amino groups about the hexose ring, show well-separated peaks. Amino acids are also retained, but with one exception, the basic amino acid histidine they elute near the void volume. Even the histidine peak is cleanly separated from the peaks of the amino sugars.[13]

The peak of lysine (not shown) is unsymmetrical and close to the peak of galactosamine. It is possible to prevent interference by amino acids by choosing the proper ammonia concentration in the eluent. The reason is that the retention of amino acids depends inversely on the square of the ammonia concentration, indicating replacement of one amino acid molecule by two ammonia molecules, whereas the retention of the amino sugars depends on the first power of the ammonia concentration. (Even though the –OH groups of the amino sugars must take part in the coordination, to account for the differences in retention, it seems that the –OH binding is sufficiently weak that one ammonia molecule can displace one amino sugar molecule from the copper ion).

In Figure 2 the absorbance in the ultraviolet is shown as the ordinate. The amino sugars and amino acids do not absorb in the ultraviolet, but their copper complexes do. As these compounds migrate along the column they take small concentrations of copper with them in the mobile phase. There is a background absorbance of the copper-ammonia complex, but at the detection wavelength, 254 nm, the copper-ammonia complex absorbs only weakly. The copper complexes of amino sugars and amino acids absorb more strongly and at higher wavelengths than the ammonia complex. This absorbance was noted in 1952 by Spies,[14] and was used for detection in amino acid chromatography by Grushka et al.[15] and later by Foucault and Rosset[16] and others. Grushka has tabulated the molar absorptivities of copper complexes of several amino acids at 230 nm, which is the preferred detection wavelength.

Masters and Leyden[17] separated amino sugars and amino acids on copper-loaded, silylated porous glass. They found the same elution order for the amino sugars as is seen in Figure 2, but the amino acids were eluted after the amino sugars, not before. The reason is not clear, but we have already emphasized that many factors influence selectivity orders in ligand-exchange chromatography.

The chromatogram shown in Figure 2 was obtained with a soft resin, a crosslinked polyacrylate, of irregular shape and particle size, 37 to 44 μm. The newer macroporous polyacrylates of 10 μm particle size, the Spherons® described in Chapter 2 would undoubtedly give sharper peaks and better performance.

IV. DIFFICULTIES IN LIGAND-EXCHANGE CHROMATOGRAPHY

All the chromatography we have described so far was done with gel-type resins. We have noted the big limitation of these resins for chromatography: the slow diffusion of ions in and out. Large ligand molecules diffuse even more slowly, and this is the main reason that ligand-exchange chromatography has not been more popular. Another reason is its complexity, the need for adding metal salts to the eluent if one is to avoid stripping the metal ions from the exchanger, and for choosing the proper concentration of metal ions in the mobile phase. Detection may be difficult. Amines and amino acids are ordinarily detected by their reaction with ninhydrin to produce red, purple, or yellow colored products, but ammonia reacts with ninhydrin too, and ammonia is the favored eluent.

To make ligand-exchange chromatography more efficient, the first step is to use surface-functional exchangers such as those that have made ion chromatography possible. However, these materials have low ion-exchange capacity. Therefore, the displacing ligand, ammonia in all the examples discussed so far, must be made more dilute. Now the problem arises that the solvent, water, is also a ligand. At low concentrations the metal ions are hydrolyzed, yielding hydroxide precipitates instead of ammonia complexes. One way to get over this difficulty is to use nonaqueous solvents in conjunction with bonded silica chelating ion exchangers. A recent paper by Takayanagi et al.[18] describes the use of copper-loaded iminodiacetate chelating exchangers in which the chelating groups are attached by carbon chains to porous silica. The mobile phase is hexane, containing methanol and ammonia (0.003 *M*) as displacing ligands. The graph of k' against $1/[NH_3]$ is not linear but curves upwards, suggesting that methanol cooperates with ammonia as a displacing ligand. Good separations of aliphatic amines were achieved; *n*-propylamine and butylamine are retained more strongly than their branched-chain isomers; however, the differentiation of isomeric amines, on which we have commented, was much less than in gel-type exchangers. Ligand exchange can also be performed in the mobile phase, and where this is done, slow diffusion is not an obstacle.

V. AMINO ACIDS AND PEPTIDES

Naturally occurring amino acids are α-amino acids, that is, they have their amino group and carboxyl group attached to the same carbon atom. They exist in the zwitterionic or dipolar-ion form:

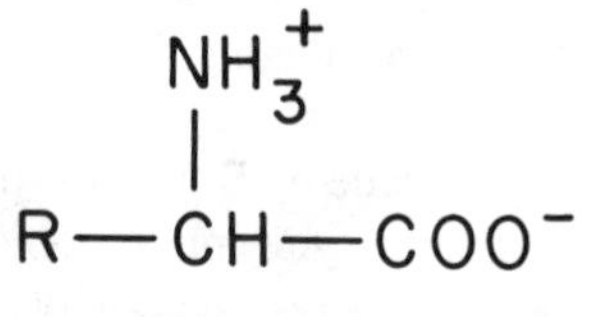

$$\begin{array}{c} NH_3^+ \\ | \\ R—CH—COO^- \end{array}$$

Coordinating with metal ions they are bidentate ligands. One amino acid ion displaces two ammonia molecules from a metal-ammonia complex ion, as we have already noted. The reaction of ligand exchange can be written thus:

$$Res_2^{2-}Cu(NH_3)_4^{2+} + L^- + B^+ = Res^-B^+ + Res^-Cu(NH_3)_2L^+ + 2NH_3$$

where L^- is the amino acid anion and B^+ is a cation, normally the cation of a buffer, Na^+ or NH_4^-. The retention of the amino acid in chromatography is influenced by the nature and concentration of the cations B, and of course the corrected retention volume depends inversely on the square of the ammonia concentration. Coordination of L^- to the metal ion reduces its positive charge from 2 to 1, but does not strip it from the resin. A very high concentration of amino acid, however, would form the uncharged 2:1 amino acid to metal complex and would strip the metal from the resin.

The first use of copper-loaded ion-exchange resins to adsorb amino acids was made by Siegel and Degens.[19] They collected traces of amino acids from sea water by passing the water through a small column of chelating resin loaded with Cu(II), then stripping the adsorbed amino acids by passing concentrated ammonia. Buist and O'Brien[20] adsorbed amino acids and peptides from urine in the same way, and separated the amino acids from the peptides because the peptides were more weakly held.

Chromatographic analysis of amino acids by ligand exchange was developed in Japan by Maeda et al.[21] and an instrument for automatic analysis was built and sold by the Hitachi Corp. The exchanger was sulfonated polystyrene, and the metal ions were zinc(II). Zinc was preferred over copper because the ligand exchange was somewhat faster. Two columns were used, one for the neutral and acid amino acids (those with one amino group and one or two carboxyls), and a separate, short column for the basic amino acids (those with one carboxyl group and two amino groups). The eluent for the first column was 0.05 *M* in sodium acetate plus acetic acid to pH 4.1, and 0.0004 *M* in zinc; that for the second column was 0.9 *M* in sodium acetate and pH 5.1, 0.001 *M* in zinc. The temperature was 55°C for both columns. The amino acids were detected by post-column reaction with pyridoxal.[22]

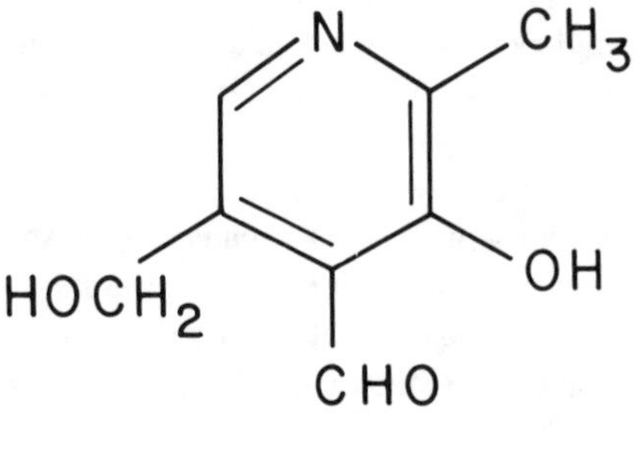

N, CH_3, OH, CHO, $HOCH_2$

This reacts in the presence of zinc ions and pyridine to give a fluorescent product. Fluorescence is excited at 390 nm, emitted at 470 nm. The reagent does not respond to ammonia, nor to tryptophane. It does respond to amino sugars.

An advantage claimed for the ligand-exchange method at the time was speed, compared with the usual two-column cation-exchange method, and in addition the pyridoxal reagent was ten times as sensitive as ninhydrin. The disadvantage that caused the method to be abandoned was the complexity; some skill was needed to get reproducible elution volumes.

Elution orders in ligand exchange are roughly the same as in cation exchange, but not exactly. A very thorough study of ligand-exchange chromatography of amino acids, first theoretical and then experimental, was made by Doury-Berthod et al.[23,24] These authors used copper and zinc ions on resins with carboxylic, phosphonic, and iminodiacetate functional groups, concentrating on copper and the polystyrene-phosphonic acid exchanger, with ammonia solution as the eluent. In their second paper they compare elution orders with the different metal-exchanger combinations and with cation exchange. Some of their results are shown diagrammatically in Figure 3. Cystine, an amino acid that contains sulfur, is much more strongly retained in ligand exchange, and so is glycine. In their first paper they show the effects of copper and ammonia concentrations on retention. Increasing the ammonia concentration lowers retention of amino acids, while increasing the copper concentration in solution first raises the retention, corresponding to raising the copper concentration in the resin, then lowers it as the copper in the solution competes for the amino acid with the copper in the resin.

A great advance in the technique of ligand-exchange chromatography was the development by Foucault et al.[25,26] of a new kind of stationary phase: copper-loaded silica gel. Foucault took acid-washed porous silica, the 7-μm Spherosil® used in high-performance liquid chromatography, put it in a column and passed a solution 1*M* in ammonia and 0.01 *M* in copper sulfate. First the ammonia combined with the hydrogen ions of the silica gel to give ammonium ions, then the copper-ammonia complex ions became attached to the silica. They remained attached to the silica and passing dilute ammonia did not wash any of the copper away. The copper-ammonia silica was an excellent stationary phase for ligand-exchange chromatography. A potential difficulty was the solubility of silica in alkaline solutions, but this was avoided, first, by passing the ammonia eluent through a precolumn of copper-loaded silica gel that could be replaced as needed, and second, by using a mobile phase that was rich in acetonitrile. Acetonitrile lowered the solubility of silica in water, and so did the copper ions; the copper-loaded silica was more resistant to alkaline attack than silica itself.

Amino acids and peptides[16] were separated chromatographically with good efficiency on this stationary phase, with water-acetonitrile ammonia eluents (see Figures 4 and 5). Figure 4 shows an isocratic run, eluent 0.5 *M* ammonia in 50% acetonitrile at 50°C; detection was by UV absorbance at 210 nm. Figure 5 includes peptides, which generally are retained more weakly than amino acids. Here a gradient was used, from 0.1 *M* ammonia in 90% acetonitrile to 0.95 *M* ammonia in 40% acetonitrile. Increasing ammonia concentration and increasing water concentration both favor the displacement of the ligands from the copper-loaded silica. Here, copper ions, 1 mg/l or 1.6×10^{-5} *M* were added to the eluents, and the ligands, amino acids, and peptides, were detected as their copper complexes by absorbance at 254 nm. A wavelength of 235 nm gave more than twice the sensitivity. Detection limits were not specified, but microgram quantities of amino acids gave good peaks.

The peak widths and resolution seen in Figures 4 and 5 are very promising. Peptides are much harder to resolve than amino acids because there are so many combinations of amino acids and so many isomers. The stationary phase, a kind of two-dimensional copper silicate, is deceptively simple. One asks what other metal ions can be attached as easily to silica. The answer is only a few; the silanol groups of silica are only weakly acidic and a

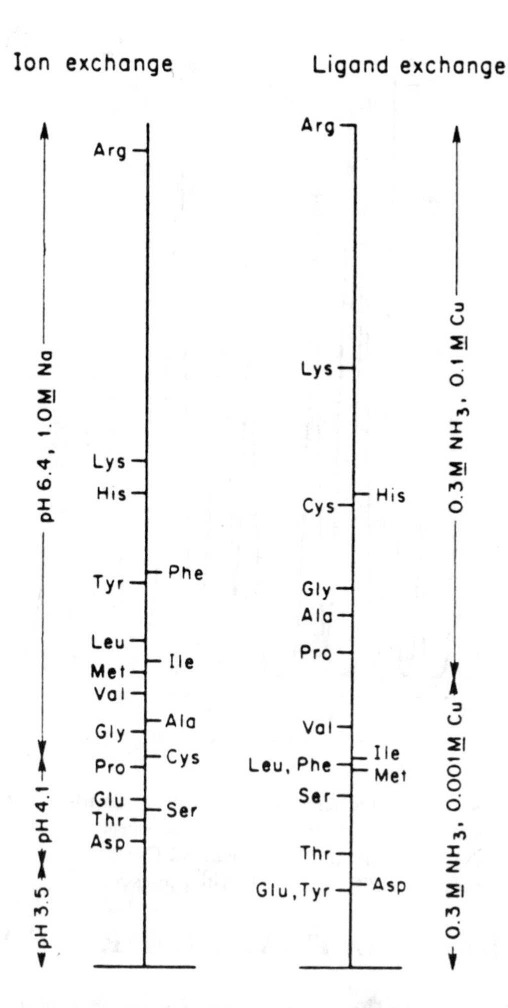

FIGURE 3. Comparison of ion-exchange and ligand-exchange elution orders of amino acids. Ordinates are elution volumes expressed as multiples of the bulk column volume. Ion-exchange column contains sulfonated polystyrene resin, 8% crosslinked; ligand-exchange column contains polyacrylate resin in Cu(II) form. (From Doury-Berthod, M., Poitrenaud, C., and Tremillon, B., *J. Chromatogr.*, 179, 37, 1979. With permission.)

high pH is necessary for their hydrogen ions to be displaced; metal ions that displace them must be cationic at high pH and this requirement limits the metal ions to those that form fairly stable ammonia complexes, like Cu, Ni, and Zn. Nevertheless one can expect to see more use of metal-loaded silica gel for ligand exchange.

The next section will include mention of ligand exchange in the mobile phase as a means of separating optical isomers of amino acids. The amino acids can be separated from one another by this means too. Grushka et al[15] used a mobile phase of aqueous copper(II) chloride, 0.0003 *M*, sometimes with an acetate buffer, and a stationary phase of bonded C18-silica. The order of retention of amino acids was like that in ligand exchange on an ion-exchange resin; the acidic amino acids were retained very weakly, the basic amino acids very strongly. What probably happens is that amino acids form uncharged CuL_2 complexes that are then retained by the hydrophobic C18 chains. The amino acids that form the most stable complexes are retained the longest. Detection was done by absorbance at 230 nm. The sensitivity was very good; detection limits were of the order of 1 ng of amino acid. Bands were very sharp.

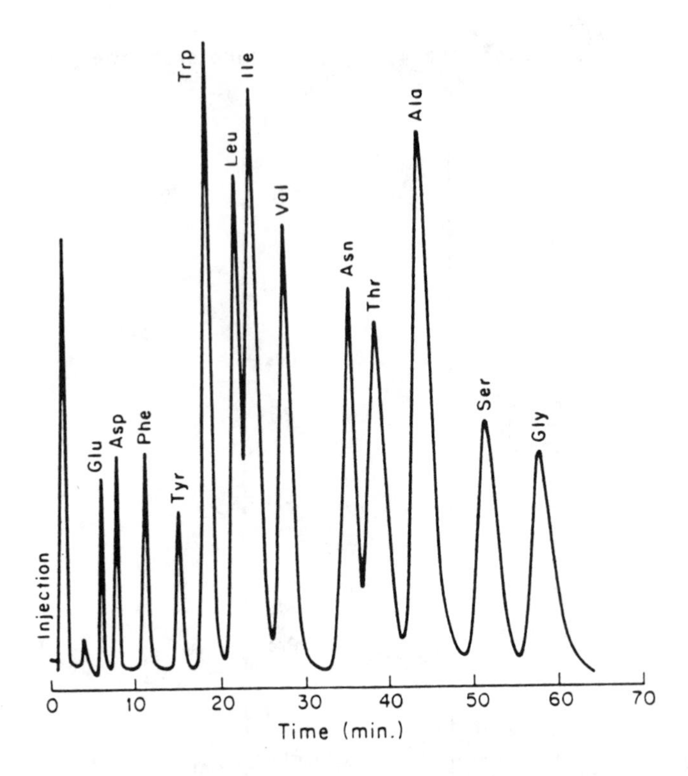

FIGURE 4. Chromatography of amino acids on Cu(II)-modified silica gel. Column, 0.5 × 15 cm; particle size, 7 μ; eluent, 0.15 *M* ammonia in 48% (v/v) acetonitrile. (From Foucault, A., Caude, M., and Oliveros, L., *J. Chromatogr.*, 185, 345, 1979. With permission.)

VI. SEPARATION OF OPTICAL ISOMERS OF AMINO ACIDS

The crowning achievement of ligand-exchange chromatography is undoubtedly the separation of optical isomers, in particular the isomers of amino acids. We have noted that ligand-exchange selectivity is sensitive to steric factors. Figure 6 shows two amino acid molecules coordinated to the same copper ion in a square planar arrangement. Around the copper ion the two nitrogen atoms are *trans*- to one another. The two side chans, R and R′, will be on the same side of the coordination plane if both amino acids have the same chiral configuration, DD or LL, but if they have different chiral configurations, a DL combination, the R groups will be on different sides of the plane and farther away. If the R groups attract one another the DD or LL combination will be more stable than DL; if the R groups repel one another, the DL combination will be more stable.

This difference forms the basis of the chromatographic separation by ligand exchange. One takes an amino acid of known configuration. If it is a natural amino acid it will have the L-configuration. Let this amino acid be attached chemically to crosslinked polystyrene, and then let it coordinate with copper ions. If the amino acid is L-proline, the result is the structure shown in Figure 7. The copper ion can now receive a second amino acid anion from the mobile phase. Either the D- or the L-form of this acid will be preferred, depending on the kind of interaction between the neighboring R groups.

In 1970 a series of papers began to appear by Davankov et al.[27] in Moscow which described the chromatographic separation of optical isomers of amino acids. The earlier reports described ligand exchange in the stationary phase. Reports after 1980 showed that ligand exchange in the mobile phase would also serve. The first task was to make a polymer that would allow the bulky amino acid ligands to move in and out with sufficient ease. Early

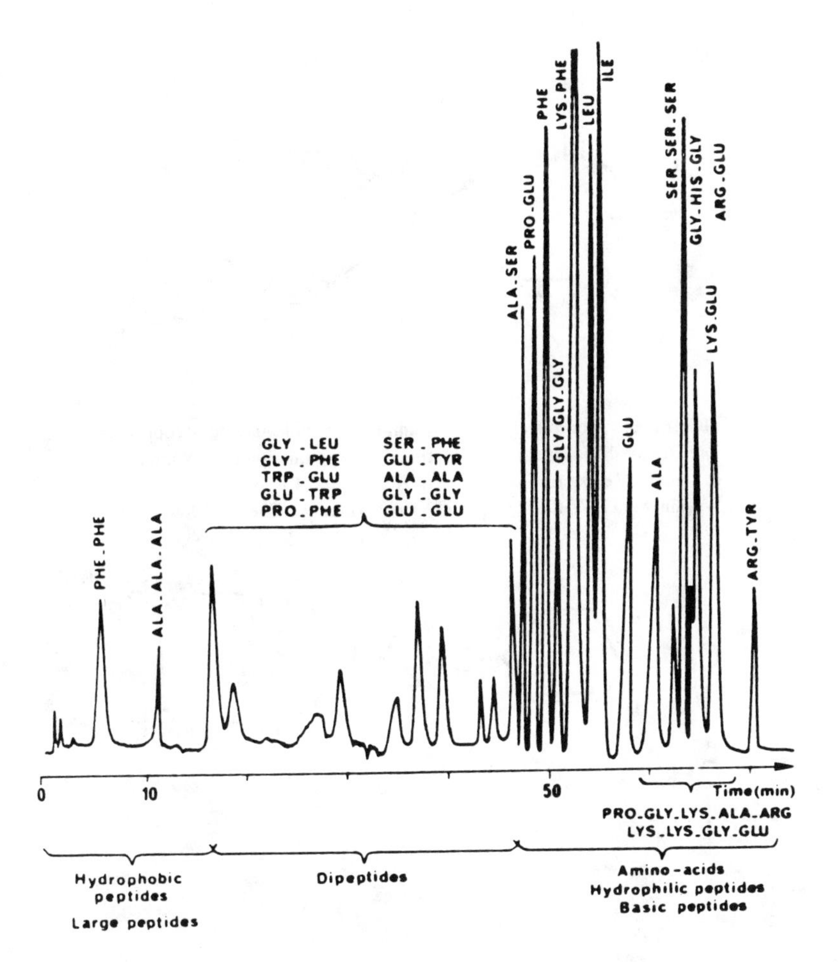

FIGURE 5. Composite chromatogram of peptides and amino acids on Cu(II)-modified silica gel. (From Foucault, A. and Rosset, R., *J. Chromatogr.*, 317, 41, 1984. With permission.)

publications[27] described the preparation of "macronet isoporous" polystyrene which was crosslinked, not with divinylbenzene, but with long rod-like molecules such as 4,4*bis*-chloromethylbiphenyl. The next step was to attach the amino acid to the polymer and this was not easy. Proline was one of the few amino acids that could readily be attached. Chloromethylated polystyrene combines with proline as it does with di- and trimethylamine (Chapter 2), but the reaction is very slow. Davankov and Rogozhin[28] found that adding sodium iodide converted the $-CH_2Cl$ group to $-CH_2I$, which reacted a lot faster.

Having made the new resin whose structure is shown in Figure 7, they packed the resin into a glass column 9 mm by 475 mm, a large and unwieldly column by present standards, converted it to the copper form, and passed 5 ml of 10% DL-proline in water; then they passed more water and collected fractions of the effluent. The fractions were tested by a polarimeter, observing the optical rotation at 436 nm. Almost immediately, L-proline appeared and before long all of it had been eluted. The 1 *M* ammonia was passed. This eluent brought out the D-proline. The isomes were cleanly separated, and of the 500 mg of DL-mixture that had been introduced, 250 mg of each isomer was recovered in pure form. (To hold back the small amounts of copper, a short postcolumn was used that contained the copper-free proline-styrene polymer).[28] Other amino acids could be separated into their D- and L-forms in the same way. Though proline was grafted to the polymer, the acid in the mobile phase need not be proline. Other amino acids were resolved too, though not as efficiently.

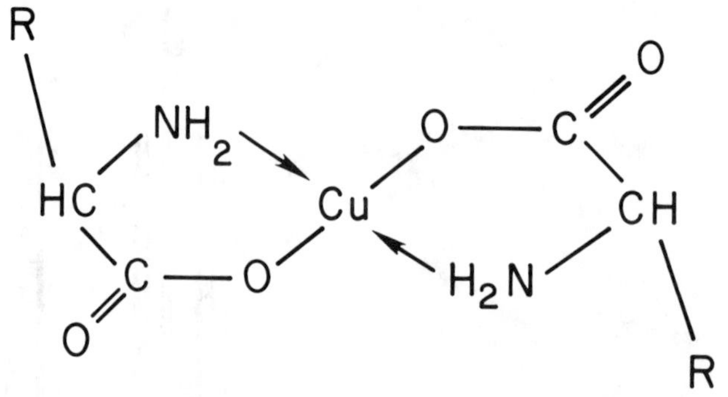

FIGURE 6. Coordination of two amino acid molecules to a copper(II) ion; DL-combination shown. (Reprinted with permission from Walton, H. F., *Ion-Exchange Chromatography* ACS Audio Courses C-93; Society: Washington, DC, 1987. Copyright 1987 American Chemical Society.)

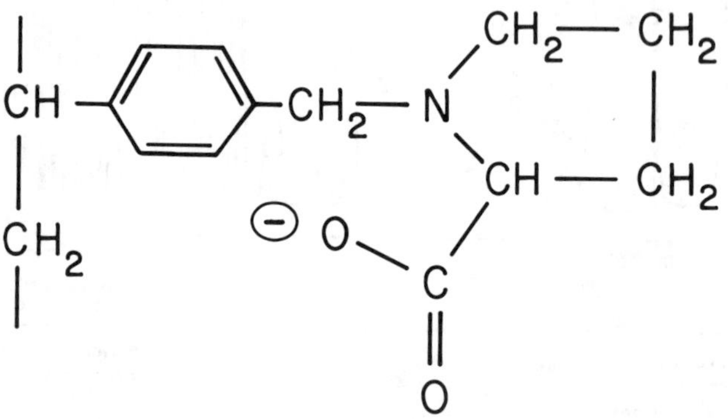

FIGURE 7. Proline molecule attached to crosslinked polystyrene.

To find the difference in stability between DL- and LL-combinations, Davankov et al.[29,30] prepared the model compound N-benzylproline in its two optically active forms and measured the formation constants of their complexes with copper ions, using the standard method of pH titration. Using the symbols D and L to represent the two forms of the ligand anion, the cumulative formation constants of the 2:1 complexes at 25°C were:

$$\beta_{2LL} = \frac{[CuL_2]}{[Cu][L]^2} = \text{antilog } 13.36$$

$$\beta_{2DL} = \frac{[CuDL]}{[Cu][D][L]} = \text{antilog } 14.36$$

The mixed complex CuDL is ten times as stable as the complex CuLL. Statistically the mixed complex is twice as probable. Repulsion between the two ligands, which are farther apart in the DL combination, may account for the remaining factor of 5; however, data of Kurganov et al.[29] show that the stabilization of CuDL is due to the entropy change, not the enthalpy.

The chromatography also indicates that the DL combination is more stable, for the L-form is eluted from the column first, and eluted by water alone. The D-form is held more strongly to the fixed L-proline in the column, and comes out afterwards with the aid of 1 *M* ammonia. It does not necessarily follow that because DL is more stable in aqueous solution that it will necessarily be more stable in the column. In the ion-exchange resin there is interaction with the polymer network. The mere fact that space inside the polymer is restricted would favor the more compact LL-arrangement. All the same, it is always the D-form of the amino acid in the mobile phase that is more strongly retained,[31] with one exception, histidine. The fact that the histidine molecule has two amino groups in position to form a chelate ring may be the reason.

Hydroxyproline, allohydroxyproline, azetidine carboxylic acid, as well as proline, can be attached to the stationary phase. (Azetidine carboxylic acid is like proline, but with one less CH_2 group, having a four-membered instead of a five-membered ring.) The different fixed ligands show somewhat different selectivities. Table 1 is taken from publications of Davankov's group[32] and shows separation factors, that is, the ratios of capacity factors, k', for the D- and L-forms of amino acids. The separation factors range up to four. Retentions and separation factors depend on the proportion of copper ions in the resin; these data correspond to 92% saturation. Enough copper salt was added to the mobile phase, up to 0.0001 *M*, to maintain this degree of saturation.

At saturation, the copper-loaded resin contains one copper atom to two proline groups. The sorption and displacement of amino acids from the mobile phase follow these equations:

$$\mathrm{RPro{-}Cu{-}ProR + Pro^- = RPro{-}Cu{-}Pro + RPro^-}$$

$$\mathrm{RPro{-}Cu{-}Pro + 2NH_3 = RPro{-}Cu(NH_3)_2^+Pro^-}$$

$$\mathrm{RPro{-}Cu(NH_3)_2^+ + RPro^- = RPro{-}Cu{-}ProR + 2NH_3}$$

In the second equation it is understood that ammonia or water molecules can coordinate the copper ions. We see the copper ions as a "zipper" to bind two separate proline units; the zipper is opened to allow dissolved proline anions to attach to one of the polymer chains and it closes after the dissolved proline is released. The closing is a slow process because the two polymer chains must come into the correct alignment.[33] As we have noted, any other α-amino acid can take the place of proline in the mobile phase.

Though this method of separating the isomers of amino acids gives very high separation factors, it is slow and inefficient by the standards of high-performance liquid chromatography. Theoretical-plate heights are of the order of 1 cm. For preparative purposes the method is very good. The column capacity is high and the amino acid isomers leave the column in virtually pure form. The ammonia is easily removed, and so is the low concentration of copper; the amino acid can be recovered in pure crystalline form.

One way to get faster ligand exchange and more rapid analysis of DL-mixtures is to fix the L-proline (or hydroxyproline) to the surface of bonded C18-silica, as a permanently coated ion exchanger (Chapter 5). Davankov and others[34] prepared N-alkyl-L-hydroxyprolines with alkyl groups of 7, 10, and 16 carbon atoms. These were soluble in methanol and could be deposited on C18-silica from a methanol or a methanol-water solution. After the packing was impregnated with alkyl hydroxyproline, a concentrated solution of cupric acetate in 15% methanol was passed. The copper-loaded packing could now be washed indefinitely with 15% methanol without "bleeding". The mobile phase for chromatography was 0.0001 *M* copper(II) acetate in 15% methanol or ethanol. The more organic solvent, the less the retention. Log k' fell linearly with the volume fraction of organic solvent, as is the normal relation in reversed-phase liquid chromatography. Again the L-amino acids were eluted before the D-forms, with the exception, once more, of histidine. Histidine has two basic nitrogen atoms separated by three carbons, and could thus form a six-membered chelate ring with the metal ion. Detection was by UV absorbance at 254 nm. In one test, seven amino acids were eluted separately, each one separated into its D- and L-forms to give 14 peaks in 40 min. Separation factors were as high as ten (for the isomers of proline). They increased slightly as the copper concentration was increased. Retention decreased as the copper concentration increased, reflecting the increasing proportion of amino acid carried as its copper complex in the mobile phase.

Most of the chiral discrimination took place in the stationary phase, where the binary complexes of Cu with (i) alkyl proline, (ii) the amino acid in solution, were formed. The exact loading of the stationary phase was not determined, but was about 0.2 mmol of alkyl proline per ml of bulk column volume; the capacity was less than that of chiral resins, but more than the capacity of surface-functional exchangers used in ion chromatography. This method is excellent for analytical purposes, but not for preparative separation; the amounts of amino acid that can be placed in the column are too small.

Optically active amino acids can be attached to silica. One way to do it is by reaction of silica with L-prolylpropyltriethoxysilane.[35] The fixed amino acid, proline in this instance, is attached to silica by the bonds $Si–(CH_2)_3–N$. With this stationary phase and a mobile phase of ammonium acetate, 0.001 to 0.1 *M*, containing 10^{-4} *M* copper acetate, pH 5, and methanol up to 30%, amino acids were resolved with rather low separation factors. The L-forms were held more weakly than the D-forms, with one or two exceptions. Another way to attach amino acids to silica, intensively studied by Gübitz et al.[36,37] is by reaction of silica with 3-glycidoxypropyl-trimethoxysilane, a favorite reagent for attaching functional groups to silica. The first-step reaction product is shaken for 2 d at room temperature with a solution of sodium prolinate, or other amino acid salt, yielding a product with this linkage to the amino acid:

$$–Si–O–Si(CH_2)_3–O–CH_2–CH(OH)–CH_2–N$$

Among the amino acid ligands attached in this way were proline, hydroxyproline, azetidine caboxylic acid, pipecolic acid, and phenylalanine. The most useful fixed ligands were proline and hydroxyproline, in the L-forms, of course. After the ligands were attached they were loaded with copper ions in the usual way. The mobile phase was 0.05 *M* phosphate, pH 4.6, or ammonium acetate, with methanol or acetonitrile added as needed to reduce retention, and 0.0001 *M* copper ions. The higher the pH the more the retention, because the proportion of amino acid anions becomes greater. Amino acids in the mobile phase were well resolved into their optically active forms, with separation factors up to three and more, but with the novel result that the D-forms came out of the columm before the L-forms; that is, the LL-combinations in the stationary phase were the more stable. A possible reason for this effect may be the –OH group in the linkage that was left by the opening of the epoxy group; it can coordinate axially with the copper ion, as shown in Figure 8. (The square planar coordination of copper(II) is only a simplification of the 6-coordinated elongated octahedron described by the Jahn-Teller effect.) Not only amino acids, but hydroxy acids are resolved into optically active forms on these stationary phases. Mandelic acid, 2- and 3-phenyllactic acids were resolved with separation factors of 1.65, 1.46, and 1.61, respectively.[38] Always the best chiral resolution is obtained with the bigger side chains (see Table 2). These stationary phases offer some promise for small-scale preparative use; 50-mg quantities of solute can be handled in the ordinary analytical columns.

A variation on Gubitz's chiral stationary phases was made by Corradini et al.[39] who reacted silica first with 3-glyidoxypropyl-trimethoxysilane and then with optically active (-)*trans*-1,2-cyclohexanediamine. The product coordinated with copper(II) in much the same way as Gubitz's proline-bound silica; the difference is that the fixed ligand holds the copper by two nitrogens rather than a nitrogen and an oxygen atom. It was able to resolve amino acids and mandelic acid.

Dansyl amino acids (see below) were separated by microcolumn HPLC by Ishii and his group;[40] they used a bonded silica carrying L-proline-Cu and ammonium acetate 0.1 m*M* in copper salt, as mobile phase with a water-to-acetonitrile gradient.

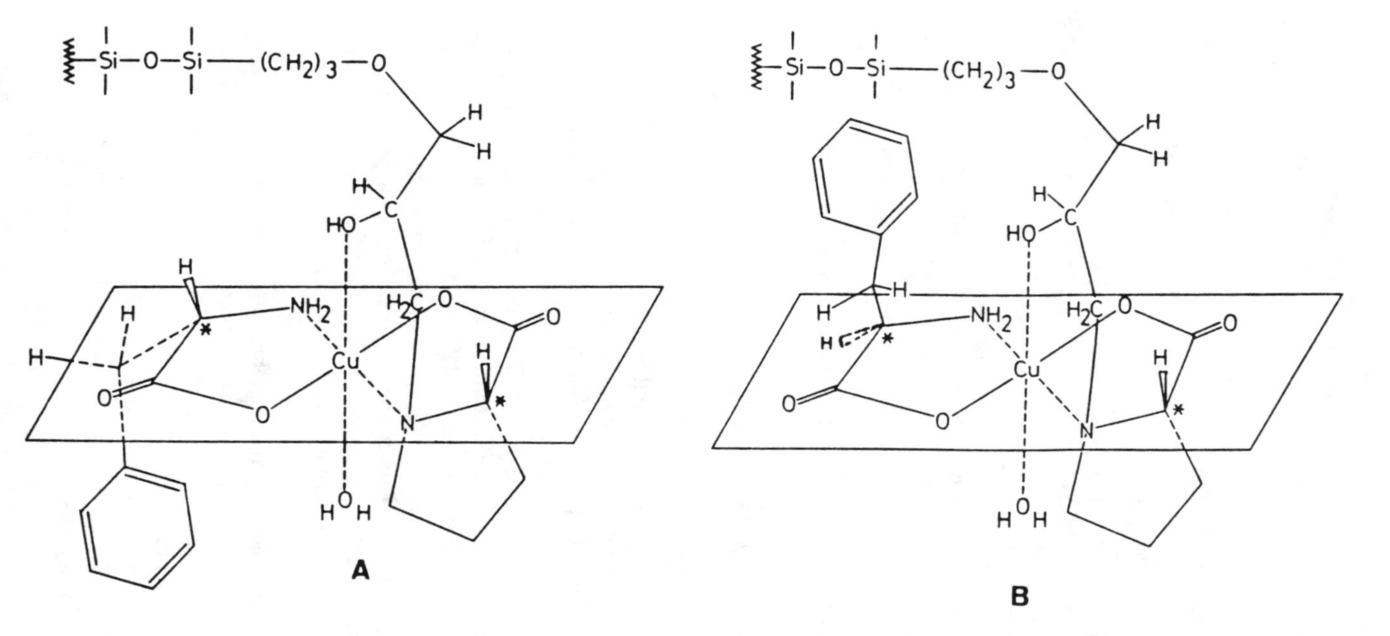

FIGURE 8. Model of a mixed-ligand copper complex: proline and phenylalanine. (From Gübitz, G., *J. Liquid Chromatogr.*, 9, 519, 1986. Courtesy of Marcel Dekker, Inc.)

TABLE 2
Capacity Factors of D- and L-Enantioners on Sorbents with Cyclic Amino Acids Bonded to Silica

	Stationary ligand								
	L-Proline			L-Hydroxyproline			L-Pipecolic acid		
Amino acid	k'_D	k'_L	Ratio	k'_D	k'_L	Ratio	k'_D	k'_L	Ratio
Alanine	1.30	1.30	1.00	1.40	1.40	1.00	1.40	1.25	0.89
Valine	2.50	3.80	1.50	2.20	2.60	1.20	1.40	2.68	1.91
Leucine	3.10	3.10	1.00	3.40	2.90	0.85	2.08	2.60	1.25
Isoleucine	3.00	3.60	1.20	3.40	3.40	1.00	1.70	2.75	1.62
Serine	2.00	3.20	1.60	2.20	3.40	1.55	1.63	2.83	1.74
Methionine	3.00	3.40	1.10	3.90	3.90	1.00	2.52	3.35	1.33
Lysine	2.00	2.20	1.10	1.60	1.60	1.00	1.10	1.55	1.41
Arginine	2.40	3.00	1.20	2.40	2.40	1.00	1.55	2.30	1.48
Histidine	6.70	12.10	1.80	7.60	17.60	2.32	5.08	7.93	1.56
Aspartic acid	3.20	4.10	1.30	5.90	7.20	1.22	2.52	3.13	1.24
Glutamic acid	2.00	2.00	1.00	4.00	4.00	1.00	1.85	2.30	1.24
Proline	2.40	1.40	0.60	3.10	1.30	0.42	1.55	1.18	0.76
Phenylalanine	3.20	9.40	2.90	5.20	12.20	2.35	4.03	8.99	2.29
DOPA	3.40	11.20	3.20	4.40	19.20	4.36	4.48	16.63	3.71
Tryptophane	7.80	27.40	3.50	9.20	39.80	4.33	12.20	31.10	2.55

From Gübitz, G., *J. Liquid Chromatogr.*, 9, 519, 1986. Courtesy of Marcel Dekker, Inc.

VII. AMINO ACIDS

A. LIGAND EXCHANGE IN THE MOBILE PHASE

To get chiral discrimination between amino acids it is necessary to bring two chiral molecules together; in ligand exchange, they are brought together with a metal ion, usually Cu, between them. This association can take place in solution as well as in a solid adsorbent.

The first people to use mobile-phase association to separate underivatized D- and L-amino acids in chromatography were Hare and Gil-Av.[41] They used as the stationary phase a gel-type sulfonated polystyrene resin in the Cu(II) form. The mobile phase was 0.05 *M* sodium acetate, pH 5.5, 0.004 *M* in copper sulfate, and 0.008 *M* in L-proline, and the column temperature was 75°C. Detection was by post-column reaction with *ortho*-phthal-aldehyde plus ethanethiol (see Chapter 4). This reagent gives fluorescent products with primary amines and amino acids, but not with secondary amines. Proline is a secondary amine and does not react, nor does hydroxyproline, but other amino acids are detected in picomole quantities.

The D-isomers were eluted before the L-. Probably this reflects the predominance of CuDL in the mobile phase. We may suppose that the amino acids are retained in the stationary phase as the 1:1 singly charged complexes, CuD^+ or CuL^+; there would be no discrimination between the D- and L-forms of these complexes.

Uncharged 2:1 amino acid to copper complexes may be retained by a nonpolar alkyl-bonded silica, and this is the basis for the most practical analytical separation of D- and L-amino acids in use today. The stationary phase is the common C18- or C8-bonded silica, and the mobile phase is a buffered aqueous solution containing copper ions and L-proline in the mole ratio 1:2. Retention is increased by raising the pH and reduced by adding a nonaqueous solvent, or by raising the copper concentration in the solution. Figure 9 shows a typical chromatogram.[41a] Separation factors are not as large as with chiral stationary phases, but the peaks are narrower and resolution is better. A drawback to the mobile-phase method is the consumption of a rather expensive chiral reagent, generally L-proline; another drawback is that, obviously, proline cannot be detected. For preparative separations one would have to remove the excess of chiral reagent, not an easy task. Summarizing, we may say that the chiral mobile phase is better for analytical purposes, the chiral stationary phase for preparative purposes.

We have described the optical resolution of underivatized amino acids. The dansyl amino acids were resolved very effectively by Karger's school,[42,43] using an optically active triamine plus zinc ions:

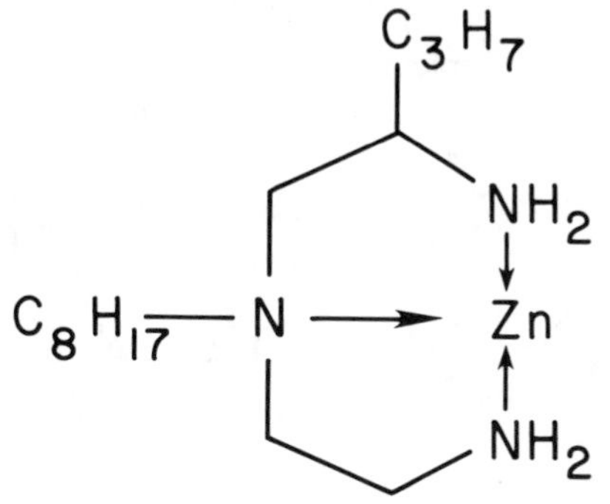

C_3H_7

NH_2

C_8H_{17}—N ⟶ Zn

NH_2

The stationary phase was C8- or C18-silica. A typical mobile phase was 0.2 *M* in ammonium acetate plus ammonia to pH 9, 0.8 m*M* in the zinc chelate shown above, and containing 35% acetonitrile. Excellent resolution of DL-amino acids was obtained in run times of 15 min with separation factors up to 2.5 (for serine). The zinc ions are believed to be 5- or 6-coordinated; three coordinate valences go to the nitrogens of the triamine, two to the carboxyl and amino groups of the dansyl amino acid. Acetonitrile coordinates with the zinc ion; thus, we can call the process "ligand exchange". Dipeptides were separated and resolved into optically active forms by this method, after first preparing their dansyl (1-dimethylamino-naphthalene- 5-sulfonyl) derivatives.

VIII. PROTEINS

A. IMMOBILIZED METAL AFFINITY CHROMATOGRAPHY

Ligand exchange was put to use by Porath and Olin[44] to separate proteins from one another. It acts in a different way from ion exchange, hydrophobic interaction, or size exclusion, and so complements these well-known methods. The ability of a protein molecule to coordinate with a metal ion depends on the presence of certain amino acids, in particular the basic amino acids, histidine and tryptophane, that have basic nitrogen atoms in their structure. Histidine is more effective in promoting attachment, as might be guessed by looking at its structure:

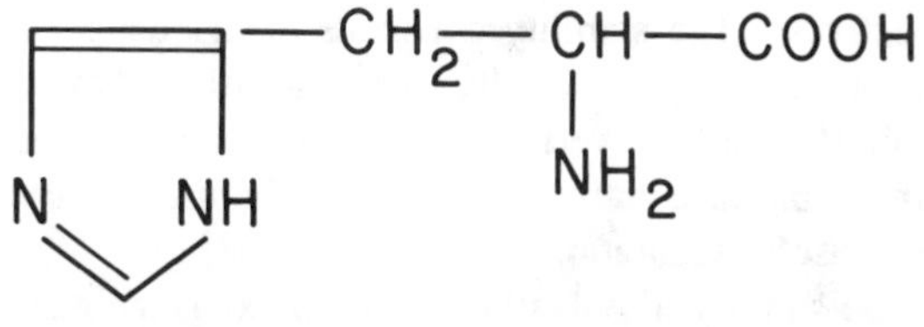

—CH_2—CH—COOH

N NH NH_2

The stationary phases used in this method are derived from agarose, a polysaccharide, by reaction with epichlorhydrin, sodium hydroxide, and sodium borohydride, followed by, in one case, the sodium salt of iminodiacetic acid. The product has the structure:

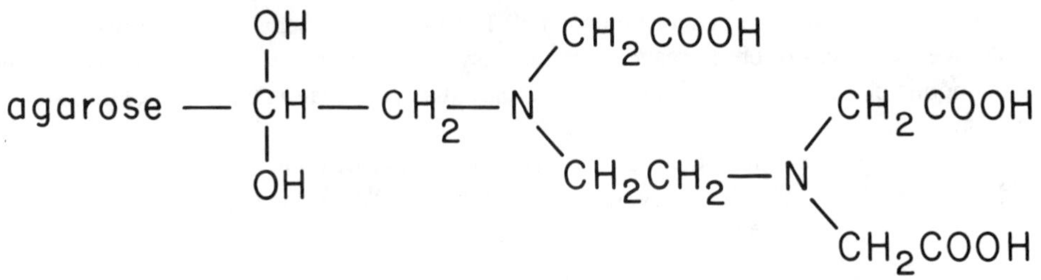

```
                OH                    CH2COOH
                |                    /
agarose  —  CH — CH2 — N                          CH2COOH
                |                    \           /
                OH                    CH2CH2 — N
                                                 \
                                                  CH2COOH
```

Or the product of reaction with epichlorhydrin is made to react with ethylenediamine, followed by sodium bromoacetate and sodium hydroxide to give this structure:

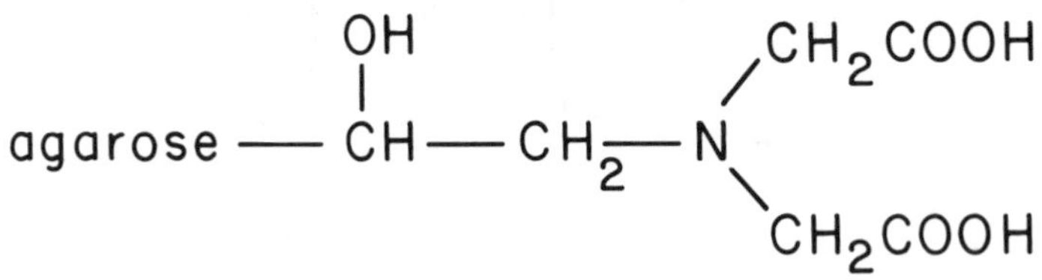

```
               OH               CH2COOH
               |               /
agarose  —  CH — CH2 — N
                               \
                                CH2COOH
```

Then the metal ion is attached to the chelating group. Porath's school has attached a great many metal ions, including Fe(II), Fe(III), Ni, Cu(II), Zn, Cd, Cd, and Tl(III). Each metal ion has its own selective binding for different proteins. Thus, two or three columns, each carrying a different metal ion, may be placed in series in the eluent stream, and the protein mixture introduced; each column will retain its own group of proteins. The loading eluent may be an acetate buffer of pH 5.5, close to the isoelectric pH of the proteins. To get the proteins off the columns one may lower the pH, converting the proteins to their cationic forms and protonating the chelating groups, or one may pass a concentrated (1 *M*) ammonium sulfate solution, or a displacing ligand such as imidazole, histidine, cysteine, or glycine. Or one may pass EDTA, which will remove the attached metal as well as the protein; or one may strip the metal off the support with a strong acid. The possibilities are many. Originally Porath called his method "metal chelate affinity chromatography", but it is not necessary for the metal ions to be held by chelation, only that they be held immobile, and so the process is called "immobilized metal-ion affinity chromatography".

IX. OXYGEN LIGANDS

A. CARBOHYDRATES

A complete contrast to the exchange of amino acid ligands on copper ions is the exchange of carbohydrate ligands on calcium ions. Copper ions are described as soft acids, nitrogen-donor ligands as soft bases, meaning that the atoms and ions are easily deformed, with overlap of electron clouds; calcium ions and oxygen-donor ligands are described as hard acids and bases, respectively. The stability constants of 1:1 copper ion-amino acid anion complexes are or the order 10^8; the stability constants of 1:1 calcium ion-carbohydrate complexes are less than 10 l/mol. Nevertheless, the power of chromatography as a multi-stage process is shown by the fact that the very weak binding of carbohydrates to calcium ions is the basis of very useful separations.

A column of gel-type sulfonated polystyrene cation-exchange resin, 4% crosslinked, in the potassium form, was used in 1968 by Saunders[45] to separate the sugars sucrose, glucose, xylose, and fructose, which were eluted in this order by passing pure water. Other workers used hydrogen-form gel-type cation-exchange resins for the same purpose; in fact, not only sugars, but low-molecular-weight alcohols, aldehydes, ketones, and other organic com-

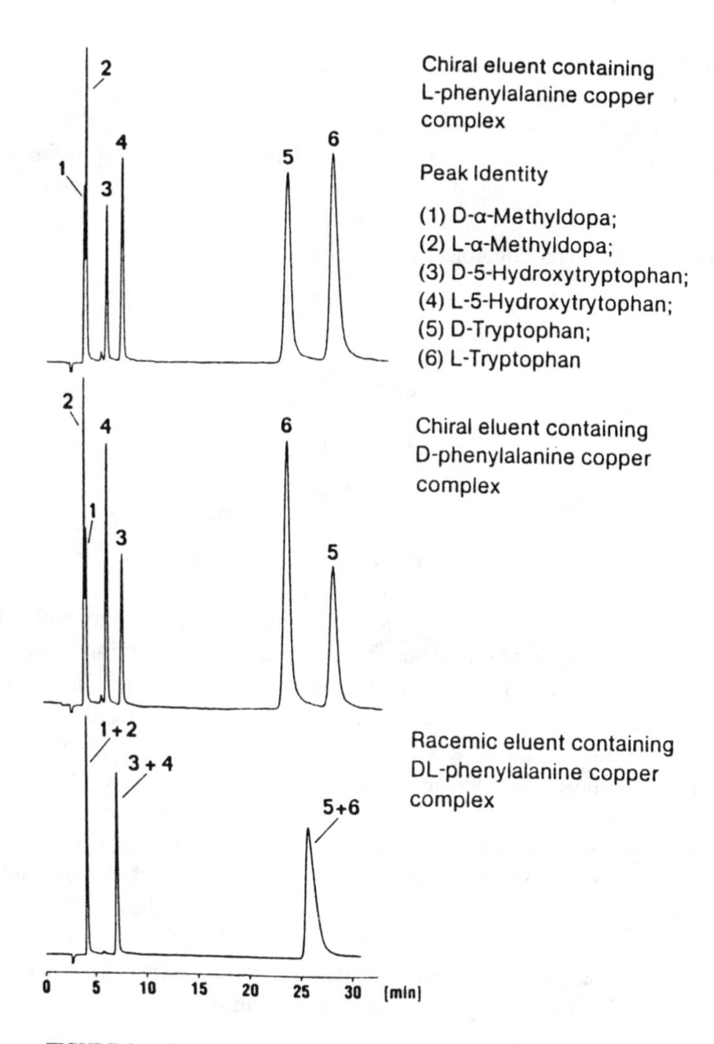

FIGURE 9. Separation of enantiomers by ligand exchange in the mobile phase. (From Oelrich, E., Preusch, H., and Wilhelm, E., *J. High-Res. Chromatogr. Chromatogr. Commun.*, 3, 269, 1980. With permission.)

pounds could be separated chromatographically, with 0.01 *M* sulfuric acid as the eluent.[46] However, the retentions were very small. The compounds eluted beyond the void volume of the column, but not much beyond, which meant that to get the necessary separation of peaks the column had to be fairly large, 30 × 0.75 cm in the references just cited. In a sense the sugars were not retained at all, but rather excluded to differing extents, for they came out before deuterium oxide, used as a tracer for water molecules, came out. The retention could be increased considerably by using acetonitrile-water mixtures[47] or ethanol-water mixtures[48] with 70% or more of the organic solvent.

In 1975 an important paper by Goulding[49] appeared in which the effect of the counter-ion was studied. With sodium as the cation in the resin and pure water as the eluent, the various sugars and sugar alcohols tested came out close together and within the overall bulk volume of the column. With calcium or lanthanum ions, however, some of the peaks emerged well beyond the bulk column volume. The peaks were spread out and well separated; some compounds were clearly being retained. Silver counter-ions gave retention too, though not as much. In all of these studies the solutes were detected by refractive index.

Goulding used twelve sugars and three polyols, glycerol, mannitol, and sorbitol. Mannitol and sorbitol were retained relatively strongly by calcium-loaded resin, as were the sugars talose and ribose. Talose is a hexose and ribose is a pentose, but both of them can exist in water as a six-membered or pyranose ring. The hydroxyl groups attached to the ring can either stick out in the plane of the ring, or nearly so, or they can point up and down roughly at right angles to the ring. These orientations are called *equatorial* and *axial*. The sugars that are most strongly retained by calcium ions are those that have three adjacent hydroxyl groups in the axial-equatorial-axial configuration, thus:

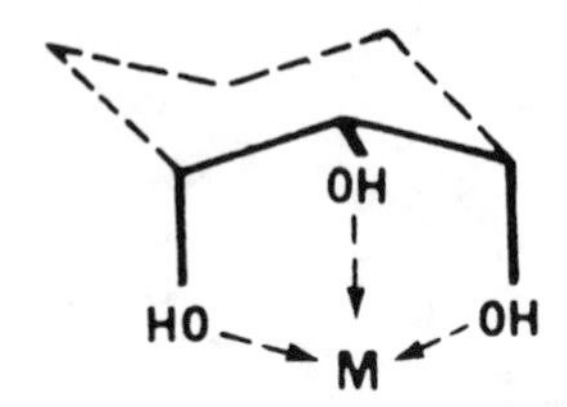

This is the configuration that gives the most effective chelation with a metal ion like Ca^{2+}. Two five-membered chelate rings are formed at once.

The mechanism of retention of these sugars is undoubtedly ligand exchange. The displacing ligand in chromatography is water. The mechanism of retention of the more weakly bound sugars, like sucrose, glucose, and fructose, is not as clear. It probably involves hydrogen bonding to the "free" water (see Chapter 3) and size exclusion. Table 3 compares the retention of various sugars by a sulfonated polystyrene resin containing four counter-ions. Retention of monosaccharides and of sucrose is greater with K^+ than with Li^+, and the potassium resin is believed to have more free water, i.e., water not bound to the ions by hydration.[50] Goulding[49] found, likewise, that rubidium ions gave more retention than sodium ions. In a classic paper (Reference 7, Chapter 3) Rueckert and Samuelson reported that the absorption of glucose into a cation-exchange resin was least with the lithium form, greater with sodium, and still greater with potassium as the exchangeable cation.

In these cases the sugars are retained less than deuterium oxide. This is to say, the ratio of sugar to water inside the gel-type resin is smaller than the ratio of sugar to water outside in the solution.

A case where exclusion from the resin on account of size seems to operate is in the chromatography of oligosaccharides, i.e., sugars formed by joining together three or more hexose units. The larger the molecules, the sooner they come out of the column. This effect is particularly well shown in a silver-loaded resin[51] (see Figure 10). The same effect is seen in a calcium-loaded resin, but to a lesser degree.[52]

The sugars of most general interest are sucrose and its hydrolysis products, glucose and fructose. A standard method for measuring the concentrations of these three sugars in corn syrup was described by Fitt et al.[53] He used a 4% crosslinked cation-exchange resin, 10 μm diameter, in the calcium form, with water as the eluent. Figure 11 shows a chromatogram of these three sugars, plus an injection of calcium chloride to mark the void volume (note: the dilute salt is excluded from the resin beads by the Donnan equilibrium), plus two sugar alcohols or polyols, mannitol and sorbitol. The polyols are retained as strongly as they are because it is easy for them to form coordinate bonds around the calcium ions; they have no ring structure to give them a fixed orientation.

It will be noted that the chromatograms of Figures 10 and 11 were run at 85°C. The reason is to hasten mutarotation, or the interconversion of anomers, so that they will elute as one single peak rather than two separate peaks. If one wants to see separate peaks for the anomers, one may do so by operating around 5°C. At intermediate temperatures the peaks may overlap. Anomers are the two isomers caused by the breaking and rejoining of the six-membered pyranose ring at the carbon-oxygen bond. This process may be catalyzed

TABLE 3
Retention of Sugars and Polyols on Sulfonated Polystyrene

Counter-ion:	Li^+	K^+	Ca^{2+}	La^{3+}
Salt peak	0.385	0.47	0.50	0.72
Sucrose	0.52	0.60	0.57	0.60
Lactose	—	—	0.56	0.60
Glucose	0.58	0.76	0.67	0.65
Xylose	—	0.82	0.77	0.71
Sorbose	—	—	0.80	0.71
Fructose	0.59	0.83	0.82	0.77
Arabinose	—	—	0.90	0.77
Pentaerythritol	0.75	0.73	0.90	1.08
Glycerol	0.77	0.82	0.96	1.10
Mannitol	0.64	0.73	1.10	1.44
Sorbitol	0.66	0.74	1.36	2.60
Ribose	0.72	0.94	1.43	2.05

Note: Deuterium oxide retention = 1.00; other retentions referred to D_2O. Retentions of D_2O were: Li, 5.81 ml; K, 5.62 ml; Ca, 5.58 ml; La, 4.95 ml. Resin was 6% crosslinked. Column temperature, 60°C.

From Walton, H. F., *J. Chromatogr.*, 332, 203, 1985. With permission.

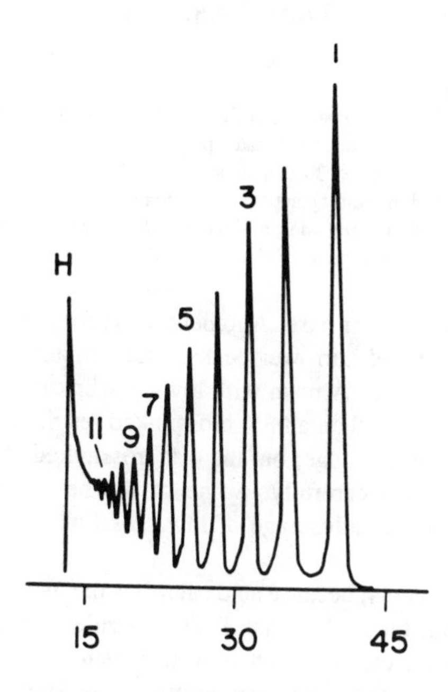

FIGURE 10. Chromatogram of polysaccharides in hydrolyzed corn syrup on a 5% crosslinked sulfonated polystyrene resin loaded with Ag and Ca. Column, 0.78 cm × 30 cm; particle size, 10 to 15 μm; temp., 85°C. Abscissae, time in minutes. Numbers show degree of polymerization. H means higher, excluded polysaccharides. (From Scobell, H. D. and Probst, K. M., *J. Chromatogr.*, 212, 51, 1981. With permission.)

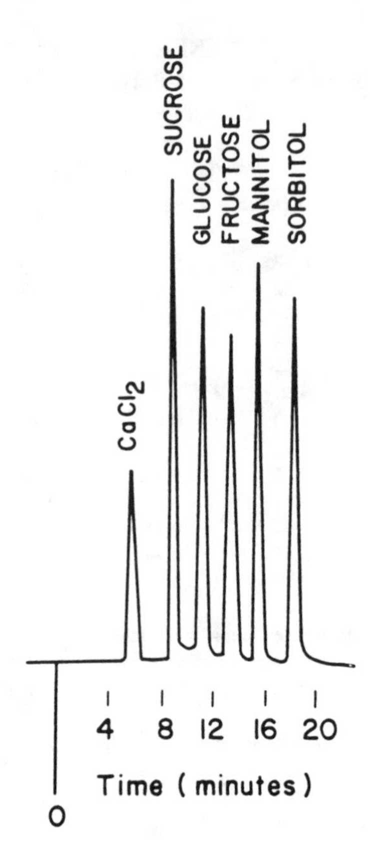

FIGURE 11. Chromatogram of common sugars and sugar alcohols on a calcium-loaded sulfonated polystyrene ion-exchange resin. Column, 0.65 cm × 30 cm; temperature, 85°C.; eluent, water; flow rate, 0.4 ml/min. Quantity of each compound, 100 μg. (From Vidalvalverde, C. and Martin-Villa, C., *J. Liquid Chromatogr.*, 5, 1941, 1982. Courtesy of Marcel Dekker, Inc.)

by adding trimethylamine or another base. Another reason for the high temperature is simply to have faster mass transport and narrower peaks. The temperature need not be as high at 85°C; 50°C is generally sufficient. A resin with low crosslinking, 4%, is preferred as it has a larger volume of imbibed water than an 8% crosslinked resin, and hence allows the sugars to be eluted over a larger volume range, but the 4% crosslinked resin is sensitive to column pressure and must be managed carefully. Some of the new solvent-modified polymeric exchangers seem to combine the advantage of low crosslinking with a greater mechanical rigidity.

One should avoid acid eluents, because acid catalyzes the hydrolysis of sucrose to glucsoe and fructose, and in general the hydrolysis of disaccharides and polysaccharides. Furthermore, hydrogen ions displace calcium ions from the resin. Natural carbohydrate mixtures from fruits usually contain organic acids and amino acids. They may be "cleaned up" by contact with an anion-exchange resin in the hydroxide form before injecting into the calcium-resin column for analysis, but it is easier to inject the samples as they are, without treatment. In this case it is a good idea to regenerate the analytical column at intervals by fushing it with a dilute calcium salt solution; or one may use as the mobile phase a very dilute calcium nitrate solution instead of pure water. Adding calcium ions to the influent will, of course, lower the retention volumes of those sugars or sugar alcohols that form complexes with calcium ions.

An attractive feature of this method of analyzing carbohydrate mixtures is that pure water is used as eluent. It is easy to recover individual sugars as they are eluted from the column; they can be used for biological testing without the need to separate them from other solutes. Detection is by refractive index, which is nondestructive and it is not as sensitive as the pulsed amperometric method described in Chapter 4, but the sensitivity required in corn syrup analysis, for instance, is not great. A drawback to the calcium-column method is that the peaks are close together, and the column must be large and well packed; care must be taken to minimize extra-column peak broadening.

Instead of calcium ions one may use lanthanum ions, another "hard acid". Lanthanum ions give greater retention of polyols and ribose than calcium ions[54] (see Table 3) but the peaks are broader because the lanthanum-loaded resin is less swollen, and it is doubtful whether the resolution is any better.

Copper ions, in the form of copper-loaded silica gel, have been used for the column chromatography of sugars.[55] The mobile phase is 75% acetonitrile in water, 1.5 *M* in ammonia and 2×10^{-5} *M* in cupric ions. Detection is by UV absorbance of the sugar-copper complexes at 280 or 254 nm, and the detection limits are a factor of ten less than by refractive index. The elution order is very different from that in the calcium-resin column; fructose, glucose, and sucrose are eluted in that order, which is just the reverse of the order in the calcium column.

X. SUGARS AS BORATE COMPLEXES

For a long time the favored method for liquid chromatography of sugar mixtures was by anion exchange of their borate complexes. Whether to call this process "ligand exchange" is dubious, for metal ions have no part; however, sugars and other hydroxy compounds having *cis*- diol groups associate with borate ions to form anionic complexes:

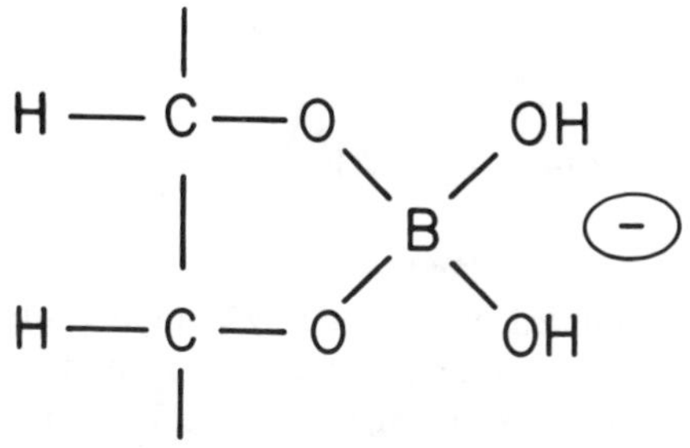

```
       |
  H — C — O         OH
       |      \    /
       |        B        (-)
       |      /    \
  H — C — O         OH
       |
```

Being anions these complexes are adsorbed on anion-exchange resins and can be displaced by borate solutions. A typical eluent is 0.4 *M* borate of pH 8. Glucose is retained more than twice as strongly as fructose; sorbitol is held more weakly and mannitol more strongly than glucose. The differences in retention are great, and gradients are commonly used. Detection is by a post-column reaction with 2-cyanoacetamide. Methods for automated analysis of carbohydrates with borate eluents are described by Honda et al.[56] and Reimerdes et al.[57]

XI. OTHER OXYGEN LIGANDS

Besides sugars, other oxygen-containing compounds have been separated by ligand exchange or removed from large volumes of water. Phenols have been so removed, with obvious benefits to pollution control. An iminodiacetate chelating resin, Chelex®-100, loaded with iron(III) recovers phenol and several chlorophenols and nitrophenols at very low concentrations, 0.1 ppm and less, from waste water at pH 3.[58] They are stripped from the resin by sodium hydroxide in 20% ethanol, and may be then extracted into methylene chloride and analyzed by gas chromatography.

Mixtures of phenols and chlorophenols may be analyzed by liquid chromatography. Instead of stripping all these compounds at once with strong alkali, they may be fractionated by passing a series of eluents of increasing pH, starting with pH 8.5 to displace only phenol, then displacing nitro-, dichloro-, and pentachlorophenol at higher pH values in that order. Using sulfonated polystyrene resins loaded with Fe(III) or Ce(III), and acid eluents, Maslowska and Pietek[59,60] separated chromatographically a number of substituted phenols. It is important to understand that coordination with the metal ions is only one of the forces that bind these compounds to the resins. The elution orders are not the orders of stabillity of metal-ligand complexes, but rather the orders of hydrophobic interaction; the same sequences are found in reversed-phase chromatography on C18-silica.

Hydroxybenzoic, hydroxynaphthoic, aminobenzoic acids, and nitrosonaphthols were separated chromatographically on exchangers loaded with iron(III), titanium(IV), copper(II), and other cations in the early days of ligand-exchange chromatography by Funasaka and his school.[61] More recently lanthanum-loaded sulfonated polystyrene has been used as a selective stationary phase for carboxylate anions, the anions of aromatic acids and hydroxy acids.[62] These methods have the potential for future development, but only if the fundamental problem of slow diffusion of the ligands in ion-exchange resins can be solved.

XII. SULFUR LIGANDS

One of the most encouraging papers on ligand-exchange chromatography to appear recently by Takayanagi et al.[63] describes the analysis of dialkyl sulfides. Chelating ligands ere grafted on to porous silica. The most successful of these was 2-amino-1-cyclopentene-1-dithiocarboxylate, or ACDA, which was attached to silica as shown:

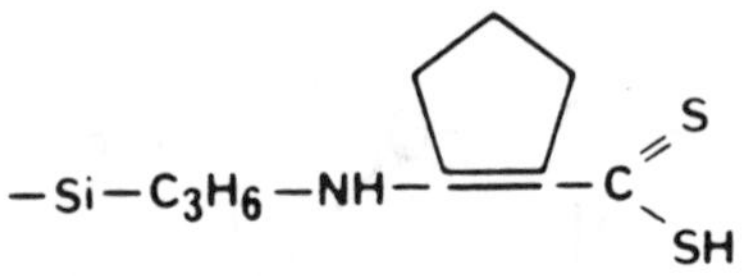

$-Si-C_3H_6-NH-$ (cyclopentene ring) $-C(=S)-SH$

This stationary phase was saturated with Cu(II) by pumping a 0.01 *M* solution of copper nitrate in methanol. For the chromatography the solvent used was hexane containing 1 to 2% methanol. The solutes were dialkyl sulfides, R_1R_2S, with R ranging from methyl to heptyl. Methanol was the displacing ligand; this was shown by the fact that the capacity factor, k′, was inversely proportional to the methanol concentration. Figure 12 shows a chromatogram.

This work is interesting for its application to nonaqueous, nonpolar solvents, and to show what might be done with ligand exchange by binding the metal ions to porous silica by long spacer chains. We have already noted the chromatography of amines by this method.

XIII. THIN-LAYER LIGAND-EXCHANGE CHROMATOGRAPHY

Ligand exchange may be used in thin-layer chromatography. Stationary phases that have been used include a chelating resin,[64] alumina and silica gel,[65] impregnated with metal ions in each case. Silica gel and alumina are simply impregnated with a solution of a metal salt and allowed to dry. The developing solvent may be hexane, benzene, or butanol; methanol may be added as the displacing ligand. Detection may be difficult. If the analytes are aromatic compounds their ultraviolet absorbance may be used; commercial silica-gel plates carring

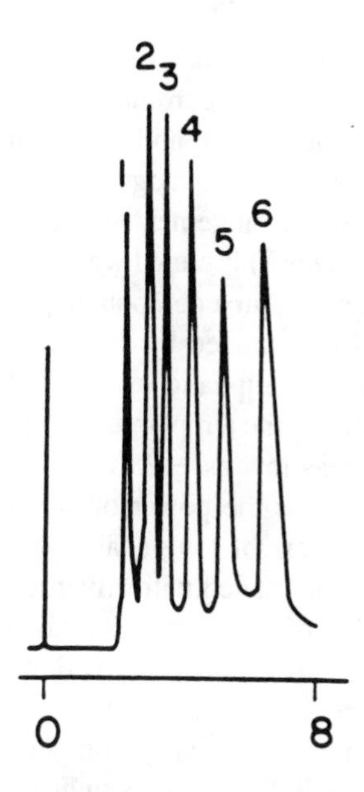

FIGURE 12. Chromatograms of dialkyl sulfides on a copper-loaded chelating bonded silica. Peaks, in order of elution: di-*n*-hexyl sulfide, di-*n*-propyl sulfide, ethyl *n*-propyl sulfide, diethyl sulfide, tetrahydrothiophene. (From Takayanagi, et al., *J. Chromatogr.*, 350, 63, 65, 1985. With permission.)

phosphorescent zinc sulfide plus a hand-held UV lamp show dark spots where UV-absorbing solutes are located.

XIV. LIGAND-EXCHANGE GAS CHROMATOGRAPHY

In his 1961 paper, Helfferich[2] wrote that ligands around a metal ion could be exchanged with other ligands from the gas phase. Little work has been done in this area, probably because capillary gas chromatography gives all the resolution that could be desired, but Fujimura and Ando[66] showed that ligand-exchange gas-solid chromatography was possible. For their stationary phase they used zirconium phosphate, an inorganic cation exchanger, loaded with ions of Cu, Zn, and Mn. For the mobile phase they used nitrogen containing ammonia. With this system they separated a large number of aliphatic and aromatic amines.

An improvement of zirconium phosphate was the use of metal stearates coated on diatomaceous earth. Manganese stearate separated many substituted aromatic amines, and its performance was much better than manganese-zirconium phosphate.[67] On copper stearate Takayanagi and co-workers[68,69] separated alkyl and aryl sulfides with good efficiency. The column temperature was 70°C or over and in every case the mobile phase was nitrogen gas containing 3 to 10% of ammonia by volume. It is interesting to note that gas-liquid chromatography on supported films of metal stearates, with a carrier gas containing ammonia, antedates the name ligand-exchange chromatography; it was used in 1959 by Barber[70] for aliphatic amines.

XV. ARGENTATION CHROMATOGRAPHY

A special area of ligand-exchange chromatography is argentation chromatography, where the stationary phase contains silver ions, and the mobile phase may be a gas or a liquid. The silver ion forms coordination complexes with ethylenic double bonds. The eluent or displacing ligand in liquid chromatography may be toluene, and if this is too weak an eluent, methanol or glycol are used. Argentation chromatography is applied to olefins, esters of unsaturated acids, lipids (including lecithins), and sterols. It will separate compounds having olefinic double bonds from their saturated analogues.

As one would expect, the steric arrangement is important; *cis–* isomers are held more strongly than *trans–*; elaidic acid (*trans–*) is eluted well before oleic acid (*cis*–9-octadecenoic acid). The more double bonds there are in a compound, the more strongly it is held; binding is stronger if the double bonds are separated by one or two carbon atoms than if they are conjugated. The length of saturated, aliphatic carbon chains has little effect. The silver-olefin complexes become less stable as the temperature is raised. Chromatography is generally done at low temperature, 15°C or as low as −25°C.

An interesting analysis that shows the power of argentation chromatography is that of insect sex attractants, which are esters of long-chain alkenols.[71] The stationary phase was a macroporous cation-exchange resin loaded with silver ions; the temperature was 13°C, the mobile phase was methanol, and a refractometric detector was used. Again the *cis–* isomers were held more strongly than the *trans–*, but the position of the double bond in the C_{11}– to C_{15} alkenols made little difference.

Another example of the separation of very complex mixtures of polyolefines is the analysis of bacterial menaquinones.[72] These compounds have 2-methyl-1,4-naphthoquinone units joined by polyisoprene chains, which have a range of numbrs of isoprene units, some of which are reduced and others of which have C=C double bonds. There is *cis–trans* isomerism around these double bonds. Separation was done on a silver-loaded ion exchanger; the mobile phase was pure methanol; the temperature was 65°C. The main factor influencing retention was the number of double bonds in the isoprene chain, but the positions of these double bonds affected retention too, and the isomers that differed in the position of the double bonds could be separated. Strangely, the *cis–* isomers were retained less than the *trans–*.

Isomeric olefines were analyzed by a form of ligand-exchange chromatography in 1971. Gil-Av and Schurig[73] used as the stationary phase a rhodium complex, $Rh(CO)_2$acac, where acac is a fluorine-substituted acetylacetone; this coordination complex was dissolved in squalane, and the column was run at 50°C. Isomeric butenes and pentenes were separated.

Reviews of argentation chromatography with long bibliographies were written by Guha and Janak[74] and Morris.[75] A comprehensive account of ligand-exchange chromatography has been given by Davankov et al.[76]

REFERENCES

1. **Tsuji, A. and Sekiguchi, K.,** Adsorption of nicotinic acid hydrazide on cation exchangers in various metal forms, *Nippon Kagaku Zasshi,* 81, 847, 1960.
2. **Helfferich, F.,** Ligand exchange: a novel separations technique, *Nature (London),* 189, 1001, 1961.
3. **Helfferich, F.,** Ligand exchange: (1) equilibrium (2) separation of ligands having different coordinative valencies, *J. Am. Chem. Soc.,* 84, 3237, 3242, 1962.
4. **Stokes, R. H. and Walton, H. F.,** Metal-amine complexes in ion exchange, *J. Am. Chem. Soc.,* 76, 3327, 1954.
5. **Cockerell, L. and Walton, H. F.,** Metal amine complexes in ion exchange. II. 2-Aminoethanol and ethylenediamine complexes, *J. Phys. Chem.,* 66, 75, 1962.
6. **Suryaraman, M. G. and Walton, H. F.,** Metal amine complexes in ion exchange. III. Diamine complexes of silver and nickel, *J. Phys. Chem.,* 66, 78, 1962.
7. **Maes, A., Peigneur, P., and Cremers, A.,** Stability of metal-uncharged ligand complexes in ion exchangers. II. The copper-ethylenediamine complex in montmorillonite and sulphonic acid resin, *J. Chem. Soc. Faraday Trans. I,* 74, 182, 1978.

8. **Maes, A., Marynen, P., and Cremers, A.,** Stability of metal-uncharged ligand complexes in ion exchangers. I. Quantitative characterization and thermodynamic basis, *J. Chem. Soc. Faraday Trans. I,* 73, 1297, 1977.
9. **Maes, A. and Cremers, A.,** Stability of metal-uncharged ligand complexes in ion exchangers. III. Complex ion stability and stepwise stability constants. IV. Hydration effects and stability changes of copper-ethylene diamine complexes in montmorillonite, *J. Chem. Soc. Faraday Trans. I,* 74, 2470, 1978 and 75, 513, 1979.
10. **Shimomura, K., Dickson, L., and Walton, H. F.,** Separation of amines by ligand exchange. IV. Chelating resins and cellulosic exchangers, *Anal. Chim. Acta,* 37, 102, 1967.
11. **Walton, H. F. and Navratil, J. D.,** *Ligand Exchange Chromatography. Recent Developments in Separation Science,* Vol. 6, Li, N. N., Ed., CRC Press, Boca Raton, FL, 1981, chap. 5.
12. **de Hernandez, C. M. and Walton, H. F.,** Ligand exchange chromatography of amphetamine drugs, *Anal. Chem.,* 44, 890, 1972.
13. **Navratil, J. D., Murgia, E., and Walton, H. F.,** Ligand exchange chromatography of amino sugars, *Anal. Chem.,* 47, 122, 1975.
14. **Spies, J. R.,** An ultraviolet spectrophotometric micromethod for studying protein hydrolysis, *J. Biol. Chem.,* 195, 65, 1952.
15. **Grushka, E., Levin, S., and Gilon, C.,** Separation of amino acids on reversed phase columns as their copper(II) complexes, *J. Chromatogr.,* 235, 401, 1982.
16. **Foucault, A. and Rosset, R.,** Ligand-exchange chromatography on copper(II)-modified silica gel, *J. Chromatogr.,* 317, 41, 1984.
17. **Masters, R. G. and Leyden, D. E.,** Ligand-exchange chromatography of amino sugars and amino acids on copper-loaded controlled-pore glass, *Anal. Chim. Acta,* 98, 9, 1978.
18. **Takayanagi, H., Tokuda, H., Uehira, H., Fujimura, K., and Ando, T.,** Ligand-exchange high-performance liquid chromatography of aliphatic amines, *J. Chromatogr.,* 356, 15, 1986.
19. **Siegal, A. and Degens, E. T.,** Concentration of dissolved amino acids from saline waters by ligand exchange chromatography, *Science,* 151, 1098, 1966.
20. **Buist, N. R. M. and O'Brien, D.,** The separation of peptides from amino acids in urine by ligand exchange chromatography, *J. Chromatogr.,* 29, 398, 1987.
21. **Maeda, M., Tusji, A., Ganno, S., and Pnishi, Y.,** Fluorometric assay of amino acids by using automated ligand-exchange chromatography, *J. Chromatogr.,* 77, 434, 1973.
22. **Maeda, M., Kinoshita, T., and Tsuji, A.,** A novel fluorimetric method for determination of hexosamines using pyridoxal and Zn(II) ion, *Anal. Biochem.,* 38, 121, 1970.
23. **Doury-Berthod, M., Poitrenaud, C., and Tremillon, B.,** Ligand-exchange separation of amino acids. I. Distribution of some amino acids between ammoniacal and copper(II) nitrate solutions and phosphonic, carboxylic and iminodiacetate ion exchangers in the Cu(II) form, *J. Chromatogr.,* 131, 73, 1977.
24. **Doury-Berthod, M., Poitrenaud, C., and Tremillon, B.,** Ligand-exchange separation of amino acids. II. Influence of eluent composition and of the nature of the ion exchanger, *J. Chromatogr.,* 179, 37, 1979.
25. **Caude, M. and Foucault, A.,** Ligand exchange chromatography of amino acids on Cu(II) modified silica gel with ultraviolet spectrophotometric detection at 210 nm, *Anal. Chem.,* 51, 459, 1979.
26. **Foucault, A., Caude, M., and Oliveros, L.,** Ligand-exchange chromatography of enantiomeric amino acids on Cu-loaded chiral bonded silica gel and of amino acids on Cu-modified silica gel, *J. Chromatogr.,* 185, 345, 1979.
27. **Davankov, V. A., Kurganov, A. A., and Bochkov, A. S.,** Resolution of racemates by high-performance liquid chromatography, in *Advances in Chromatography,* Vol. 22, Giddings, J. C., Grushka, E., Cazes, J., and Brown, P. R., Eds., Marcel Dekker, New York, 1984, chap. 3.
28. **Davankov, V. A. and Rogozhin, S. V.,** Chromtography of ligands; a new method for studying mixed complexes, *Dok. Akad. Nauk SSSR Chem. Proc.,* 193, 94, 1971; English translation, Consultants Bureau, p. 460.
29. **Kurganov, A. A., Zhuchkova, L., Ya., and Davankov, V. A.,** Stereochemistry in *bis*-(α-amino acid) copper(II) complexes. VII. Thermodynamics of N-benzylproline coordination to copper(II), *J. Inorg. Nucl. Chem.,* 40, 1081, 1978.
30. **Davankov, V. A., Rogozhin, S. V., and Kurganov, A. A.,** First case of stereoselectivity in complexes of copper with amino acids, *Isv. Akad. Nauk SSSR Ser. Khim.,* 204, 1971; English translation, Consultants Bureau, p. 193.
31. **Davankov, V. A., Zolotarev, Yu. A., and Kurganov, A. A.,** Ligand exchange chromatography of racemates. XI. Complete resolution of some chelating racemic compounds and nature of sorption enantioselectivity, *J. Liquid Chromatogr.,* 2, 1191, 1979.
32. **Davankov, V. A. and Zolotarev, Yu. A.,** Ligand exchange chromatography of racemates, V, VI, VII, *J. Chromatogr.,* 155, 285, 1978.
33. **Semechkin, A. V., Rogozhin, S. V., and Davankov, V. A.,** Ligand exchange chromatography of racemates, IV. Influence of stationary-complex structure on the mechanism of ligand exchange, *J. Chromatogr.,* 131, 65, 1977.

34. **Davankov, V. A., Bochkov, A. S., Kurganov, A. A., Roumeliotis, P., and Unger, K. K.,** Separation of unmodified α-amino acid enantiomers by reversed-phase HPLC, *Chromatographia,* 13, 677, 1980.
35. **Roumeliotis, P., Unger, K. K., Kurganov, A. A., and Davankov, V. A.,** HPLC of α-amino acid enantiomers: studies on bonded 3-(L-prolyl)- and 3-(L-hydroxyprolyl) propyl silicas, *J. Chromatogr.,* 255, 51, 1983.
36. **Gübitz, G., Juffman, F., and Jellenz, W.,** Direct separation of amino acid enantiomers by high-performance ligand-exchange chromatography on chemically bonded chiral phases, *Chromatographia,* 16, 103, 1982.
37. **Gübitz, G.,** Direct separation of enantiomers by high performance ligand exchange chromatography on chemically bonded chiral phases, *J. Liquid Chromatogr.,* 9, 519, 1986.
38. **Gübitz, G. and Mihellyes, S.,** Direct separation of 2-hydroxyacid enantiomers by high performance liquid chromatography on chemically bonded chiral phases, *Chromatographia,* 19, 257, 1984.
39. **Corradini, C., Federici, F., Sinibaldi, M., and Messina, A.,** High-performance ligand-exchange chromatography on diamine-bonded silica, *Chromatographia,* 23, 118, 1987.
40. **Takeuchi, T., Asai, H., and Ishii, D.,** Enantiomeric separation of amino acids by micro HPLC on an L-proline-bonded stationary phase, *J. Chromatgr.,* 407, 151, 1987.
41. **Hare, P. E. and Gil-Av, E.,** Separation of D- and L-amino acids by liquid chromatography — use of chiral eluents, *Science,* 204, 1226, 1979.
41a. **Oelrich, E., Preusch, H., and Wilhelm, E.,** Separation of enantiomers by high-performance liquid chromatography using chiral eluents, *J. High-resolut. Chromatogr. Chromatogr. Commun.,* 3, 269, 1980.
42. **Lindner, W., LePage, J. N., Davies, G., Seitz, D. E., and Karger, B. L.,** Reversed-phase separations of optical isomers of Dns-amino acids and peptides using chiral metal chelate additives, *J. Chromatogr.,* 185, 323, 1979.
43. **LePage, J. N., Lindner, W., Davies, G., Seitz, D. E., and Karger, B. L.,** Resolution of the optical isomers of dansyl amino acids by reversed phase liquid chromatography with optically active metal chelate additives, *Anal. Chem.,* 51, 433, 1979.
44. **Porath, J. and Olin, B.,** Immobilized metal ion affinity adsorption and affinity chromatography of biomaterials, *Biochemistry,* 22, 1621, 1983.
45. **Saunders, R. M.,** Separation of sugars on an ion-exchange resin, *Carbohydr. Res.,* 7, 76, 1968.
46. **Pecina, R., Bonn, G., Burtscher, E., and Bobleter, O.,** HPLC elution behaviour of alcohols, aldehydes, ketones, organic acids and carbohydrates on a strong cation-exchange stationary phase, *J. Chromatogr.,* 287, 245, 1984.
47. **Kuwamoto, T. and Okada, E.,** Separation of mono- and disaccharides by HPLC with a strong cation-exchange resin and an acetonitrile-rich eluent, *J. Chromatogr.,* 258, 284, 1983.
48. **Hobbs, J. S. and Lawrence, J. G.,** Separation of carbohydrates on cation-exchange resin columns having organic counter-ions, *J. Chromatogr.,* 72, 311, 1972.
49. **Goulding, R. W.,** Liquid chromatography of sugars and related polyhydric alcohols on cation exchangers: effect of cation variation, *J. Chromatogr.,* 103, 229, 1975.
50. **Walton, H. F.,** Counter-ion effects in partition chromatography, *J. Chromatogr.,* 332, 203, 1985.
51. **Scobell, H. D. and Probst, K. M.,** Rapid high-resolution separation of oligosaccharides on silver-form cation-exchange resins, *J. Chromatogr.,* 212, 51, 1981.
52. **Schmidt, J., John, M., and Wandrey, C.,** Rapid separation of malto-, xylo- and cello-oligosaccharides, *J. Chromatogr.,* 213, 151, 1981.
53. **Fitt, L. E., Hassler, W., and Just, D. E.,** A rapid method to determine the composition of corn syrup by liquid chromatography, *J. Chromatogr.,* 187, 381, 1980.
54. **Petrus, L., Bilik, V., Kuniak, L., and Stankovic, L.,** Chromatographic separation of alditols on a cation-exchange resin in lanthanum form, *Chem. Zvesti,* 34, 530, 1980.
55. **Leonard, J. L., Guyon, F., and Fabiani, P.,** HPLC of sugars on Cu(II)-modified silica gel, *Chromatographia,* 18, 600, 1984.
56. **Honda, S., Takahashi, M., Kakehi, K., and Ganno, S.,** Rapid automated analysis of monosaccharides, *Anal. Biochem.,* 113, 130, 1981.
57. **Reimerdes, E. H., Rothkitt, K. D., and Schauer, R.,** Determination of carbohydrates by anion exchange of their borate complexes, *Fresenius Z. Anal. Chem.,* 318, 285, 1884.
58. **Petronio, B. M., Lagana, A., and Russo, M. V.,** Some applications of ligand exchange: recovery of phenolic compounds from water, *Talanta,* 28, 215, 1981.
59. **Maslowska, J. and Pietik, W.,** Effect of complex formation in the separation of phenols by ion-exchange chromatography, *J. Chromatogr.,* 201, 293, 1980.
60. **Maslowska, J. and Pietik, W.,** Separation of chlorophenols on a cation exchanger in cerium(III) form, *Chromatographia,* 18, 704, 1984; 20, 46, 1985.
61. **Fujimura, K., Koyama, T., Tanigawa, T., and Funasaka, W.,** Ligand-exchange chromatography separation of hydroxybenzoic and hydroxynaphthoic acid isomers, *J. Chromatogr.,* 85, 101, 1973.
62. **Otto, J., de Hernandez, C. M., and Walton, H. F.,** Chromatography of aromatic acids on lanthanum-loaded ion-exchange resins, *J. Chromatogr.,* 247, 91, 1982.

63. **Takayanagi, H., Hatano, O., Fujimura, K., and Ando, T.,** Ligand-exchange high-performance liquid chromatography of dialkyl sulfides, *Anal. Chem.*, 57, 1840, 1985.
64. **Antonelli, M. L., Marino, M., Messina, A., and Petronio, B. M.,** A proposal for the application of ligand-exchange chromatography in thin layers, *Chromatographia,* 13, 167, 1980.
65. **Shimomura, K. and Walton, H. F.,** Thin-layer chromatography of amines by ligand exchange, *Separation Sci.*, 3, 493, 1968.
66. **Fujimura, K. and Ando, T.,** Studies on ligand-exchange chromatography. V. Gas chromatographic separation of lower aliphatic amines by ligand exchange, *J. Chromatogr.*, 114, 15, 1975.
67. **Fujimura, K., Kitanaka, M., and Ando, T.,** Ligand-exchange gas chromatographic separation of aniline bases, *J. Chromatogr.*, 241, 295, 1982.
68. **Fujimura, K., Kitanaka, M., Takayanagi, H., and Ando, T.,** Ligand-exchange gas chromatography of lower aliphatic amines on solid and liquid crystalline stationary phases, *Anal. Chem.*, 54, 918, 1982.
69. **Takayanagi, H., Hashizuma, M., Fujimura, K., and Ando, T.,** Ligand-exchange gas chromatography of dialkyl sulfides, *J. Chromatogr.*, 350, 63, 65, 1985.
70. **Barber, D. W., Phillips, C. S. G., Tusa, F. F., and Berdin, A.,** The chromatography of gases and vapours. VI. Use of the stearates of bivalent Mn, Co, Ni, Cu and Zn as column liquids in gas-liquid chromatography, *J. Chem. Soc.*, 18, 1959.
71. **Houx, N. W. H., Voerman, S., and Jongen, W. M. F.,** Purification and analysis of synthetic insect sex attractants by liquid chromatography on a silver-loaded resin, *J. Chromatogr.*, 96, 25, 1974.
72. **Kroppenstedt, R. M.,** Separation of bacterial menaquinones by HPLC, using reverse-phase and a silver-loaded ion exchanger as stationary phases, *J. Liquid Chromatogr.*, 5, 2359, 1982.
73. **Gil-Av, E. and Schurig, V.,** Gas chromatography of olefins, *Anal. Chem.*, 43, 2030, 1971.
74. **Guha, O. K. and Janak, J.,** Charge-transfer complexes of metals in the chromatographic separation of organic compounds, *J. Chromatogr.*, 68, 325, 1972.
75. **Morris, L. J.,** Separation of lipids by silver-ion chromatography, *J. Lipid Res.*, 7, 717, 1966.
76. **Davankov, V. A., Navratil, J. D., and Walton, H. F.,** *Ligand Exchange Chromatography,* CRC Press, Boca Raton, FL, 1988.

Chapter 7

AMINO ACIDS, PEPTIDES, PROTEINS, AND CARBOHYDRATES

I. AMINO ACIDS

The separation and quantitative analysis of mixtures of amino acids was one of the early triumphs of ion-exchange chromatography, following soon after the separation of the lanthanide ions. The classic work of Moore and Stein,[1] published in 1951, earned these investigators the Nobel prize for chemistry in 1972.

Amino acids are amphiprotic. They can lose protons to become anions, and they can gain protons to become cations. Thus, they can be separated by both anion and cation exchange. Cation exchange was used by Moore and Stein, and has been overwhelmingly preferred since then. The main reason for preferring cation exchange is that strong-acid sulfonated polystyrene exchangers are chemically more homogeneous and more reproducible than quaternary-base anion exchangers.

Amino acids are the units of which proteins are formed. Some 20 amino acids occur normally in proteins, and they are all α– amino acids having the general formula $R{-}CH(NH_3^+)COO^-$. They exist in the dipolar-ion or Zwitterion form, both in aqueous solution and in solid crystals, but for ease in writing they are usually shown as $R{-}CH(NH_2)COOH$. The R groups are different in different amino acids. Table 1 lists the common amino acids with the formulas of their R groups. Note that proline and hydroxyproline are secondary amines and that their entire formulas are written here. The amino acids fall into three classes, neutral, acidic and basic. The acidic amino acids have an extra carboxyl group in their R groups; the basic amino acids have an extra amino group, while the neutral amino acids have neither. Peptides and proteins are formed from α–amino acids by condensation and loss of water molecules to form peptide links, –CO–NH–.

At a sufficiently low pH, about 3, all the acids shown in Table 1 form cations and can be adsorbed on cation-exchange resins. Moore and Stein separated the cations by chromatography on a column of the sulfonated polystyrene resin Dowex®-50, particle size 250 to 500 mesh (50 to 100 μm), crosslinking 8%. The bed size was 0.9 × 100 cm, and the column was water jacketed to allow temperature control. Flow was under gravity. A series of eluents was used, starting with a buffer of pH 3.4, 0.1 *M* in citrate ions. The buffer pH was raised to 6.7 in a series of steps as the elution proceeded, and at the same time the temperature was raised from 37 to 75°C. The acidic amino acid aspartic acid came out first, followed by the neutral amino acids, but the basic amino acids remained on the column and had to be eluted with bicarbonate-carbonate buffers of pH 8.3 to 11.0. The flow rate was 4 ml/h. To detect and measure the amino acids as they came out of the column, fractions of 1 ml were collected and adjusted to pH 5, then the reagent ninhydrin was added and the solution was heated to 100°C for 20 min. A purple-colored compound was formed by this reaction:[2]

$$\mathrm{R-\underset{NH_2}{\overset{H}{C}}-COOH + 2\ C_6H_4(CO)_2C(OH)_2 \longrightarrow C_6H_4(CO)C(OH){=}C{-}N{=}C(CO)_2C_6H_4 + RCHO}$$

Its concentration was measured spectrophotometrically. The maximum light absorption was

TABLE 1
Common Amino Acids from Proteins, General Formula: $RCH(NH_3)^+COO^-$

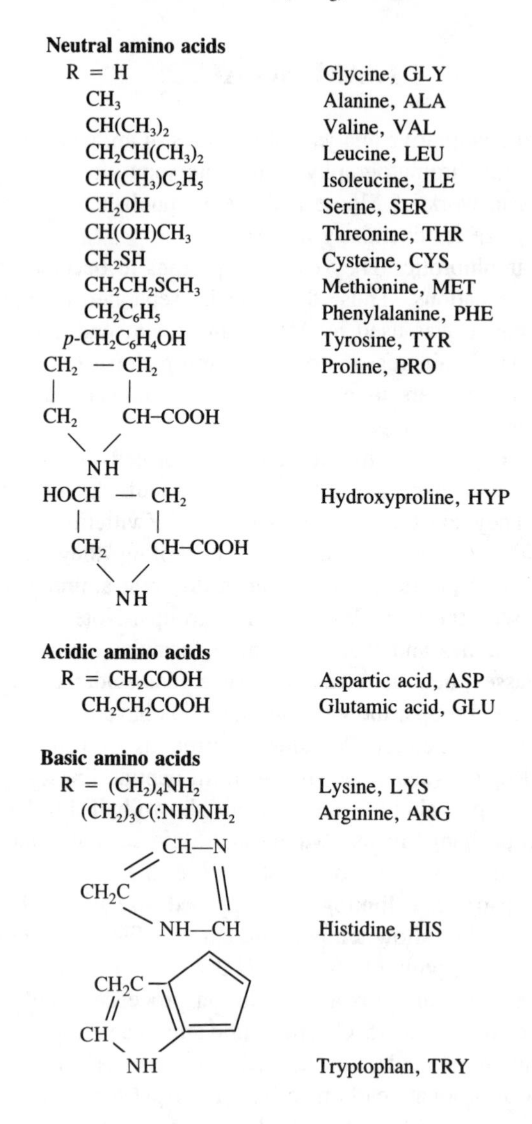

R	Amino acid
Neutral amino acids	
R = H	Glycine, GLY
CH_3	Alanine, ALA
$CH(CH_3)_2$	Valine, VAL
$CH_2CH(CH_3)_2$	Leucine, LEU
$CH(CH_3)C_2H_5$	Isoleucine, ILE
CH_2OH	Serine, SER
$CH(OH)CH_3$	Threonine, THR
CH_2SH	Cysteine, CYS
$CH_2CH_2SCH_3$	Methionine, MET
$CH_2C_6H_5$	Phenylalanine, PHE
p-$CH_2C_6H_4OH$	Tyrosine, TYR
CH_2 — CH_2 | | CH_2 CH–COOH \\ / NH	Proline, PRO
HOCH — CH_2 | | CH_2 CH–COOH \\ / NH	Hydroxyproline, HYP
Acidic amino acids	
R = CH_2COOH	Aspartic acid, ASP
CH_2CH_2COOH	Glutamic acid, GLU
Basic amino acids	
R = $(CH_2)_4NH_2$	Lysine, LYS
$(CH_2)_3C(:NH)NH_2$	Arginine, ARG
CH—N // || CH_2C \\ NH—CH	Histidine, HIS
CH_2C—(benzene ring) // CH \\ NH	Tryptophan, TRY

570 nm. (Proline and hydroxyproline, being secondary amines, formed different compounds having maximum absorbance at 440 nm.)

To elute the acidic and neutral amino acids required 100 h; to elute the four basic amino acids required 75 h more, and their recovery was only 70%, contrasted with nearly 100% recovery for the other amino acids. Because the basic amino acids took so long to come out and because their recovery was incomplete, Moore and Stein set up a separate, shorter column to analyze the basic amino acids. A separate sample was placed on this column and more concentrated buffers, citrate and phosphate of pH 6.5 to 6.8, were used; the higher concentrations gave greater displacing power. The acidic and neutral amino acids came out almost immediately and were discarded; then the basic amino acids, tryptophane, histidine, lysine, and arginine, were eluted in that order, with recoveries near 100%. Running the two columns in parallel, Moore and Stein were able to analyze the amino acids in protein hydrolyzates in 5 d. Precision was about 2% relative standard deviation.

Today the same analysis can be made in less than an hour without loss of precision. Cation-exchange chromatograms can be run in 20 min. The same chemistry is used today that was used by Moore and Stein, but with the following improvements.

A. DETECTION

Instead of collecting fractions and analyzing them one by one, the color-producing reagent and the buffer are fed into the effluent solution after it leaves the column, then the mixed solution is made to flow through a long coil of capillary tubing (the "delay coil") at an elevated temperature to allow time for reaction. At 130°C the delay time need only be 2 min. With the color fully developed, the solution now flows through the cell of a spectrophotometer and the light absorbance is measured and recorded continuously.

Ninhydrin is the reagent commonly used in commercial amino-acid analyzers, but another reagent much used for post-column (and also pre-column) derivatization is *ortho*-phthalaldehyde plus ethanethiol or 2-mercaptoethanol, $HO{\cdot}C_2H_4SH$.[3,4] These compounds form fluorescent products with primary amines but not secondary amines or ammonia. The solution must be alkaline, pH 9.5 to 10.5 and a borate buffer is used. The thiol must be added simultaneously with the aldehyde or before, not after. The reaction is probably:

CHO, CHO (on benzene ring) + $HS{-}C_2H_5$ + $H_2N{-}CHR{-}COOH$ ⟶ (benzene ring fused) C(SC_2H_5)–N–CHR–COOH, CH

Fluorescence is excited at 335 nm, emitted at 425 nm. Detection limits are below 50 pmol, 10 to a hundred times less than the limits with ninhydrin.

Being secondary amines, proline and hydroxyproline are not detected. However, they may be made detectable by adding sodium hypochlorite to the column effluent before pumping in the ethanethiol-*o*-phthalaldehyde mixture.

A new method is pulsed amperometric detection, described in Chapter 4. It works well for carbohydrates and will be mentioned later in this chapter, but its application to amino acids is not yet fully tested. For it to be effective the solution must be alkaline, for hydroxyl ions take part in the electrode reaction. Polta and Johnson[5] separated amino acids by anion exchange with 0.25 *M* sodium hydroxide eluent and used pulsed amperometric detection.

B. THE MOBILE PHASE

Sodium citrate buffers are commonly used. To elute the basic amino acids in the alkaline range, sodium borate-boric acid and sodium carbonate-sodium bicarbonate buffers may be used. As to citrate, one should recall that citric acid is made commercially by bacterial fermentation of corn syrup, and is therefore liable to contain traces of amino acids. Citric acid used to prepare buffers for amino-acid analysis must be carefully purified, or else avoided altogether by using other weak acids.

The function of the buffer is twofold. By regulating the pH it regulates the ionic charge of the amino acids, and the fraction of amino acids in its various ionic forms. The lower the pH, the greater the fraction in the cationic form, which, having a net positive charge, is able to be held on a cation-exchange resin. The fraction also depends on the ionization constant of the acid; the stronger the acid, other things being equal, the more weakly it is held by the exchanger at a given pH. The second function of the buffer is to supply cations that compete with the amino acid cations and displace these cations from the exchanger by ion exchange. Thus, the elution of an amino acid from the column can be hastened in two ways; (1) by raising the pH; and (2) by raising the sodium-ion concentration. There is enough difference in the strengths of binding of the different amino acids that the eluent strength must be increased in the course of a chromatographic run. This may be done in a series of steps with buffer changes, as was done by Moore and Stein,[1] or it may be done through a continuous gradient. The gradient may be one of increasing pH at constant sodium-ion concentration, increasing sodium-ion concentration at constant pH, or a combination of the two. The temperature is another parameter that can be varied. Binding of amino acids is weakened as the temperature rises. Changing the temperature may also affect peak separation; Moore and Stein found that methionine and isoleucine, also tyrosine and phenylalanine, were resolved at 60° but not at 25°C. Raising the temperature also sharpens the peaks because it makes diffusion faster. Today's practice is to use a temperature of 60 to 70°C but not to change it during the run.

Lithium salts are often used as buffers instead of sodium salts.[6-8] Lithium ions are bound more weakly than sodium ions by sulfonated polystyrene cation exchangers, but this effect can be offset by using higher concentrations. The advantage of lithium buffers is that they give sharper peaks and allow certain peaks to be separated that overlap with sodium buffers, e.g., glutamine and asparagine. Thus, lithium buffers are preferred for the analysis of physiological fluids, which are much more complex than the mixtures of amino acids formed by protein hydrolysis. The reason for the lithium effect is not clear; however, it is found that when sulfonated polystyrene resins are used as supports for reversed-phase chromatography of neutral compounds, the ionic form of the resin makes a difference, and peaks are narrower and better separated on lithium-form resin than on sodium or potassium-form resin. Theophylline and caffeine, e.g., are well resolved on lithium-form resin with an aqueous lithium buffer but give only one peak with sodium-form resin.[9]

We must emphasize again that organic compounds are attached to organic polymeric ion exchangers by mixed mechanisms. Electrostatic attraction and ion exchange are not the only forces operating. Hydrophobic interaction is a factor too. Ion-exchanging polymers have the quality of reversed-phase supports, and adsorb and retain organic compounds, particularly those with aromatic structures, even when they have no ionic charge.

In conclusion, we note that hydrogen ions have been used as the displacing ions for chromatography of amino-acid cations. Moore and Stein[1] reported on their use. Recently 0.001 *M* hydrochloric acid has been used to elute acidic and neutral amino acids from a low-capacity latex-coated cation exchanger.

C. THE STATIONARY PHASE

Stein and Moore used 8% crosslinked sulfonated polystyrene as beads 100 μm in diameter. It is no wonder that their chromatography took 5 d. The first step to improving this performance was to use smaller resin particles. In 1958 Hamilton devised a hydraulic technique for refining commercial bead-type resins according to size. He isolated fractions having diameters 22 ± 3 μm (these were the swollen diameters in water) and found, as expected, that this resin gave greatly improved column performance. Later, when commercial amino acid analyzers became available, the production of sulfonated polystyrene resins with closely controlled particle diameters, 10 μm and less, and closely controlled crosslinking was developed into a fine art. Analysis times for protein hydrolyzates were brought down to 8 h, then to 4 h.

The next advance was the production of macroporous resins.[8] Most of these are based on polystyrene, but Spheron® and other acrylate polymers were also developed. Many chromatograms have been published to show the performance of these resins, and these are impressive; 20-min analyses are possible for protein hydrolyzates, though longer analysis

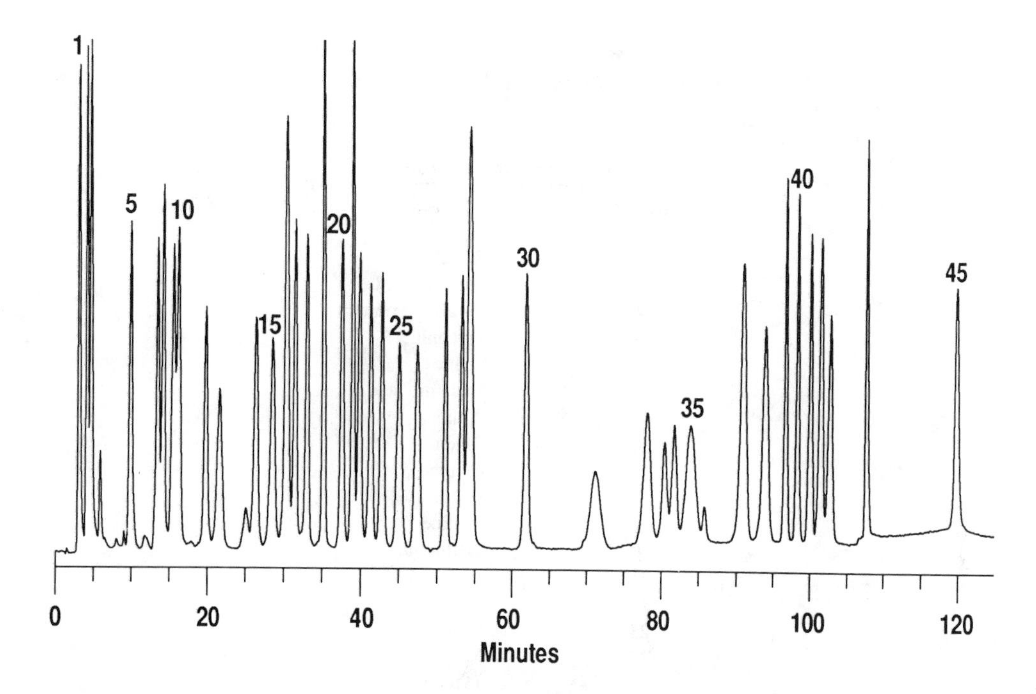

FIGURE 1. Chromatogram of physiological amino acids. Stationary phase, fully sulfonated polystyrene-DVB, 5 μm; mobile phase, lithium citrate buffers. Detection by ninhydrin; 2.5 nmol of each amino acid. For peak identity, see Table 2. (By courtesy of Dionex Corporation.)

times are needed for physiological fluids which are much more complex. (Estimating overall analysis times, one must of course include the time needed to reverse the gradient at the end of the run and restore the column to equilibrium with the first eluent.) The exact nature and method of preparation of the macroporous resins is, of course, kept confidential by the commercial laboratories that have developed them.

Figure 1 shows a chromatogram of amino acids in a physiological fluid, on a 5-μm fully sulfonated (high-capacity) polystyrene-DVB resin with lithium citrate buffers as eluents. The identity of the amino acids and the order of their elution is given in Table 2. Chromatograms of amino acids from protein hydrolyzates are, we repeat, much simpler; they can be run in 20 to 30 min, compared with the 120 min run time shown here.

Until recently, only high-capacity resins, gel-type or macroporous, were used for amino acid analysis. Now the techniques of high-performance ion chromatography have been adapted to this purpose. Latex-coated ion exchangers have been used, as we mentioned above, and experiments have been made with anion exchangers as well as cation exchangers. In anion exchange the order of elution of the amino acids is roughly the reverse of that in cation exchange. Low-capacity exchangers require small sample sizes, hence sensitive detection. Post-column reaction with *o*-phthalaldehyde gives the necessary sensitivity. If anion exchange is used for separation, pulsed amperometric detection may be used, as noted above.

D. ELUTION ORDERS

As a rough generalization we have said that the acidic amino acids are eluted first in cation exchange, then the neutrals, then the basic amino acids. Table 2 gives the order of elution of the amino acids shown in Figure 1, which is for a physiological fluid. A study of the table shows that this statement is not quite true. The first amino acids to be eluted from a protein hydrolyzate are, in this order, aspartic, threonine, serine, glutamic. Glutamic acid is an acidic amino acid with two carboxyl groups, and so is aspartic acid, but threonine and serine are neutral. Amino acids stick to a cation exchanger because of their positive

TABLE 2
Elution Order of Amino Acids

Amino acids in protein hydrolyzates	
Peak No.	**Acid**
4	(Urea)
5	Aspartic acid
6	Hydroxyproline
7	Threonine
8	Serine
10	Glutamic acid
13	Proline
14	Glycine
15	Alanine
18	Valine
19	Cystine
20	Methionine
21	Cysteine
22	Isoleucine
23	Leucine
24	Norleucine
25	Tyrosine
26	Phenylalanine
31	Tryptophane
35	(Ammonium)
38	Lysine
39	Histidine
45	Arginine

charge. Therefore, the fraction of acid in the fully protonated, positively charged form should determine the strength of binding to the exchanger. The pH of the buffer used to begin the elution is generally 3.0. At this pH only a small fraction of each amino acid is in its fully protonated, positively charged form. The first ionization constants of the fully protonated ionic forms, H_3L^+ for the acidic amino acids. H_2L^+ for the neutral amino acids, where L is the deprotonated amino acid ligand, are as follows, expressed as their negative logarithms:

> Aspartic acid, 1.87; threonine, 2.17; serine, 2.55; glutamic acid, 2.30. (For comparison, glycine and alanine have pK_1 values of 2.35). The corresponding fractions of each amino acid in the positively charged form at pH 3.00 are: aspartic acid, 0.07; threonine, 0.13; serine, 0.25; glutamic acid, 0.17; glycine and alanine, 0.18.

The smaller this fraction, the faster the amino acid should be eluted. This relationship holds for the first three acids, but not for the others. It is the polar interaction of –COOH and –OH with water that seems to determine the binding, rather than the fraction that is positively charged.

Comparing glycine, alanine, valine, and leucine, where the number of carbon atoms in the R– side chain are 0, 1, 3, and 4 and polar groups in the side chain are absent, we see that the elution order follows the number of carbon atoms, or the increasing order of hydrophobicity. Isoleucine, leucine, and norleucine (which has the straight-chain R, $CH_3CH_2CH_2CH_2$) are eluted in that order; it is the straight chain that causes the most disruption of the water structure and is most adsorbed by the exchanger. A similar effect is seen in ligand-exchange chromatography of isomeric butylamines (see Chapter 6). Amino acids having aromatic rings are preferentially bound by polystyrene-based exchangers; tyrosine and phenylalanine are more strongly bound that lysine, and tyrosine, having a polar phenolic –OH, is more weakly bound than phenylalanine.

While the order of elution of amino acids from polystyrene-based cation exchangers is generally the same for different buffer systems, the buffer does make a difference and so does temperature. Changing the temperature or the buffer may change the separation of neighboring peaks and can change the elution order. Much trial and error goes into a successful amino acid analysis, and elution variables must be carefully controlled, especially in analyzing complex physiological fluids. One must always bear in mind that ion exchange is only one of the mechanisms for retention of organic compounds, and that ion-exchanging polymers can act as reversed-phase adsorbents and can retain uncharged, non-ionic organic solutes, particularly if they have aromatic character.

II. PROTEINS

Proteins are made from amino acids by eliminating water molecules and joining the amino acid units through peptide links:

$$-CHR_1-CO-NH-CHR_2-CO-NH-CHR_3-$$

The nature of the side chains, R_1, R_2, and so on, their order in the chain and the length of the chain all go to determine the kind of protein. The chain is twisted and bent upon itself as different R units are attached to one another through hydrogen bonds. Another kind of secondary linkage is the disulfide bond, –S–S–, formed between two cysteine molecules by two –SH groups being oxidized and losing two hydrogen atoms. When any of these bonds is broken the protein is changed, often irreversibly. This change is called *denaturation*.

Of the different R units, some, as we have seen, carry carboxyl groups (the units of aspartic and glutamic acid) and some carry amino groups (the units of arginine, lysine, histidine, and tryptophane). These groups can ionize and make the protein either positively or negatively charged, depending on pH. Through these charges proteins are adsorbed on both cation and anion exchangers and can be separated by anion- and cation-exchange chromatography. An important characteristic of a protein is its *isoelectric-point,* the pH at which the net charge on the protein molecule is zero. When the net charge is zero the protein does not move in an electric field. It is through electrophoretic measurements that the isoelectric points of proteins are determined.

Mixtures of proteins may be analyzed by electrophoresis, anion exchange, cation exchange, ligand exchange, hydrophobic interaction, and reversed-phase chromatography. In this chapter we are concerned with anion and cation exchange.

Proteins have large molecules, with molecular weights in the tens of thousands. Moreover they are hydrophilic. Removal of water may change their structure and denature them. Thus, the ion exchangers used in the chromatography of proteins must have special characteristics. They must be very porous on the molecular scale and they must be hydrophilic. The conventional gel-type polystyrene-based ion exchangers fail on both these counts.

A. ION EXCHANGERS FOR PROTEIN CHROMATOGRAPHY

The first successful ion exchangers for use in chromatography of proteins were the Glycophases made by Regnier and his group.[10-13] The starting point is controlled-pore glass or porous silica, which is made to react with a silane carrying the epoxide or oxirane group. Reaction with a 10% aqueous solution of glycidoxypropyl-trimethoxysilane gives this product:

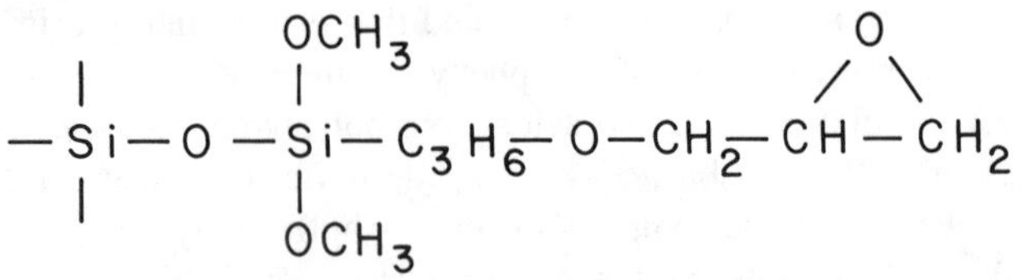

$$
-\overset{|}{\underset{|}{Si}}-O-\overset{OCH_3}{\underset{OCH_3}{Si}}-C_3H_6-O-CH_2-\overset{O}{CH-CH_2}
$$

Under the conditions used in the reaction the epoxy group adds a molecule of water to give the terminal group $-CH(OH)-CH_2OH$. This product is called "Glycophase G" and is used for size-exclusion chromatography. The next step is to couple this product, using boron trifluoride as catalyst, with a compound that has two or three epoxy groups in its molecule, namely, di- or triglycidylglycerol. Diglycidylglycerol has this formula:

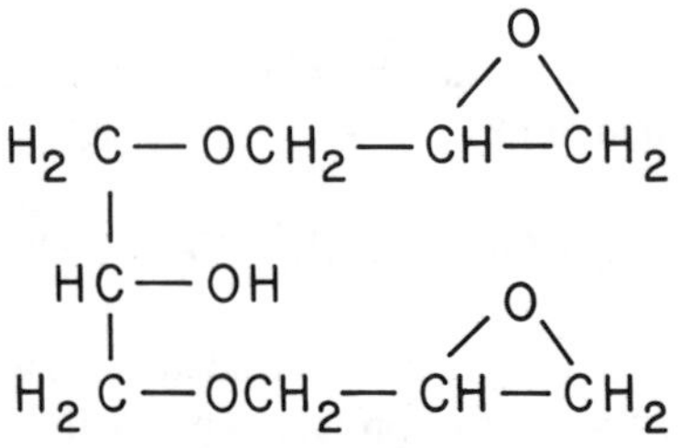

$$
\begin{array}{l}
H_2C-OCH_2-CH\overset{O}{-}CH_2 \\
\;\;| \\
HC-OH \\
\;\;| \\
H_2C-OCH_2-CH\overset{O}{-}CH_2
\end{array}
$$

These molecules introduce crosslinking and coat the silica or glass with what is virtually glycerol, chemically bound, and very hydrophilic.

To introduce ionic functional groups a suitable reagent is added during the crosslinking step. This may be diethylaminoethanol (DEAE). A weakly basic anion exchanger then results, having the structure:

$$
-OCH_2CH(OH)CH_2-OCH_2CH_2N(C_2H_5)_2
$$

This product can be turned into a strong base by reaction with methyl iodide to give a quaternary ammonium groups. A weakly acidic cation exchanger is made by adding allyl glycidyl ether.

$$
CH_2{=}CHCH_2OCH_2CH\overset{O}{—}CH_2
$$

The product contains the group $-CH_2CH{=}CH_2$ which is then oxidized to $-CH_2-COOH$ with alkaline periodate or permanganate. (Perbenzoic acid oxidizes the allyl group to the oxirane, $-CH_2-CH(O)CH_2$.) If sodium bisulfite is used in this step the sulfonic acid group is introduced, giving a strong-acid cation exchanger. Direct reaction of the epoxy group with sodium bisulfite also introduces the sulfonic acid group, thus:

$$
-O-CH_2CH(OH)-CH_2-SO_2OH
$$

A different series of reactions was used by Alpert[14,15] to bind polyaspartic acid to silica, making a weak-acid cation exchanger. The starting point was aminopropyl silica, $Si-O-Si(CH_2)_3-NH_2$. This was made to react with polysuccinimide,

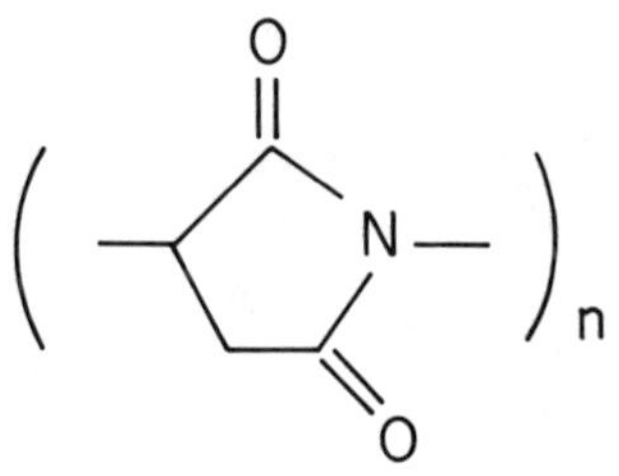

```
        O
        ||
   /  /   \    \
  ( —<     N—   )
   \  \___/    / n
          \\
           O
```

which had been made by heating aspartic acid. The polysuccinimide-silica was hydrolyzed to give polyaspartic acid-silica, which is a weak-acid cation exchanger:

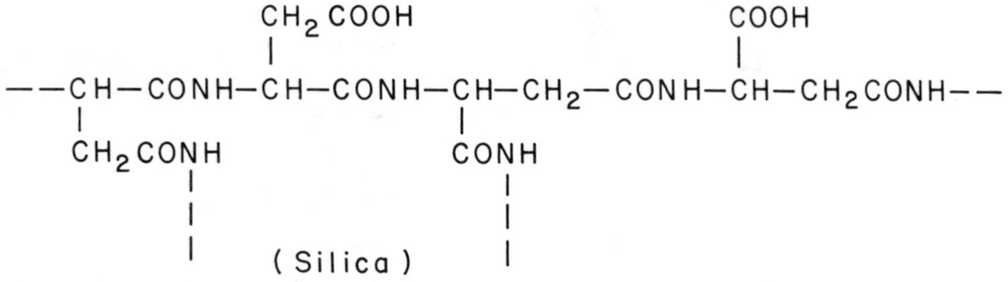

```
                CH2COOH                               COOH
                 |                                     |
——CH—CONH—CH—CONH—CH—CH2—CONH—CH—CH2CONH——
  |                    |
 CH2CONH              CONH
      |                 |
      |                 |
      |     ( Silica )  |
```

A strong-acid cation exchanger was prepared by incorporating taurine, $H_2N{\cdot}C_2H_4{\cdot}SO_3H$.[16]

Porous silica may be coated with polyethyleneimine, $-(CH_2CH_2NH)_n$, which is then crosslinked in various ways. Di- and triglycidyl glycerol, mentioned above, can be used. Another crosslinking agent is 1,4-butanediol diglycidyl ether. The result is a weakly basic anion exchanger.

Exchangers are also made with a polymeric base instead of silica. One approach is to take a surface-sulfonated macroporous polystyrene and attach low-molecular-weight polyethyleneimine (say, PEI-6, with average molecular weight 600) electrostatically, then hold the film in place by crosslinking with di- or triglycidylglycerol.[17] The bound polyethylene imine may be quaternized and made into a strongly basic anion exchanger by reaction with methyl iodide. Rounds and Regnier[18] have started with a 3 μm gel-type polystyrene fully sulfonated, and introduced polyethylene imine (molecular weight 600) by cation exchange, followed by crosslinking with 1,4-butanediol diglycidyl ether and quaternization with methyl iodide, to produce a very efficient strong-base anion exchanger.

It will be seen that there is no lack of ion-exchanging stationary phases suitable for protein chroatography. Proprietary columns are available, like Mono Q® and Mono S® (Pharmacia Fine Chemicals). These are polymer based, with quaternary ammonium and sulfonate groups, respectively. Their particle diameter is 10 μm. Other stationary phases, silica-based, are sold by Toyo Soda Corporation, PolyLC, and SynChrom, Inc. Stationary phases without ionic groups are used for size-exclusion and hydrophobic-interaction chromatography. A new development is the use of latex-coated ion exchangers for the cromatography of proteins and peptides (see Figures 6, 7, and 9, below).

B. BINDING OF PROTEINS BY ION EXCHANGERS

Like the amino acids from which they are made, proteins can have a positive or a negative charge, and can be adsorbed on an exchanger by cation or anion exchange. The strength of chromatographic retention depends on pH. It also depends on the nature and concentration of the displacing ion, as we expect in ion exchange. It is very sensitive to the concentration of the displacing ions, because it takes several small ions like sodium or chloride ions to displace one large protein ion.

Consider first the dependence on pH. At the isoelectric point, the pH at which the protein

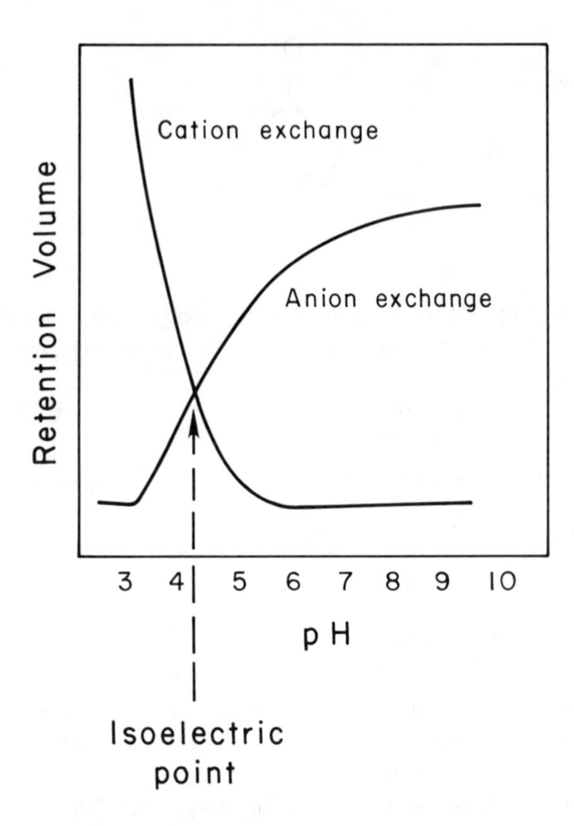

FIGURE 2. Retention map of a hypothetical protein.

molecule does not move in an electric field and has no net charge, one might expect that the protein would not be adsorbed by a cation exchanger, nor by an anion exchanger. For a few proteins this is true. Most proteins, however, are adsorbed to some extent by cation exchangers and by anion exchangers at the isoelectric point.[19] This fact suggests that the distribution of ionic charges on the protein molecule is not uniform. There may be a cluster of positive charges in one area of the big protein molecule, and a cluster of negative charges in another area, in accordance with the amino acid sequence and the shape of the amino acid chain. The total net charge on the molecule determines the movement in an electric field; the local concentrations of ions determine the binding to ion exchangers.

It is also possible that nonionic forces such as hydrophobic interaction contribute to binding. With exchangers as polar and hydrophilic as the Glycophases this is unlikely, but under special circumstances nonionic binding may occur.

The isoelectric pH may certainly be used as a guide to whether to use anion or cation exchange in the chromatography of a mixture of proteins. Acidic proteins, those that have isoelectric points at low pH, are best separated by anion exchange, while basic proteins, with high isoelectric pH, are best handled by cation exchange. Retention by cation exchange gets stronger as the pH is lowered away from the isoelectric point; retention by anion exchange gets stronger as the pH is raised above the isoelectric point.

The dependence of retention on pH is shown by retention maps such as that shown schematically in Figure 2.[20] The ordinate here is the retention time or volume in a chromatographic column *using a specified* and *exactly reproduced salt - concentration gradient.* Retention depends very strongly on salt concentration, as we shall see. With the help of retention maps for the various proteins one wishes to separate one can decide whether to use anion or cation exchange, and also one can choose the best pH.

Retention maps are usually drawn for ion exchangers that are strong acids or strong bases. If the exchangers are weak acids or bases, retention will first increase as one moves the pH away from the isoelectric point, then decrease as the fixed ionic groups lose their charge. For example, a weakly acidic cation exchanger like one that has functional carboxyl ions, COO^-, gains protons at low pH to form uncharged COOH which does not bind protein cations.

Another way to choose between anion and cation exchange and to find the best pH uses electrophoresis.[21] A glass plate is coated with a gel, and a pH gradient from 3 to 10 is set up across the plate from left to right. Then a sample of the protein mixture is spread across the middle of the plate from left to right in a narrow band or line. Now an electrical potential is applied at right angles to this line, from the top to the bottom of the plate. The different proteins in the mixture migrate up or down and produce, as it were, graphs of net charge vs. pH for each of the constituents. A glance at these traces tells immediately whether the proteins are best separated as cations or as anions and what is the pH that gives the best separation. This method shows the net charge, not the localized charge that is effective in ion exchange; moreover, the electrophoretic velocity depends on the molecular size and shape as well as on the charge; nevertheless, this method is a good guide to chromatographic conditions and it is fast.

Now let us consider the effect of salt concentration. It is usual to run protein chromatograms at a constant pH, maintained by a dilute buffer, say 0.05 *M* phosphate, whose concentration remains constant through the run. The concentration of salt, which is usually sodium chloride, is increased in a linear gradient from zero up to 1.0 *M* or thereabouts. Somewhere along this gradient is a narrow range of concentrations below which the protein is firmly attached to the exchanger, and above which it is not attached at all. Within this narrow range it is possible to measure corrected elution volumes (proportional to the capacity factor, k′) as a function of salt concentration. Graphs like that shown in Figure 3 are obtained, which is taken from an important and much quoted paper by Kopaciewicz et al.[19] The log-log plots are linear and very steep. The slope gives the number of small ions that must cooperate to dislodge one protein molecule. The lines to the left in Figure 3 have slopes somewhat greater than eight.

If this were simple ion exchange, the process could be represented thus, considering the protein, P, to be a cation with charge +n, displaced by n sodium ions:

$$PEx + nNa^+ = P^{n+} + Na_nEx$$

where Ex stands for the ion exchanger. From Figure 3 one would conclude that n = 8. Kopaciewicz et al.[19] make a different interpretation. They consider that the protein, too, is an ion exchanger and that the charges on the protein ion must be balanced by small counter-ions. They represent the displacement of the protein thus:

$$PEx + nNa^+ + nCl^- - P^{n+}Cl_n^- + Na_nEx$$

In this interpretation, n = 4. It takes four sodium ions and four chloride ions, acting jointly, to displace one protein ion from the surface of the exchanger. There are four points of attachment between the protein ion and the exchanger surface. The net charge of the protein ion depends on pH and is a lot more than four.

Figure 3 shows data for anion exchange of β-lactoglobulin, whose isoelectric point is 5.2. At pH 5 the slope of the line in Figure 3 is 3.6, indicating n = 1.8. Close to the isoelectric point, the charges on the protein are fewer and farther apart, so that the number of charges interacting with the protein are fewer. For proteins in general the slopes of these graphs decrease as one approaches the isoelectric points.

Lest one get the idea that the linear plots of Figure 3 are the rule, we include some

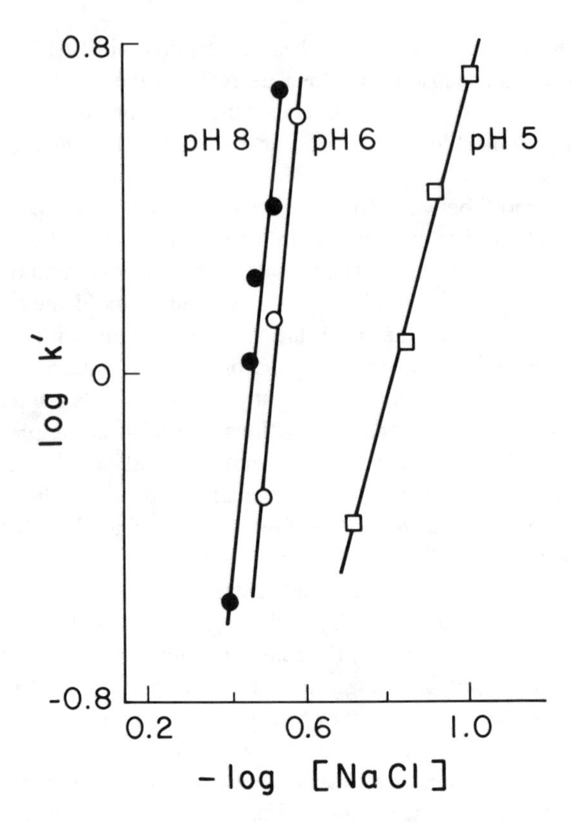

FIGURE 3. Retention of β-lactoglobulin vs. salt concentration on a strong-base bonded-silica anion exchanger. (After Kopaciewicz et al., *J. Chromatogr.*, 226, 3, 1983.)

graphs of retention vs. salt concentration published by Hearn et al.[22] (see Figure 4). When k′ falls below 1 the graphs are far from linear. One must not read too much importance into these graphs, however, for the calculation of low k′ values depends greatly on the value chosen for the void volume of the column. Exact evaluation of the void volume is never easy, and for the porous exchangers used in protein chromatography it is hard to define what one means by the void volume. Anyway, the lesson of Figures 3 and 4 for the practical chromatographer is that he should work at a pH well away from the isoelectric point wherever possible.

Support for the idea that counter-ions for the protein must be supplied during desorption is given by the fact that the magnitude of k′ depends not only on the nature of the displacing ion, the ion that has the same charge sign as the protein, but also on the nature of the counter-ion. Kopaciewicz et al.[19] reported that the relative retentions of cytochrome C on a strong cation exchanger were, for the following displacing salts: Na citrate, 1.00, NaF 0.79, NaCl 0.68; LiCl 1.00, NaCl 0.58, KCl 0.57.

The idea that the displaced protein molecule must acquire its own counter-ions when it leaves the surface of the exchanger is disputed by Velayudhan and Horvath.[23] The exact mechanism of displacement is not clear. The density of ionic charges on the protein and on the stationary phase, and the thickness of the electrical double layer must affect the situation. The fact is certain that the partition of protein between the exchanger and the solution depends on a high power of the salt concentration. The same is true of hydrophobic interaction chromatography, where a more hydrophobic stationary phase is used and a salt gradient of *decreasing* concentration is used for desorption. The practical result of this high-power dependence is that the peaks in protein chromatography are extremely sharp. Figure 5 shows

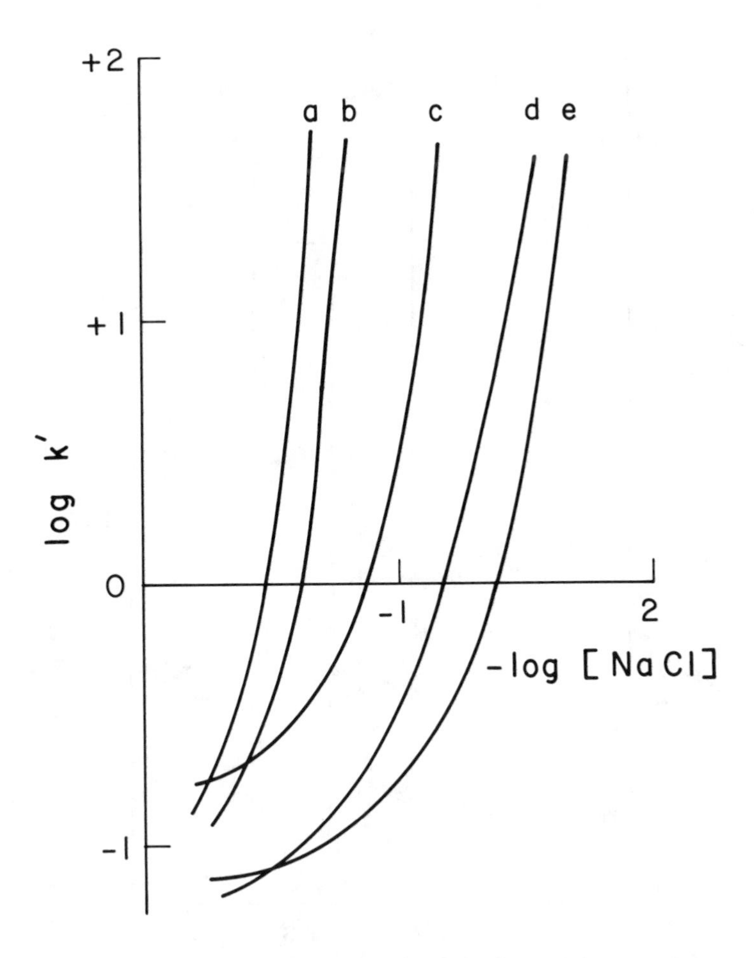

FIGURE 4. Retention vs. salt concentration for various proteins on a strong-base polymer-type anion exchanger, Mono-Q®. Buffer, piperazine, pH 9.6. Proteins: (a) serum albumin, pI 5.85; (b) ovalbumin, pI 4.70; (c) hemoglobin, pI 6.80; (d) carbonic anhydrase, pI 5.89; and (e) myoglobin, pI 7.68. (After Hearn et al., *Chromatographia,* 24, 769, 1987.)

a chromatogram of purified protein standards on a polyaspartic acid-silica column run by cation exchange. The sharpness of these peaks may be a surprise to non-biochemists who think of proteins as ill-defined macromolecules. Proteins have large molecules, but these molecules are highly organized and exactly defined.

Figures 6 and 7 show chromatograms of proteins run by anion exchange on a latex-coated exchanger. Figure 7 is a fast chromatogram of six standard proteins run in less than 3 min.

It remains to add that proteins are adsorbed and desorbed without denaturation or loss of their intricate tertiary structure. The evidence is that enzymes retain their activity after adsorption and desorption. This being so, ion exchange can be used for the preparative chromatography of proteins, as well as for analysis.

In Chapter 6 we described immobilized metal-ion affinity chromatography of proteins, which is a form of ligand exchange. Until recently, resins like Chelex®-100 (crosslinked polystyrene with iminodiacetate groups) were used for this purpose. Resolution was poor because of slow mass transfer, and the method was more useful for gross separations than for fine analytical chromatography. The advent of new macroporous polymers that carry chelating groups in their structure has changed this situation. A recent paper by Porath[24]

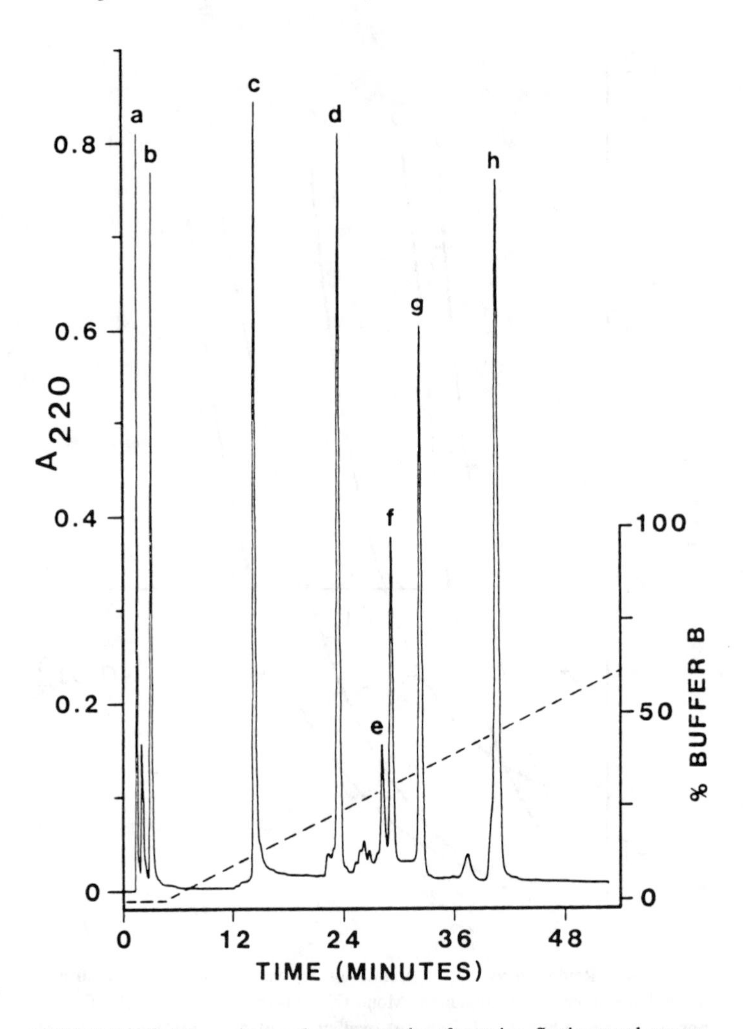

FIGURE 5. Cation-exchange chromatography of proteins. Stationary phase, polyaspartic acid bonded to silica, 5 μm. Mobile phase, 0.05 *M* phosphate, pH 6.0; Buffer B is 0.6 *M* in NaCl but has the same pH and phosphate concentration as the first buffer. Proteins, in order of elution: ovalbumin, bacitracin, myoglobin, chymotrypsinogen A, cytochrome C, reduced, ribonuclease A, cytochrome C, oxidized, lysozyme. After Alpert.[14] (By courtesy of PolyLC, Inc.)

describes high-resolution chromatography of proteins and peptides by ligand-exchange chromatography on chelating polymers loaded with ions of Cu, Zn, and Ni.

III. PEPTIDES

Peptide molecules are chains of amino acids linked by peptide bonds, –CONH–. They may be only 2 amino-acid units long or they may have 50 units or more, but the chains are not long enough to turn and twist and form the hydrogen-bonded tertiary structure of proteins. Like amino acids and proteins, however, they can form both cations and anions and can be separated by cation and anion exchange. The smallest peptides, two or three units long, can be separated by chromatography on ion-exchange resins, gel-type or macroporous, using the same methodology as for amino acids. The longer peptides are generally separated by reversed-phase chromatography. Because they do not have the well-defined three-dimensional structure of proteins, denaturation on the reversed-phase support is not a problem, and mixed aqeuous-nonaqueous solvents can be used as mobile phases. The normal procedure

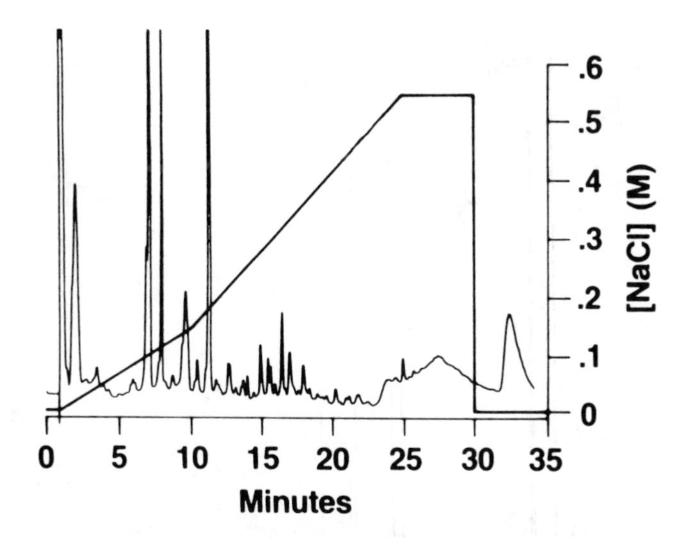

FIGURE 6. Anion-exchange chromatography of spinach proteins. Stationary phase, polystyrene-DVB latex, Dionex ProPac® PA1. Mobile phase, Tris buffer, pH 8.4, with NaCl gradient as indicated. Sample was a spinach extract precipitated with 90% ammonium sulfate and desalted. (By courtesy of Dionex Corporation.)

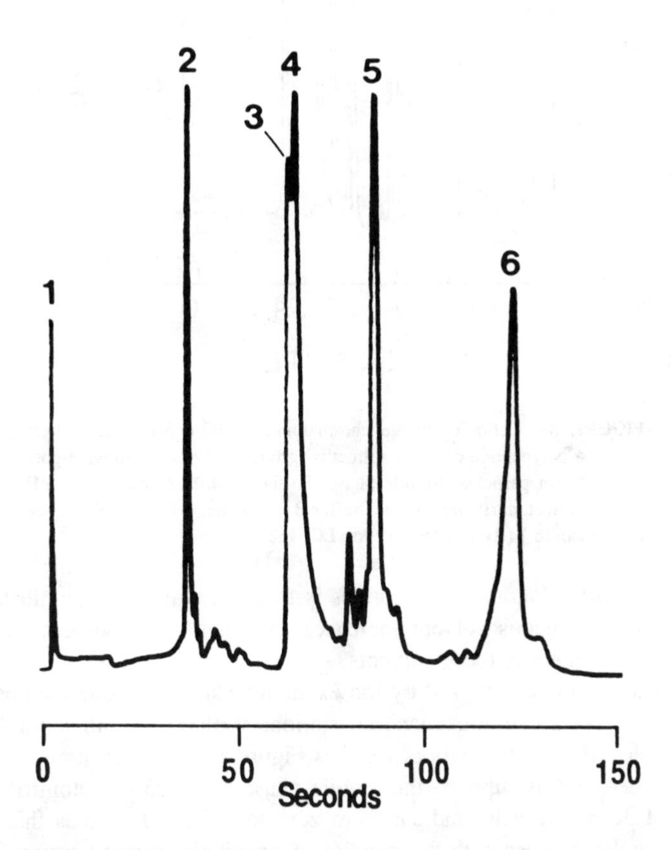

FIGURE 7. Fast protein separation. Stationary phase, Dionex ProPac® PA1. latex coated, 10 μm. Mobile phase, Tris buffer, pH 8.1, with increasing NaCl gradient. Proteins, in order of elution: (1) myoglobin (peak 2 a contaminant), (3) and (4) conalbumin; (5) ovalbumin; and (6) soybean trypsin inhibitor. (By courtesy of Dionex Corporation.)

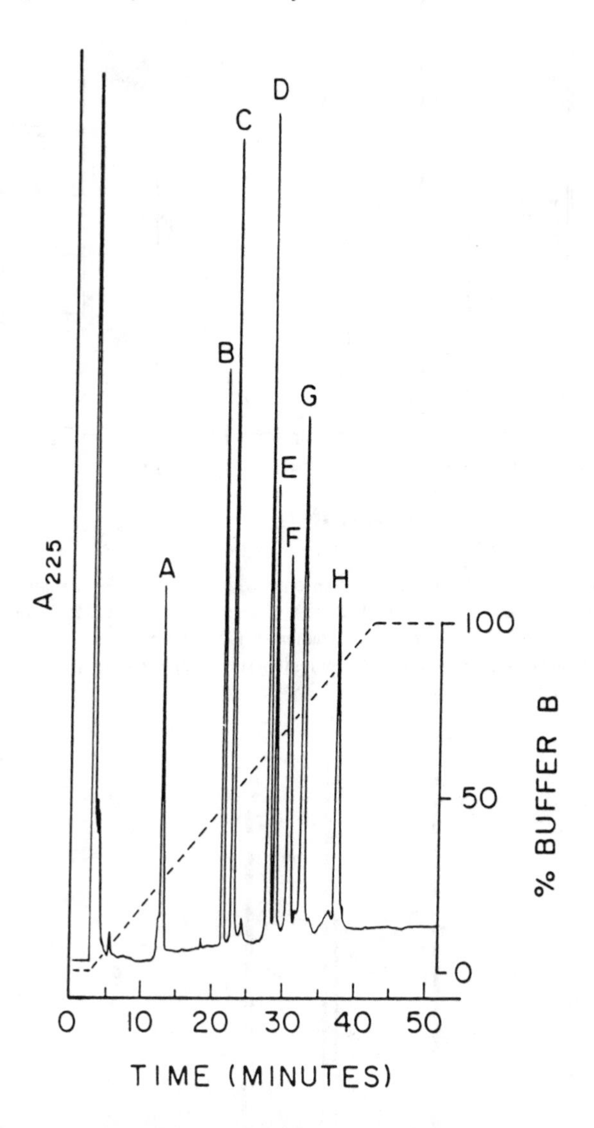

FIGURE 8. Cation-exchange chromatography of peptides. Stationary phase, a polypeptide carrying sulfoethyl groups and bonded to wide-pore silica, 5-μm particles. Mobile phase, buffer A, 0.05 *M* phosphate, pH 3.0, 25% acetonitrile by volume; buffer B, the same, 0.25 *M* in KCl. (See Reference 26.) (By courtesy of PolyLC, Inc.)

in reversed-phase chromatography of peptides is to use a water-to-acetonitrile gradient; as the proportion of nonaqueous solvent increases, peptides are desorbed. Large peptides, however, tend to aggregate in these solvents.

Peptide mixtures can be analyzed by ion exchange using the same methodology that is used for proteins.[25] As a cation-exchanging stationary phase the silica-bonded sulfoethyl aspartamide developed by Alpert works well, as Figure 8 shows. Crimmins[26] lists retention times of 54 peptides on this support, the mobile phase being 25% acetonitrile, 0.005 *M* in phosphate at pH 3, with a salt gradient from zero to 0.5 *M*. He notes that the retention increased in a regular manner with the number of positively charged amino acid residues. Cation exchange was able to separate pairs of very similar peptides where reversed-phase chromatography could not.

Dizdroglu et al.[27] has used a silica-based weak-base anion exchanger to separate mixtures

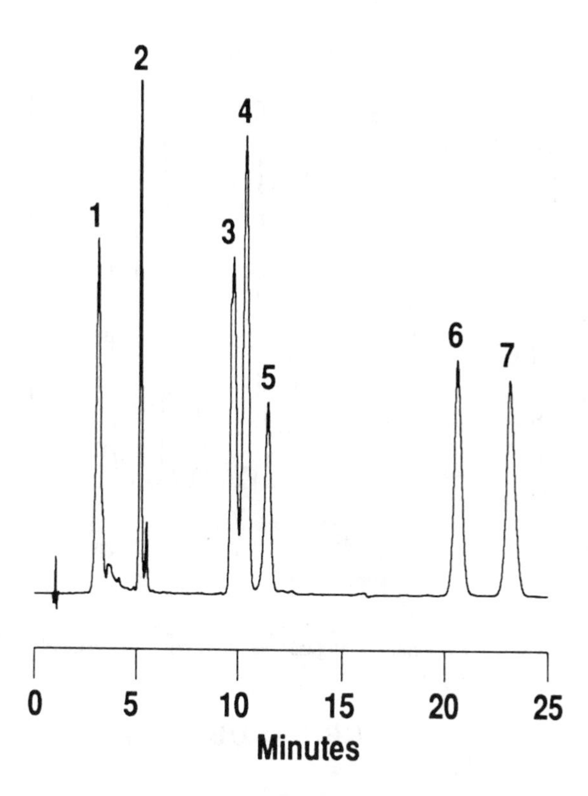

FIGURE 9. Chromatography of peptides on a latex-coated cation exchanger. Peptides and amino-acid sequences are (1) Angiotensin-III, RVYIHPF; (2) Des-Tyr-Met-Enkephalin, GGFM; (3) Val^5-Angiotensin-II, DRVYVHPF; (4) Angiotensin-II, DRVYIHPF; (5) Angiotensin-I, DRVYIHPFHL; (6) Leu-Enkephalin, YGGFL; and (7) Met-Enkephalin YGGFM, Code: (D) aspartic acid; (F) phenylalanine; (G) glycine; (H) histidine; (I) isoleucine; (L) leucine; (M) methionine; (P) proline; (R) arginine; (V) valine; and (Y) tyrosine. Note similarity of peptides 3 and 4, also 6 and 7. (By courtesy of Dionex Corporation.)

of angiotensins, peptides 7 to 10 amino acid units long that are hormones. Different angiotensins may differ in only one amino acid unit, yet they can be separated by anion exchange. A gradient of diminishing acetonitrile concentration is used for anion exchange, along with a dilute buffer, triethylammonium acetate of pH 6.0, 0.01 *M*. This author has compared the ion-exchange method with reversed-phase chromatography and has shown that the two methods can complement one another; compounds that are eluted together in reversed-phase chromatography can be transferred to an anion-exchange column and then separated from one another.[28]

Figure 9 shows a chromatogram of seven peptides, angiotensins and enkephalins, on a latex-coated polymer-based ion-exchange resin. Pairs of peptides that differ in only one amino acid unit are nicely separated.

Peptides are successfully separated by ligand-exchange chromatography. The work of Foucault and Caude with Cu(II)-loaded silica, eluting with solutions of ammonia in aqueous acetonitrile, was described in Chapter 6. We just mentioned Porath's[24] technique of "high-performance immobilized metal ion affinity chromatography" and its use for peptides and proteins. Porath used Ni(II), Cu(II), and Zn(II) as the coordinating metal ions; elution was done by gradients of either pH, salt concentration or of the displacing ligand imidazole.

The general subject of ion-exchange chromatography of peptides and proteins has been reviewed by Hearn et al.,[22] Regnier,[20] Ritchey[21] and others.

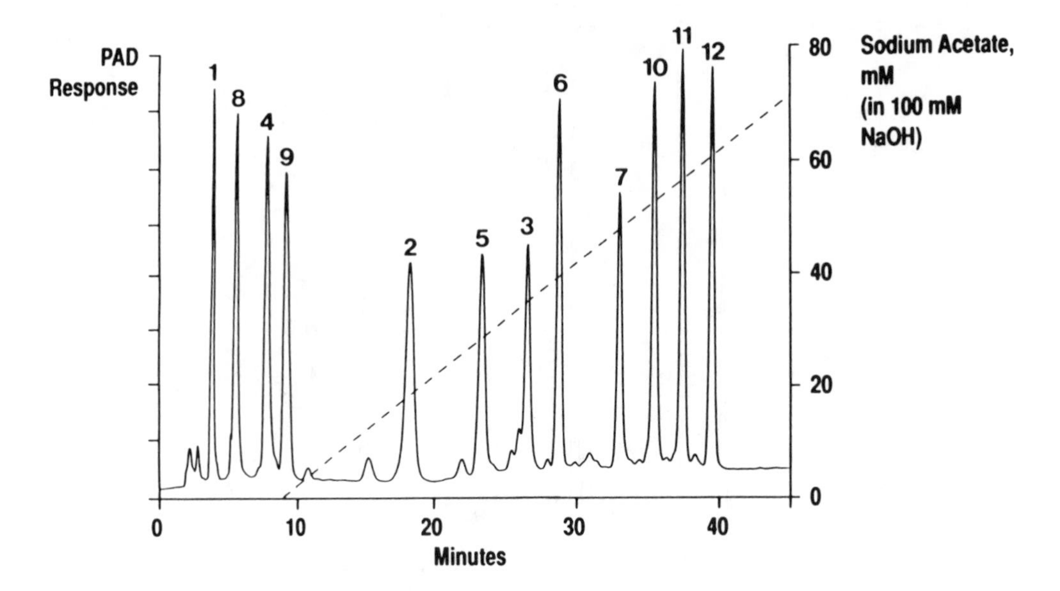

FIGURE 10. Anion-exchange chromatogram of oligosaccharides from glycoproteins. with pulsed amperometric detection. (By courtesy of Dionex Corporation.)

IV. CARBOHYDRATES

The chromatography of sugars and other carbohydrates by ligand exchange on calcium-form cation-exchange resins, with water eluent and refractive index detection, was described in Chapter 6, and a brief mention was made of the anion exchange of sugar-borate complexes. It has lately come to be appreciated that carbohydrates are very weak acids with pKa values around 12, and that they may be separated as their anions on high-efficiency crosslinked polystyrene anion exchangers in mobile phases of pH 11 to 13. Pulsed amperometric detection (described in Chapter 4) is a highly sensitive and selective method for detecting carbohydrates at high pH. The high pH is necessary because hydroxyl ions are consumed in oxidizing carbohydrate molecules to CO_3^{2-} and H_2O. Since both the anion exchange separation and pulsed amperometric detection of these compounds can only be done at high pH, these separation and detection modes combine to give a successful new method for determining carbohydrates.

Eluents for separating mono- and disaccharides are 0.005 to 0.15 *M* sodium hydroxide. The column temperature is from ambient to about 35°C. Oligo- and polysaccharides are separated by gradient elution. A constant sodium hydroxide concentration (0.05 to 0.1 *M*) is used with an increasing concentration of sodium acetate, usually up to about 0.5 *M* (see Figure 10).

The elution order of carbohydrates using anion exchange is similar to the elution order using borate-complex separation on weak-base silica columns. Carbohydrates of low molecular weight elute first. Higher molecular weight oligo- and polysaccharides elute roughly in the order of their molecular weight. However, the structure of the carbohydrate has a strong influence on retention. This elution order is the reverse of that on high-capacity sulfonated resins. Because of this, selectivity for oligo- and polysaccharides is much better than that obtained on sulfonated resins. An example is the separation of oligo- and polysaccharides from hydrolyzed corn starch shown in Figure 21, Chapter 4. Polysaccharides having up to 22 glucose units (DP 22; DP = degree of polymerization) can be separated. High-capacity gel-type sulfonated polystyrene resins cannot separate polysaccharides beyond DP 10.

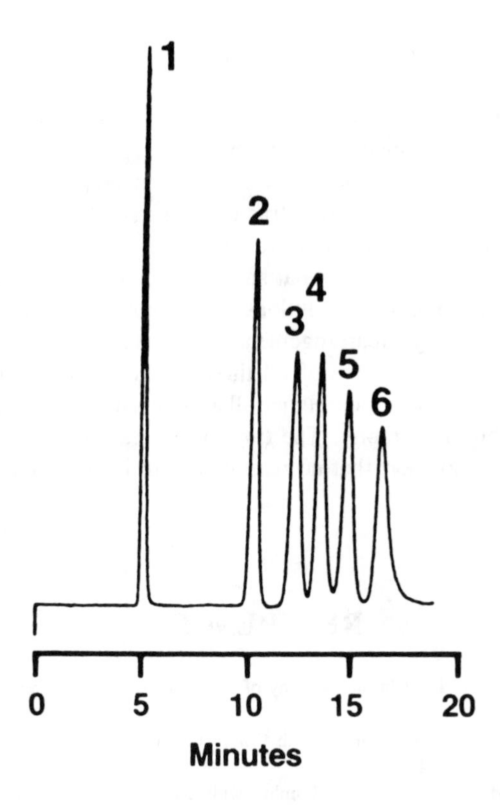

FIGURE 11. Anion-exchange separation of neutral and basic monosaccharides. Mobile phase, 0.015 *M* NaOH. Pulsed amperometric detection with post-column addition of 0.3 *M* NaOH. Peaks: (1) fucose; (2) galactosamine; (3) glucosamine; (4) galactose; (5) glucose; (6) mannose. There are 5 nmol of each. (By courtesy of Dionex Corporation.)

Major applications for anion exchange with pulsed amperometric detection are the analysis of foods and beverages and the study of complex oligosaccharides from glycoconjugates. Most glycoconjugates consist of oligosaccharides linked to either lipids or proteins, forming glycolipids, glycoproteins, or proteoglycans. For glycoprotein enzymes the oligosaccharides apparently do not affect the enzymatic function of the protein, since they are not located at the protein's active site, but they greatly affect the protein's three-dimensional structure. Changes in the outer structure of the protein have a major effect on its recognition by the immune system, or by the organism for its function to be turned on.

Because the purification of large quantities of protein is a difficult and time-consuming process, highly sensitive analytical techniques are required for protein analysis. Anion exchange followed by pulsed amperometric detection has recently been successfully used for this purpose.[29,30]

The oligosaccharide portions of the glycoprotein are first removed from the protein by enzymatic or chemical means, then analyzed by injecting directly. Identification is facilitated by matching retention times with standards. A separation of twelve standards is shown in Figure 10. This separation required gradient elution with acetate as the displacer ion.

In addition to the direct injection of the oligosaccharides, they can also be hydrolyzed to their constituent monosaccharides. These are neutral sugars like mannose and glucose, and also amino sugars like galactosamine and sialic acids. These compounds are separated by using a much more dilute eluent, typically 10 to 20 m*M* NaOH (Figure 11). It will be noted that the dilute eluent was not sufficiently alkaline to allow detection at the anode potential used, and therefore a more concentrated sodium hydroxide solution was fed into the effluent before it entered the detector.

The combination of anion-exchange separation with pulsed amperometric detection is a powerful method for the determination of carbohydrates. Although there may be interference from amines and certain sulfur species, the technique is very sensitive and selective. It is much more sensitive than detection by refractive index, but unlike ligand exchange with water eluent, it is not suitable for preparative separations.

It has been known for a long time that sugar molecules are degraded and changed by alkali, and this sensitivity to alkaline solutions is a potential difficulty in anion chromatography. In the "Lobry de Bruyn rearrangement", glucose rearranges to form fructose and mannose, and an equilibrium mixture of all three of these compounds results. However, this rearrangement is slow, and needs higher alkali concentrations and higher temperatures than are used in ion chromatography. The time that sugars spend in the column between injection and detection is so short that no rearrangement is observed.

REFERENCES

1. **Moore, S. and Stein, W. H.,** Chromatography of amino acids on sulfonated polystyrene resins, *J. Biol. Chem.*, 192, 663, 1951.
2. **Stein, W. H. and Moore, S.,** Photometric ninhydrin method for use in the chromatography of amino acids, *J. Biol. Chem.*, 176, 337, 1948.
3. **Pfeiffer, R. F. and Hill, D. W.,** HPLC of amino acids; ion exchange and reversed-phase strategies, *Adv. Chromatogr.*, 22, 44, 1983.
4. **Benson, J. R. and Hare, P. W.,** Ortho-phthalaldehyde: fluorogenic detection of primary amines in the picomole range; comparison with fluorescamine and ninhydrin, *Proc. Natl. Acad. Sci. U.S.A.*, 72, 619, 1975.
5. **Polta, J. A. and Johnson, D. C.,** The direct electrochemical detection of amino acids at a platinum electrode in an alkaline chromatographic effluent, *J. Liquid Chromatogr.*, 6, 1727, 1983.
6. **Benson, J. V., Gordon, M. J., and Patterson, J. A.,** Accelerated chromatographic analysis of amino acids in physiological fluids containing glutamine and asparagine, *Anal. Biochem.*, 18, 228, 1967.
7. **Mondino, A., Bongiovanni, G., and Fumero, S.,** Effect of pH of the test solution on amino-acid ion-exchange chromatography in a lithium cycle, *J. Chromatogr.*, 71, 363, 1972.
8. **Benson, J. V. and Woo, D. J.,** Polymeric columns for liquid chromatography, *J. Chromatogr. Sci.*, 22, 386, 1984.
9. **Dieter, D. S. and Walton, H. F.,** Counterion effects in ion-exchange partition chromatography, *Anal. Chem.*, 55, 2109, 1983.
10. **Regnier, F. E. and Noel, R.,** Glycerolpropylsilane bonded phases in the steric exclusion chromatography of biological macromolecules, *J. Chromatogr. Sci.*, 14, 316, 1976.
11. **Chang, S. H., Noel, R., and Regnier, F. E.,** High-speed ion-exchange chromatography of proteins, *Anal. Chem.*, 48, 1839, 1976.
12. **Chang, S. H., Gooding, K. M., and Regnier, F. E.,** Use of oxiranes in the preparation of bonded phase supports, *J. Chromatogr.*, 120, 321, 1976.
13. **Regnier, F. E. and Gooding, K. M.,** Review: HPLC of proteins, *Anal. Biochem.*, 103, 1, 1980.
14. **Alpert, A. J.,** Cation-exchange HPLC of proteins on poly(aspartic acid) silica, *J. Chromatogr.*, 266, 23, 1983.
15. **Alpert, A. J.,** Hydrophobic-interaction chromatography of proteins, *J. Chromatogr.*, 359, 85, 1986.
16. **Alpert, A. J. and Andrews, P. C.,** Cation-exchange chromatography of peptides on poly(2-sulfoethyl aspartamide)-silica, *J. Chromatogr.*, 443, 85, 1988.
17. **Vanacek, G. and Regnier, F. E.,** Macroporous high-performance anion exchange of proteins, *Anal. Biochem.*, 121, 156, 1982.
18. **Rounds, M. A. and Regnier, F. E.,** Synthesis of a nonporous polystyrene-based strong anion-exchange packing material and its application to fast HPLC of proteins, *J. Chromatogr.*, 443, 73, 1988.
19. **Kopaciewicz, W., Rounds, M. A., Fausnach, J., and Regnier, F. E.,** Retention model for high-performance ion-exchange chromatography, *J. Chromatogr.*, 266, 3, 1983.

20. **Regnier, F. E.,** High-performance ion-exchange chromatography, *Methods Enzymol.*, 104, 170, 1984.
21. **Ritchey, J. S.,** Optimal pH conditions for ion exchangers on macroporous supports, *Methods Enzymol.*, 104, 223, 1984.
22. **Hearn, M. T. W., Hodder, A. N., Stanton, P.G., and Aguilar, M. I.,** High-performance liquid chromatography of amino acid peptides and proteins. LXXXIII. Evaluation of retention and bandwidth relationships for proteins separated by isocratic anion-exchange chromatography, *Chromatographia*, 24, 769, 1987.
23. **Velayudhan, A. and Horvath, Cs.,** Preparative chromatography of proteins: analysis of the multivalent ion-exchange formalism, *J. Chromatogr.*, 443, 13, 1988.
24. **Porath, J.,** High-performance immobilized-metal-ion affinity chromatography of peptides and proteins, *J. Chromatogr.*, 443, 3, 1988.
25. **Bradshaw, R. A., Bates, O. J., and Benson, J. R.,** Peptide separations on substututed polystrene resin: effect of crosslinkage, *J. Chromatogr.*, 187, 27, 1980.
26. **Crimmins, D. J., Gorka, J., and Schwartz, B. D.,** Peptide characterization with a sulfoethyl aspartamide column, *J. Chromatogr.*, 443, 63, 1988.
27. **Dizdaroglu, M., Krutzch, H. C., and Simic, M. G.,** Separation of angiotensins by HPLC on a weak anion-exchange bonded phase, *Anal. Biochem.*, 123, 190, 1982.
28. **Dizdaroglu, M. and Krutzch, H.,** A comparison of reversed-phase and weak-anion exchange HPLC methods for peptide separation, *J. Chromatogr.*, 264, 223, 1983.
29. **Hardy, M. R. and Townsend, R. R.,** Separation of positional isomers of oligosaccharides and glycopeptides by high-performance anion-exchange chromatography with pulsed amperometric detection, *Proc. Natl. Acad. Sci. U.S.A.*, 85, 3289, 1988.
30. **Hardy, M. R., Townsend, R. R., and Lee, Y. C.,** Monosaccharide analysis of glycoconjugates by anion exchange with pulsed amperometric detection, *Anal. Biochem.*, 170, 54, 1988.

Chapter 8

"STOP AND GO" SEPARATIONS OF METAL IONS

I. INTRODUCTION

In this chapter we describe a group of analytical methods for separating metal ions that use conventional high-capacity ion-exchange resins, anion exchangers as well as cation exchangers, and eluents that are relatively concentrated, like 9 *M* hydrochloric acid. We call them "stop and go" or "all or nothing" methods[1] because by choosing the nature and concentration of the eluent we can bring out one element at a time and bring it out completely, leaving the other elements on the column. The separation factors are very high, sometimes several powers of ten. In general the methods rely on the formation and dissociation of metal complexes that can be anionic or cationic, and their incorporation in the resin by ion exchange. They were developed before ion chromatography and have to some extent been replaced by ion chromatography, but they have a place in modern chemical analysis. They are primarily methods of separation, not measurement. They are useful in analyzing complex materials, like ores and minerals and special alloys, where they make it possible to recover every element, or nearly every element, in pure form, each in its separate container apart from the others, where the quantity can be measured by any method of choice.

These procedures are "analysis" in the truest sense of the word. They may be used to check and calibrate the faster and easier methods used in routine analysis, like atomic absorption and emission spectrometry or even ion chromatography. They may be used to separate small amounts of elements sought from much larger amounts of main constituents. They are used in radiochemistry for examining cyclotron targets and for post-irradiation separations in activation analysis.

Because these methods were developed before ion chromatography, they use small, open glass columns and ordinary grades of ion-exchange resins, grades that are chemically clean but have relatively large particles, compared with the fine, uniform resins used in high-performance chromatography.

Nearly always the selectivity depends on complex-ion formation. The first methods we shall describe involve the formation and exchange of negatively charged chloride complexes.

II. ANION EXCHANGE OF METAL CHLORIDE COMPLEXES

Metal ions in solution can combine with negatively charged ligands to form complex ions that, when fully coordinated, carry a negative charge. As such they can be adsorbed by anion-exchange resins. The strength of binding depends not only on the concentration of the ligand and the stability of the complex in solution, but also on the affinity of the resin for this complex; in other words, the stability of the anionic complex in the resin is generally not the same as its stability in aqueous solution.

Anion exchange of chloride complexes in aqueous hydrochloric acid solutions was studied intensively in the early 1950s by Kraus and Nelson[2] at Oak Ridge National Laboratory. Their work was presented at the first United Nations Conference on Peaceful Uses of Atomic Energy in 1955, and the publication is widely quoted.[2] Similar studies, but narrower in scope, were published in Germany by Jentzch.[3] Distribution ratios of metals betweeen aqueous hydrochloric acid solutions and quaternary-base anion-exchange resins were measured with the help of radioactive tracers and found to range over several powers of ten, depending on the metal and the hydrochloric acid concentration. For each metal a graph was drawn of (log D) against (molar conc. of HCl). Figure 1 shows these graphs. It

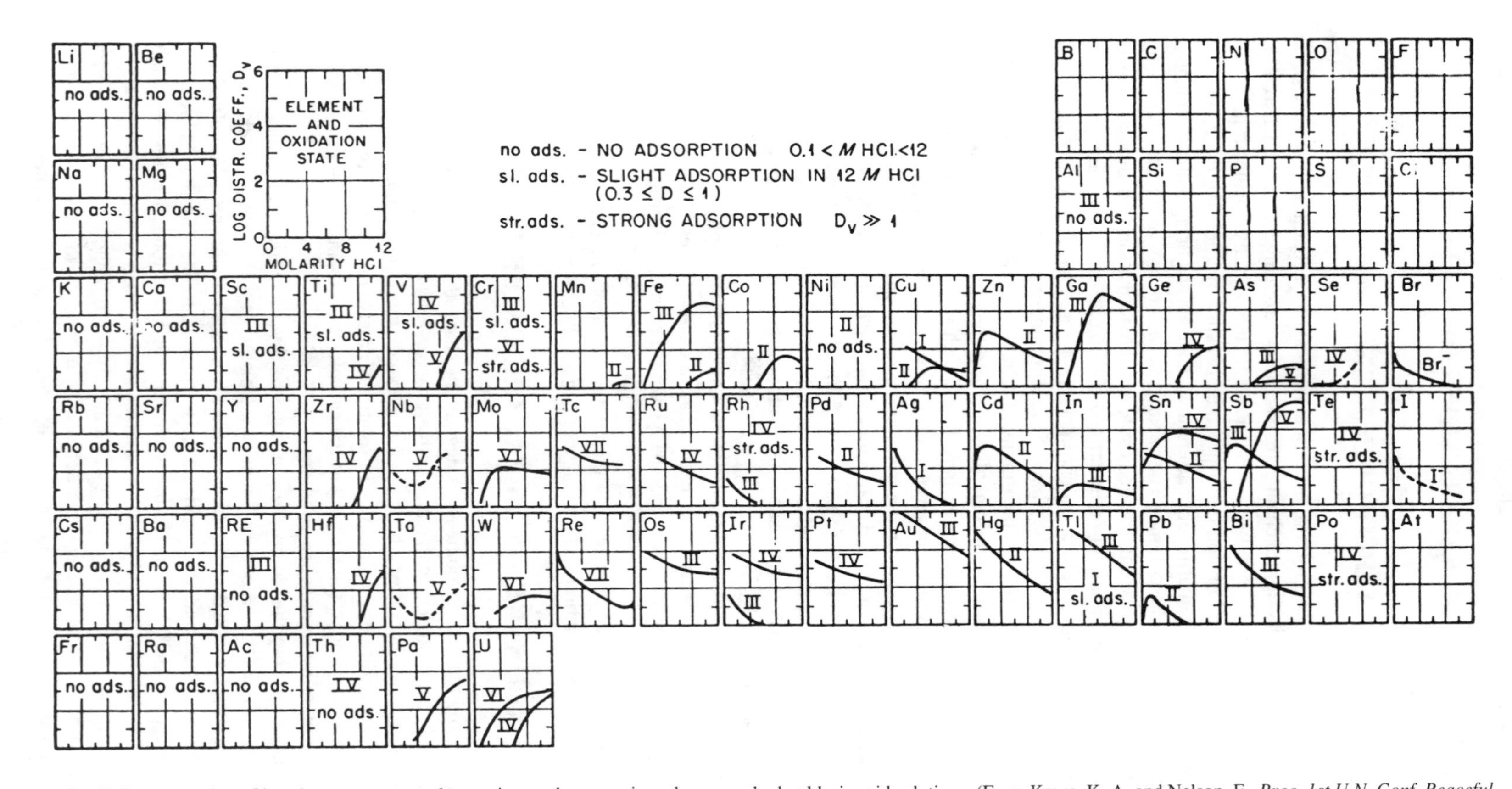

FIGURE 1. Distribution of ions between a strong-base anion-exchange resin and aqueous hydrochloric acid solutions. (From Kraus, K. A. and Nelson, F., *Proc. 1st U.N. Conf. Peaceful Uses Atomic Energy,* 1955, 113. With Permission.)

is immediately evident that an enormous number of separations are possible, with very large separation factors. Generally speaking a metal is bound more strongly, the higher the hydrochloric acid concentration; thus, a group of metals could be adsorbed on a resin at a high concentration, then released one at a time by lowering the hydrochloric acid concentration in a series of steps. The operations can be performed in open glass columns.

A typical separation, that of a mixture of iron(III), cobalt(II), and nickel(II), goes as follows.

A glass column, 1 cm internal diameter, is filled with a strong-base anion-exchange resin, particle size 100 to 200 mesh (75 to 150 μm) to a depth of 10 to 15 cm; the resin is washed with 9 *M* hydrochloric acid, and the acid is drained down to the level of the top of the bed. (Surface tension will prevent the level from dropping more than a millimeter or two below the top of the bed, with resin of this particle size). On to the resin bed is poured a small volume, 2 to 5 ml, of a solution containing 1 mmol each of $FeCl_3$, $CoCl_2$, and $NiCl_2$. Some 20 ml of 9 *M* HCl are passed. Soon a green solution will appear at the column outlet, which contains nickel chloride. When the effluent is no longer green, 3 *M* HCl is passed and the receiver is changed. The colored band at the top of the column, which was originally dark brown with a blue fringe at the bottom, will separate into two bands, the brown iron(III) band remaining at the top while a deep blue band, containing the blue tetrahedrally coordinated $CoCl_4^{2-}$ ion, moves down the column. As it moves the blue zone becomes narrower as the top of the band turns pink, the color of the aquo-complex or the hydrated, octahedral cobalt(II) ion, $Co(OH_2)_6^{2+}$. When the band reaches the bottom of the column the solution flowing out of the column is at first deep blue, then rose pink; the pink solution floats on top of the blue, and when the contents of the receiver are swirled to mix them, the color of the whole effluent changes to pink. (This makes a nice lecture demonstration!) After the cobalt has been collected, the receiver is changed and water is passed through the column. The brown band at the top of the column moves along, lightening as it goes, and emerges as a yellow solution of iron(III) chloride. (The yellow color is not that of hydrated Fe(III), which is almost colorless; it is the color of cationic $FeCl^{2+}$ and $FeCl_2^+$).

There are now three receivers, one containing only nickel, one containing only cobalt, and one containing only iron. The separation, which takes about 20 min, is virtually complete, the cross-contamination virtually nil. The solutions may be evaporated to remove the hydrochloric acid, and the quantities of the metals may be measured in any manner one wishes. Titration with EDTA is probably the most accurate.

Had the solution contained zinc as well, one would have eluted the iron with 1 *M* hydrochloric acid, at which concentration zinc is still strongly retained; D = 100 or more. To remove the zinc, one would pass 0.1 *M* HCl, dilute (1 *M*) HNO_3 or water. Wilkins[4] described a method for the analysis of a mixture of Fe, Co, Ni, and Zn in which the final eluent was 3 *M* nitric acid. Zinc ions are colorless, but may be detected by the reagent dithizone at pH 5, or Zincon® at pH 9.

Many separation schemes like this have been worked out to meet analytical problems, especially in the analysis of complex minerals or special alloys. A published analysis of silver solder[5] describes dissolving the solder in nitric acid, evaporating excess nitric acid, and titrating the silver ions potentiometrically with standard hydrochloric acid, filtering the silver chloride, then separating the remaining metal ions, Ni, Cu, Zn, and Cd, thus:

> Make the solution 6 *M* in HCl. Pass through anion-exchange resin: 6 *M* HCl elutes Ni; 1 *M* HCl elutes Cu; 0.01 *M* HCl elutes Zn; water elutes Cd.

Again, each element is recovered separately and may be measured by any desired method, which in this case was EDTA titration.

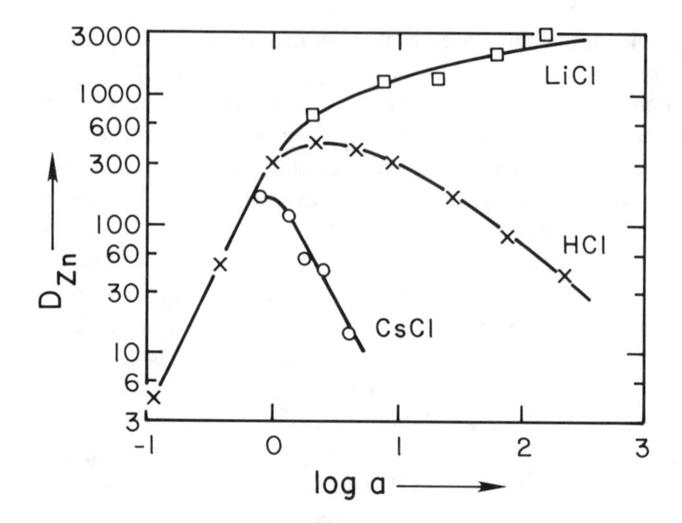

FIGURE 2. Distribution of tracer zinc between a strong-base anion-exchange resin and aqueous chloride solutions; "a" is the activity of the electrolyte in solution, not its concentration. From Schindewolf, U., Z. *Elektrochem.*, 162, 335, 1958. With permission.)

III. ELEMENTS THAT FORM CHLORIDE COMPLEXES

Inspection of Figure 1 shows that most of the metals in the periodic table form chloride complexes that are taken up by the resin, with the exception of the alkali and alkaline-earth metals, aluminum, yttrium, and the lanthanides, thorium and nickel. Except for nickel, these are the hard-acid metal ions, those that behave as hard charged spheres. One would not expect these ions to associate with chloride ion, which is a somewhat soft base. Aluminum ions form $AlCl_4^-$, but not in the presence of water; hydrated $Al(OH_2)_6^{3+}$ is much more stable. Nickel does associate with chloride ions to some extent, forming positively charged $NiCl^+$; the green color of the solution becomes deeper as hydrochloric acid is added; and in water-organic solvent mixtures of lower dielectric constant than water, there is evidence that anionic complexes do form; see below.

Certain metals are very strongly adsorbed. Those having distribution coefficients greater than 10^4 are: (Fe(III), Ga(III), Au(III), Tl(III), Bi(III), Sb(V), and Hg(II). All of these ions except the last two form singly charged complex anions of the formula MCl_4^-. We have noted that such ions form ion pairs with large singly charged positive ions; the functional group of strong-base anion-exchange resins is a large singly charged positive ion.

A feature of most of the graphs of Figure 1 is that the distribution coefficient rises with increasing HCl concentration, passes through a maximum, and then falls. The cause of the rise is obvious enough; the decrease at higher concentrations probably is due to displacement of the fully coordinated complex anion by the chloride ions. However, at concentrations as high as 6 to 12 *M* there is considerable invasion of the swollen resin by hydrochloric acid; Donnan exclusion is swamped, and the laws of dilute solution certainly do not apply. The polymer network is probably taking part as a nonaqueous solvent. It may be well here to note that a sulfonated polystyrene cation exchanger adsorbs iron(III) strongly from 6 to 9 *M* HCl, in spite of the fact that the iron may not be in the cationic form; see below.

The maximum adsorption of zinc(II) occurs near 1 *M* hydrochloric acid, a concentration that is low enough for ideal solution behavior to be at least approximated. Figure 2 plots the logarithm of the distribution coefficient of a zinc tracer between 4% crosslinked Dowex®-1 and chloride solutions against the logarithm of the activity (not the concentration) of the three electrolytes hydrochloric acid, cesium chloride, and lithium chloride.[6] At activities below 1 the graph is a straight line with slope 2, corresponding to the equilibrium.

$$2ResCl + Zn^{2+} + 2Cl^- = Res_2ZnCl_4$$

At higher concentrations (and activities) the line for CsCl is what we would expect, a straight line with slope −2, corresponding to the ion exchange of one $ZnCl_4^{-2}$ ion for two chloride ions. The sorption of zinc from hydrochloric acid and from lithium chloride solutions, however, is a good deal greater. This effect is attributed[7] to the higher activity coefficients of HCl and LiCl. In 5 *m* solutions at 25°C the mean activity coefficients of CsCl, HCl, and LiCl are respectively 0.475, 2.38, and 2.02.[8] A higher activity coefficient implies greater invasion of the resin by the electrolytes present. The position of HCl is anomalous, and the anomaly is perhaps explainable by the fact that there is a considerable fraction of undissociated, molecular HCl at higher concentrations.[9]

A remarkable fact observed by Schindewolf[6] is that curves almost exactly like those of Figure 2 are obtained when a liquid ion exchanger, methyl dioctylamine dissolved in trichloroethylene, is substituted for the anion-exchange resin. Thus, the shapes of the curves are due to interactions of the dissolved ions with the substituted ammonium cations, which may be described as ion-pair formation, and to the properties of the solutions, and not to any work of stretching the polymer network of the resin. Another interesting feature of Schindewolf's data is the effect of the concentration of the tertiary amine in the organic phase. The distribution coefficient of tracer zinc was directly proportional to the square of the amine concentration, other variables being equal, and this was true over the entire range of concentrations of electrolytes in solution. This proportionality is expected if the ion-exchanging phase is a homogeneous phase of variable capacity (see Chapter 3).

IV. COMPLEXES WITH OTHER ANIONS

Other anions besides chloride form complexes with metal ions and cause them to be adsorbed by anion exchangers. Bromide ions form complexes that in general are somewhat more stable than the chloride complexes, but there is not enough difference between chloride and bromide to be of much interest, save for one element: lead. Lead(II) is bound more strongly by an anion exchanger, Dowex®-1, from hydrobromic acid solutions than from hydrochloric acid.[10] The distribution coefficients fall with increasing acid concentration, and their ratio is largest near 2 *M* acid, where the coefficient is 130 in HBr, 10 in HCl. Thus, lead can be separated from other elements in aqueous solution by making the solution 2 *M* in hydrobromic acid and passing it through a short column of anion-exchange resin in the bromide form, which has been brought to equilibrium with 2 *M* HBr. Lead is retained, along with Zn, Cd, Bi, In, Au, Pt, and Pd. Other metal ions, including Cu, Fe, Co, Ni, Mn, Al, Mg, and Ca, pass on. The lead is eluted with 6 *M* HCl. The other ions that were retained remain on the column. They may be eluted later with dilute nitric acid.

For many separations fluoride complexes are useful. Fluoride is the smallest of the halide ions and forms octahedral six-coordinated complexes where chloride forms only tetrahedral four-coordinated complexes. This fact leads to significant differences in chemical behavior. Also, fluoride is a hard-acid ligand and coordinates preferentially with hard-base metal ions, highly charged, and small ions like Ti(IV) and Be(II).

In 1960 Nelson, et al.[11] published a detailed study of anion-exchange adsorption of metals in mixed HCl-HF solutions. Solutions containing only hydrochloric acid of various concentrations were compared with the same solutions that were also 1 *M* in hydrofluoric acid. For many metals the hydrofluoric acid made little difference, and at high acid concentrations the differences were hard to rationalize, but certain regularities were seen in the behavior at low hydrochloric acid concentrations, 2 *M* and below. Hydrofluoric acid increased the adsorption of Be, Sc, Ti, Zr, Hf, Ge, and Sb(V). It decreased the adsorption of Sn(II), Sn(IV), Fe(III), Ga(III), and Sb(III).

In general, fluoride complexes are more stable than chloride complexes in aqueous solution, but they may be more weakly adsorbed by the anion-exchange resin. Consider the case of iron(III). Adding fluoride ions to a dilute solution of ferric chloride causes the yellow color to disappear, for FeF_6^{-3} is colorless, yet the ion-exchange distribution coefficient, D, for iron(III) in 1 *M* HF-2 *M* HCl is 1/50 that in 2 *M* HCl alone. We may presume that the very strong adsorption of Fe(III) from hydrochloric acid solutions is due to ion pairing between the large, singly charged quaternary ammonium group of the resin and the large, singly charged anion $FeCl_4^-$, for the ion $FeCl_4^-$ in aqueous solution is very unstable; it hardly forms at all. The ion FeF_6^{-3} is stable in aqueous solution, but its high charge would prevent it from forming ion pairs with the quaternary ammonium groups of the resin.

This interpretation is speculative, but the effects are there, and they make a number of useful separations possible. Nelson et al.[11] show some of them. Aluminum and beryllium are separated by placing them on the column and passing, first, a solution 1 *M* in HF and 0.01 *M* in HCl. Aluminum is immediately eluted, but the fluoride ions keep beryllium on the column as stable BeF_4^{-2}. Beryllium is eluted by passing 1 *M* HCl. Then vanadium(IV) and titanium(IV) are separated by passing, first, 0.1 *M* HCl-1 *M* HF; vanadium is eluted; then 6 *M* HCl-1 *M* HF elutes titanium. (In cation exchange with the same eluents the order is reversed; the more dilute eluent removes Ti, the more concentrated removes V.) Another interesting separation that these authors describe is that of zinc, gallium, germanium, and arsenic, a separation that has obvious use in the semiconductor field. The separation can be made by anion exchange in hydrochloric acid solutions, but then germanium is eluted in 5 *M* HCl in the form of germanium tetrachloride, which is very volatile and easily lost. An alternative procedure using hydrofluoric and hydrochloric acids is this:

> First add a little chlorine to oxidize arsenic to As(V); then pass 0.3 *M* HCl-1 *M* HF. Gallium appears at the void volume, followed (at five void volumes) by As(V). Now pass 0.01 *M* HCl-1 *M* HF; zinc is eluted. Finally pass 6 *M* HCl-1 *M* HF; germanium is eluted as a fluoride complex that is stable in solution but weakly adsorbed by the resin.

Sorption of metal ions by anion-exchange resins in solutions of hydrofluoric acid alone has been studied by Faris,[12] who has published a set of graphs in periodic table form, like those for hydrochloric acid that are shown in Figure 1. The graphs for hydrofluoric acid are less varied, however. Nearly all show a steady drop with increasing acid concentration, and the differences between individual elements are less pronounced. The elements that show large adsorption (D greater than 100) are: Be, B, Sc, Ti, As, Zr, Nb, Mo, Pd, Hg, Sn, Sb, Te, Hf, Ta, W, Re, Pt, and U(VI). Several others, aluminum, for example, have D values between 10 and 100. The main importance of anion exchange in hydrofluoric acid solutions (or mixed HCl-HF solutions) is for the hard-acid metallic species that hydrolyze easily in water, namely, ions of titanium, zirconium, hafnium, niobium, and tantalum. Almost the only way to make stable aqueous solutions of these species is to form their fluoride complexes. Zirconium and hafnium were first separated by Huffman and Lilly[13] in 1949 on a column of anion-exchange resin with an eluent 0.2 *M* in HCl, 0.01 *M* in HF; zirconium was eluted first. This was a major triumph of ion exchange, for these two elements are as similar to one another as are two adjacent rare earths, and for the same reason: their ionic radii are nearly the same. (Anion exchange in sulfuric acid solutions is more efficient; see below.)

The separation of niobium and tantalum is easier. A method for the separation of the constituents of a high-temperature alloy was published by Hague and Machlan;[14] it goes as follows:

> The alloy is dissolved in aqueous HF and the solution placed on a column of strong-

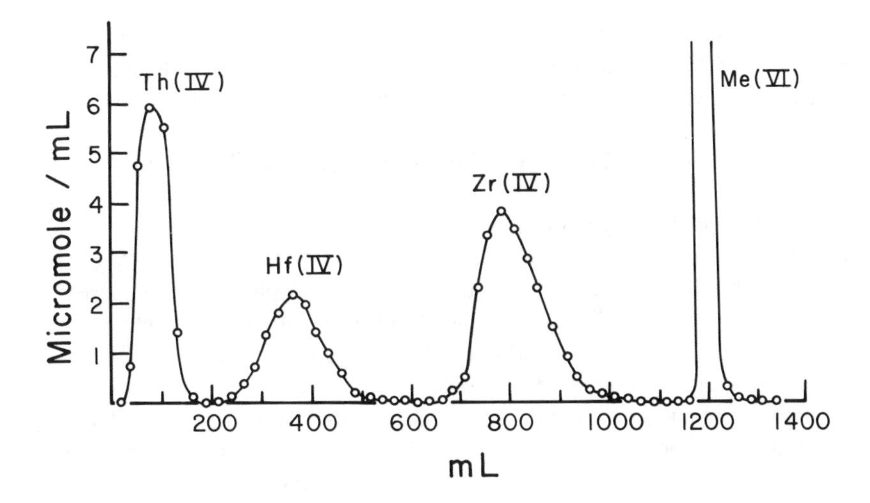

FIGURE 3. Elution of elements from a strong-base anion-exchange resin. Thorium is not adsorbed; halfnium is eluted with 1.25 *N* H_2SO_4 containing 0.1% H_2O_2; zirconium is eluted with 2.0 *N* H_2SO_4 containing 0.1% H_2O_2; the column is then washed with 2 *N* H_2SO_4, and molybdenum is eluted with 2.0 *N* NH_4NO_3-0.5 *N* NH_4OH. Flow rate, 1.2 ml/min. (From Strelow, F. W. E. and Bothma, C. J. C., *Anal. Chem.*, 39, 595, 1967. Copyright, American Chemical Society, 1967. With permission.)

base anion-exchange resin that has been washed with 1.25 *M* hydrofluoric acid. Passing more of this solution removes Al, Fe, Co, and Ni. Then 8 *M* HCl elutes Ti; a solution 7 *M* in HCl and 2.5 *M* in HF elutes W; a solution 3 *M* in HCl and 5 *M* in HF elutes Mo; then 2.5 *M* ammonium chloride, 1 *M* in HF, elutes Nb; then 2.5 *M* ammonium chloride, 1 *M* in NH_4F, elutes Ta.

Note that because HF is a weak acid, ammonium fluoride provides more fluoride ions than hydrofluoric acid of equal concentration. Hafnium is separated from tantalum on a column of anion-exchange resin by eluting Hf with 5 *M* HF-0.25 *M* HCl, then Ta with 1 *M* NH_4F-2.5 *M* NH_4Cl.[15]

One should observe that the fluorides of certain elements are very insoluble, and have to be removed by filtration if they are formed. These include fluorides of calcium, yttrium, and the lanthanides.

Anion exchange in nitric acid solutions was also studied by Faris.[16] Only a few metals show high adsorption, with $D > 100$. One of these is thorium, whose adsorption starts near zero in very dilute nitric acid, then rises to a maximum of $D = 300$ in 8 *M* nitric acid. Other metals that are strongly adsorbed are Tc, Re, Au(III), Pd(IV), Np(IV), and Pu(IV).

In sulfuric acid, uranium(VI) is very strongly adsorbed from 0.01 *M* solution; $D = 1100$. Other metals that are strongly adsorbed are Cr(VI), Mo(VI), W(VI), Zr, Hf, and Ta. A table of D values at different acid concentrations is given by Strelow and Bothma,[17] who also show several separation schemes. One is shown in Figure 3. Hafnium is eluted from the strong-base anion exchanger Bio-Rad® Ag1 × 8 by 1.25 *N* H_2SO_4, zirconium by 2.0 *N* acid; the separation is good, though the bands are broad. So is the separation from thorium, which is eluted almost at void volume. Now, there is a precaution necessary when handling solutions of zirconium and hafnium that Strelow points out. These ions are strongly hydrolyzed in water, even in the presence of much acid. The hydrolysis is slow, and leads to polymerization through –Zr–O–Zr– bonds, eventually giving a hydrous oxide precipitate. A solution of zirconyl sulfate, prepared and left on the shelf, becomes turbid after a day or two. For the ion-exchange separation to be successful the sample containing Zr or Hf must be dissolved in hot concentrated sulfuric acid, then cooled and diluted to the desired acid concentration (1.25 *N*), then applied to the column and eluted without delay, at least within an hour or two. Perhaps the broadness of the bands in Figure 3 is due to hydrolysis.

V. CATION-EXCHANGE SEPARATIONS

Chromatographic separations of metal ions on cation exchangers are now made by high-performance ion chromatography, described in Chapter 4. Dilute eluents are used to get adequate retention, and to provide adequate separation, complexing eluents are used, generally containing organic acids. Dynamic ion exchange with ion-interaction agents gives excellent separation. Before the advent of ion chromatography, Strelow, Nelson, Kraus, and others showed that cation exchange can be made as selective and versatile as anion exchange by using simple inorganic acids at high concentrations. Strelow et al.[18] studied the distribution of 45 metal ions in nitric acid and sulfuric acid solutions, at concentrations from 0.1 *N* up to 4.0 *N*, between the solutions and the sulfonated polystyrene resin AG-50W × 8. The results were presented in the form of extensive tables. Typically the distribution coefficients were high in very dilute acid solutions and fell as the acid concentration increased, corresponding to displacement of the cations from the resin by ion exchange. The authors took care to measure the distributions at as nearly as possible the same loading, i.e., with the metal ions occupying 40% of the exchange capacity from 0.5 to 12 *M*, so their selectivity orders may not be the same as those determined by other authors who used radioactive tracers. The selectivity orders depended on the kind of acid and on its concentration. Thus, a number of separation schemes could be worked out.

Nelson et al.[19] published an extensive study of cation exchange of metal ions in solutions of hydrochloric and perchloric acids of concentrations ranging from 0.5 to 12 *M*. The stationary phase was 4% crosslinked sulfonated polystyrene, Dowex®-50 × 4. They chose the 4% crosslinking to get rapid equilibration, but found that the resin contracted considerably as the acid concentration was raised, so that ion exchange was still slow. To measure distribution coefficients they used radioactive tracers, and the resin loading was below 1%.

In the low concentration range, up to 2 to 3 *M* acid, the distribution coefficient, D, fell with increasing acid concentration, as is expected with hydrogen ions displacing metal cations. Even in this range, D was higher in perchloric acid than in hydrochloric. As the concentrations were raised to higher values, the typical course of events was that, for perchloric acid, D went through a minimum and then rose to very high values; for hydrochloric acid, D dropped lower and went through a flat minimum, or continued to drop as the acid concentration rose. Figures 4 and 5 show representative curves. The ideal curve in Figure 4 is for simple cation exchange, assuming a constant ratio of activity coefficients and no Donnan invasion of the resin by the electrolyte. Not all the metals showed this pattern; a number of cations, such as the alkali metals, were not adsorbed at all. An interesting exception to the pattern is iron(III) (see Figure 6). We note the sharp minimum of adsorption in 4 *M* HCl and the very high values of D that are reached in 6 to 9 *M* HCl.

The reasons for this behavior are not clear. Probably the very strong hydration of the proton in perchloric acid has something to do with it, for it reduces the activity of the remaining water. Hydrochloric acid is not nearly as strongly hydrated. Rather, a considerable proportion of the acid in solutions of 6 *M* and more is present as molecular HCl. Regardless of the reasons for the behavior, the possibilities for analytical separations are many. Several separations are illustrated in the reference cited.

A long paper by Nelson and Michaelson[20] describes cation exchange of metal ions in aqueous hydrobromic acid solutions of concentrations up to 12 *M*. A striking effect was the order of elution of alkaline-earth metal ions. Beryllium and magnesium were eluted first with 12 *M* HBr, followed by radium and barium; then strontium and calcium were eluted, well separated, by 5 *M* HBr.

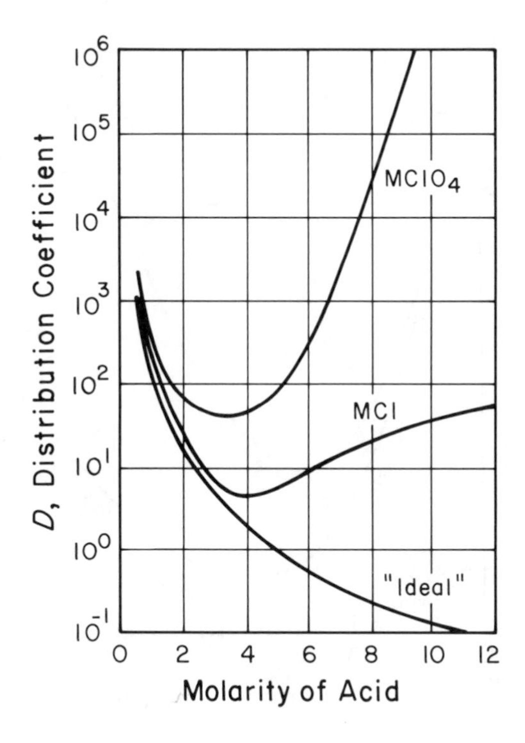

FIGURE 4. Distribution of scandium ions between a cation-exchange resin, Dowex®-50 × 4, and aqueous acids. The "ideal" curve assumes simple ion exchange at trace levels with constant activity coefficients and no electrolyte invasion. (From Nelson, F., Murase, T., and Kraus, K. A., *J. Chromatogr.*, 13, 503, 1964. With permission.)

The use of hydrobromic acid in cation-exchange separations in mixed solvents will be described below.

The effects seen in concentrated solutions are difficult to interpret, but are potentially useful in chemical separations, for the separation factors are very high. As a practical matter, though, the use of concentrated acids is inconvenient and ill-adapted to the equipment of high-performance liquid chromatography.

VI. USE OF NONAQUEOUS SOLVENTS

Mixing organic solvents with water lowers the dielectric constant. Representative dielectric constants at 25°C are water 80; methanol 32.6; ethanol 24.3; 1-propanol 20.1; 2-propanol 18.3; acetone 20.7; and tetrahydrofuran (THF) 8.

Lowering the dielectric constant increases the electrostatic forces between ions and stabilizes metal-chloride complexes and other complexes of metal ions with negatively charged ligands. Thus, the adsorption of metal ions by anion-exchange resins may be expected to increase when an organic solvent is added; or, lower hydrochloric acid concentrations should be needed to obtain a given distribution if an organic solvent is present. If a cation-exchange resin is used in the presence of an anionic complex former, the binding of metal ions should decrease when the organic solvent is added.

Fritz and co-workers[21,22] studied the effect of acetone on the binding of certain metal ions by a cation-exchange resin in the presence of varying proportions of acetone. Figure 7 shows how log D for Cu(II) varies with the fraction of acetone at constant acid concentrations. 0.5 *M* perchloric acid and 0.5 *M* hydrochloric acid. With perchloric acid, D rises somewhat with increasing acetone concentration. Probably this is because the resin takes up water

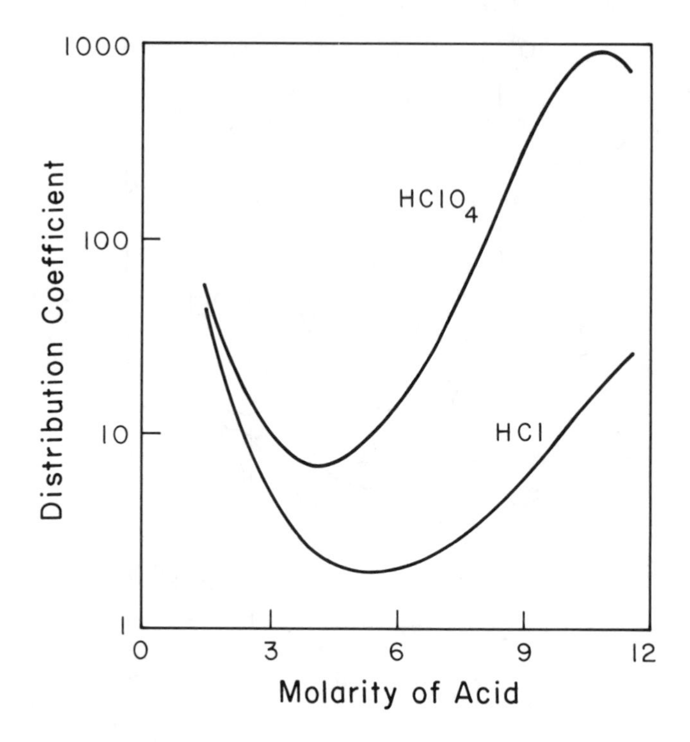

FIGURE 5. Distribution of calcium ions between a cation-exchange resin and aqueous hydrochloric and perchloric acids. (From Nelson, F., Murase, T., and Kraus, K. A., *J. Chromatogr.*, 13, 503, 1964. With permission.)

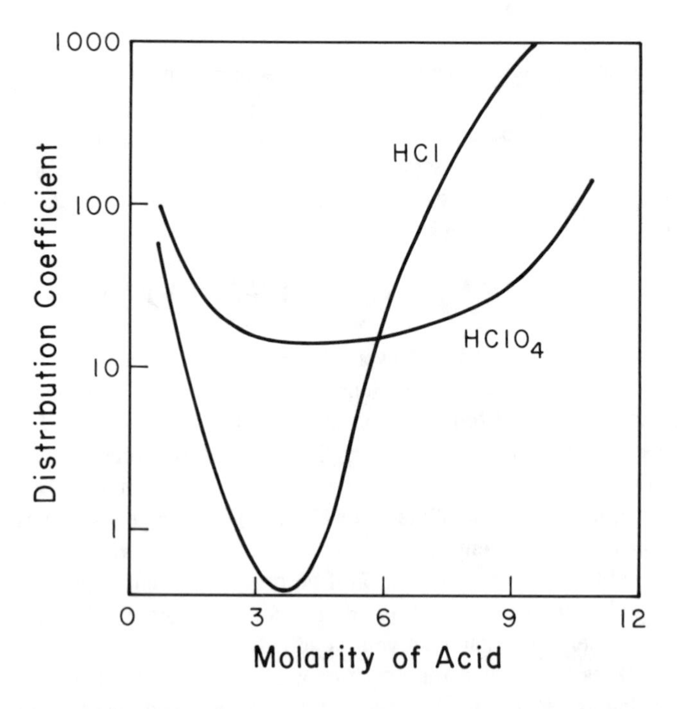

FIGURE 6. Distribution of iron(III) between a cation-exchange resin and aqueous hydrochloric and perchloric acids. (From Nelson, F., Murase, T., and Kraus, K. A., *J. Chromatogr.*, 13, 503, 1964. With permission.)

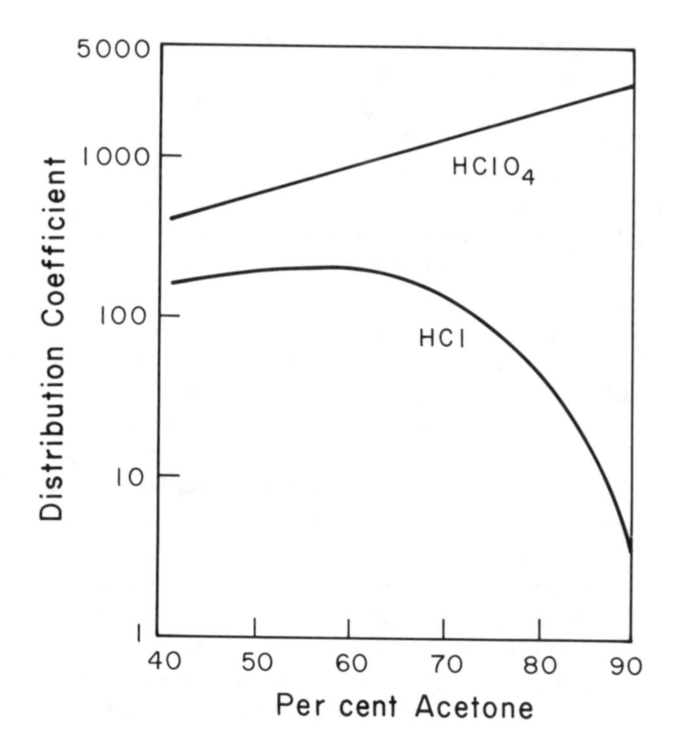

FIGURE 7. Distribution of copper(II) between a cation-exchange resin and 0.5 *M* hydrochloric and perchloric acids. vs. volume per cent of acetone. (From Fritz, J. S. and Rettig, T. A., *Anal. Chem.*, 34, 1562, 1962. Copyright, American Chemical Society, 1962. With permission.)

preferentially (Chapter 3; see also Davies and Owen[22a]) and the cupric ion is more hydrated inside the resin. There is no indication of association between Cu^{2+} and ClO_4^-. With hydrochloric acid at higher acetone concentrations D falls as the acetone concentration rises, probably because the complex ion $CuCl_4^{2-}$ becomes more stable. This is the normal behavior, shown by Mn, Co(II), Fe(III), Zn, Cd, and Bi.

Fritz and Pietrzyk[23] showed that in anion exchange in hydrochloric acid, distribution coefficients were increased by adding the alcohols, methanol, ethanol, and 2-propanol; the effect on D was greatest with propanol and least with methanol, as we would expect from the dielectric constants (see above). The effect was quite large; for iron(III) in 0.3 *M* HCl, D = 1 in 60% ethanol, 1000 in 96% ethanol. Various separation schemes were described, and all had the advantage over the separations in aqueous solutions described by Kraus and Nelson that the concentrations of acid required were much lower.

VII. ANION-EXCHANGE SEPARATIONS IN AQUEOUS ACETONE

A great deal of work on ion-exchange separations in mixed solvents has been done by Korkisch and his school in Vienna. One of the first publications describes the separation of iron, cobalt, and nickel.[24] It is to be compared with the separation described early in this chapter. A column of the same dimensions may be used, containing the same resin. One should only note that the surface tension of acetone is much less than that of water, and the acetone-water solvents will not be held up by surface tension in the column packed with 100 to 200 mesh resin, as was the case with aqueous solutions. On the other hand, the lower viscosity of acetone allows faster flow under gravity. The procedure is as follows:

> Wash the bed of Dowex®-1 × 8 resin with a solvent made by mixing 10 volumes of 6 *M* aqueous hydrochloric acid with 90 volumes of acetone. Add a small volume of a mixed solution containing the chlorides of iron(III), cobalt(II), and nickel(II) as before. Pass more of the same solvent. Iron is eluted immediately, but the solution is almost colorless; test for complete removal of iron by a spot test with thiocyanate. (Iron gives a red color). Now pass a solvent made by mixing 30 volumes of 2 *M* hydrochloric acid with 70 volumes of acetone. Nickel is eluted, and its apple-green color is easily seen. Continued elution with the same solvent will bring out cobalt, which is deep blue, but the cobalt can be eluted faster by passing aqueous 1 *M* hydrochloric acid. It comes out then as the pink aquo-complex. (For lecture purposes one may simplify the procedure and speed it up, with a very slight loss of efficiency, by eluting nickel with the 90% acetone-10% HCl solvent diluted 1:1 with water, then eluting the cobalt with water made slightly acid with HCl).

Some interesting points arise here. First, the behavior of iron: it is not adsorbed, and it comes out as an almost colorless solution, contrasted with the deep yellow solution in aqueous hydrochloric acid. The only conceivable explanation is that it forms an ion pair, $H^+FeCl_4^-$, with the proton solvated by acetone. Acetone behaves differently from ethanol, investigated by Fritz (see above), in spite of an almost equal dielectric constant; ethanol strengthens the binding of iron(III) to an anion-exchange resin in hydrochloric acid solutions. The :CO group of acetone is probably involved, becoming protonated to $:COH^+$. Tetrahydrofuran, with its ether oxygen atom, acts in the same way; it prevents iron from being adsorbed by the anion exchanger.

The difference between acetone and alcohol at high mole fractions is seen in Figure 8. For cobalt the attachment to a cation-exchange resin falls rapidly with increasing proportion of acetone above 60%, while with increasing proportion of alcohol it continues to rise.

Second, the behavior of nickel: the low dielectric constant stabilizes the nickel-chloride complex, which is then attached to the anion-exchange resin, yet the results of Fritz and Rettig[21] show that acetone makes nickel(II) stick more tightly to a cation-exchange resin in 0.5 *M* HCl. One must beware of making too-simple interpretations of effects that involve several kinds of interactions.

Nevertheless, the practical value of the scheme of Hazan and Korkisch[24] is evident. One is more likely to need to measure small amounts of cobalt and nickel in the presence of large quantities of iron than vice versa. Using aqueous hydrochloric acid the column must be big enough to hold back large amounts of iron, and then, "fronting" of iron may contaminate the small amounts of nickel and cobalt. With acetone-HCl the column need only be large enough to hold the nickel and/or cobalt, and the operation will be much faster. We have already mentioned the advantage of having lower hydrochloric acid concentrations and less excess HCl to remove before making quantitative measurements of eluted metal ions.

Hazan and Korkisch[24] describe an application of this method to steel analysis. They note that manganese, aluminum, and chromium(III) accompany the nickel, and that copper accompanies the cobalt. They do not mention a possible source of error: the reduction of Fe(III) to Fe(II). Impurities in the anion-exchange resin could reduce traces of Fe(III). To prevent this from happening, some authors advocate adding hydrogen peroxide or a few parts per million of chlorine.

Acetone and tetrahydrofuran in high concentration, 80 to 90%, prevent the adsorption or iron(III) by an anion-exchange resin in the presence of hydrochloric acid. They also prevent its adsorption by a cation-exchange resin. Gallium(III), thallium(III), and gold(III), which in aqueous hydrochloric acid are bound very strongly to an anion-exchange resin, are likewise prevented from binding to anion or cation-exchange resins by these solvents.[25,26]

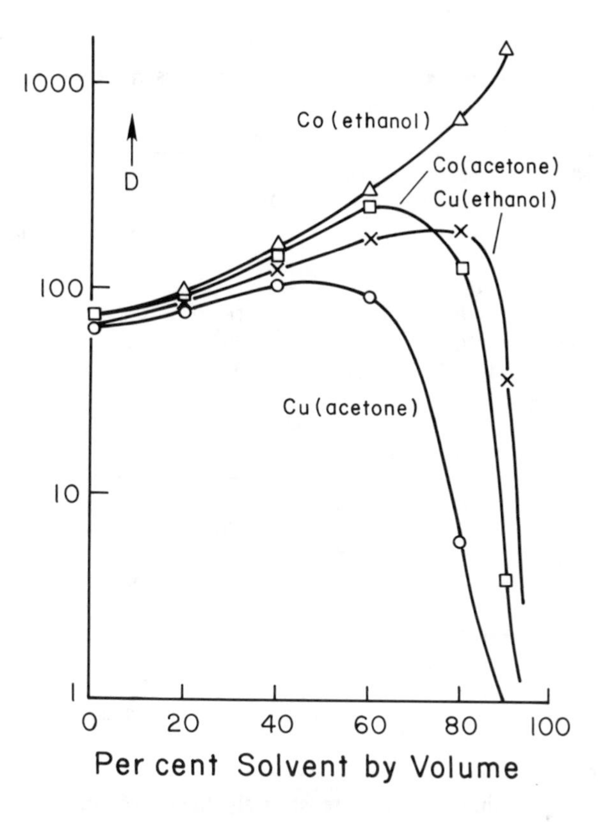

FIGURE 8. Distribution of copper(II) and cobalt(II) between a cation-exchange resin and mixtures of water with alcohol and acetone, all 0.5 *M* in hydrochloric acid. Resin was half-saturated with metal ions. Wt. resin, 2.5 g; vol. solution, 250 ml. (From Strelow, F. W. E., *Ion Exchange and Solvent Extraction,* Vol. 5, Marinsky, J. A. and Marcus, Y., Eds., Marcel Dekker, New York, 1973. Chap. 2. Courtesy of Marcel Dekker, Inc.)

The conclusion seems reasonable that these electron-donating, oxygen-containing solvents coordinate with the protons and stabilize the ion pairs $H^+ \; MCl_4^-$, where M is the trivalent metal. The ion pairs remain in the solution phase.

Strelow and co-workers[27] measured the adsorption of many metals by a cation-exchange resin in acetone-water mixtures containing hydrobromic acid. Table 1 presents a small selection of their voluminous data. Certain trends are evident: (1) raising the acid concentration lowers the adsorption, in general accord with the mass-action law; (2) metals that can form singly charged, 4-coordinated bromide complexes are adsorbed weakly or not at all at high acetone concentrations; (3) hard-acid cations like lanthanide ions are very strongly adsorbed; and (4) the effect of acetone concentration depends on the cation type. For alkali and alkaline-earth ions the adsorption increases with increasing acetone concentration; for highly charged ions, zirconium and the lanthanides, the adsorption decreases; for ions that form bromide complexes of intermediate stability, such as Cu, Zn, Cd, and Pb, the adsorption first rises with increasing acetone concentration, then falls quite rapidly as the concentration goes above 60 to 80%.

Figure 9 is an example of the separations that Strelow[27] was able to make, based on these distribution ratios. The zinc-cadmium separation is very effective, and so is the zirconium-hafnium separation, which is not shown in this figure. (Another zirconium-hafnium separation is shown in Figure 3). Zinc is bound much more strongly than cadmium

TABLE 1
Cation Exchange in HBr-Aqueous Acetone

	Acetone%						
	60	60	60	60	60	40	80
	HBr, M						
	0.1	0.2	0.5	1.0	2.0	0.5	0.5
D, ml/g							
Na	230	150	44	26	8	26	78
K	440	230	100	50	22	50	190
Mg	6000	1800	320	100	21	160	520
Ca	large	4400	825	240	80	340	2000
Cu	4500	1200	140	37	1.2	150	3
Zn	970	200	9	2	1.2	83	2
Cd	7	1.7	1	—	—	1	0.8
Pb	3500	100	9	6	—	11	6
Al	large	large	3100	380	60	1100	5600
Fe(III)	large	large	1400	110	6	930	6
U(VI)	—	—	370	130	50	180	500

Note: Very strongly adsorbed: Zr, Th, lanthanides, Ga. Very weakly adsorbed: Au, platinum metals, Hg, Ge. Large = over 5000.

Condensed from Strelow et al., *Anal. Chim. Acta,* 76, 377, 1975, with numbers rounded from data in six tables.

in 60% acetone, and hafnium much more strongly than zirconium in 90% acetone. The separation of cadmium and indium shown in Figure 9 is also noteworthy.

VIII. COMBINED ION EXCHANGE AND SOLVENT EXTRACTION

This name was given by Korkisch[28] to describe methods for separating metal ions that use ion exchange in conjunction with organic solvents. The separation just described are of this kind.

Three parameters can be chosen: nature of the ion exchanger (cation or anion exchanging); nature and concentration of the inorganic acid (usually nitric or hydrochloric); and nature and composition of the organic solvent mixture. By manipulating these parameters a great variety of metal separations may be realized, generally with high separation factors, and trace metals may be recovered from water, rocks, and minerals. Much of the work described by Korkisch and his school was "mission oriented", related to the nuclear power industry, and concerned uranium. As we shall see, the ion UO_2^{2+} has special properties.

The effect of various solvents on the distribution of uranium(VI) between a cation-exchange resin and dilute nitric acid is shown in Table 2. This table is condensed from a more extensive table published by Korkisch[28,29] and the numbers are rounded off for clarity. With alcohols and acetic acid the distribution coefficient, D, rises continuously as the fraction of organic solvent increases, and rises very rapidly as this fraction approaches one. With tetrahydrofuran and acetone D rises with the fraction of organic solvent up to about 0.6, or 60% by volume. As the fraction is raised beyond 0.6 D falls rapidly. Tetrahydrofuran (THF) is an ether; acetone is a ketone. Methyl isobutyl ketone (MIBK, 2-methyl-4-pentanone) shows the same effect, though this compound is not itself miscible with water, it may be made miscible by adding a third component, like acetone, THF, or methyl glycol. The behavior of methyl glycol is seen in Table 2. Here, too, D rises to a maximum around 60%

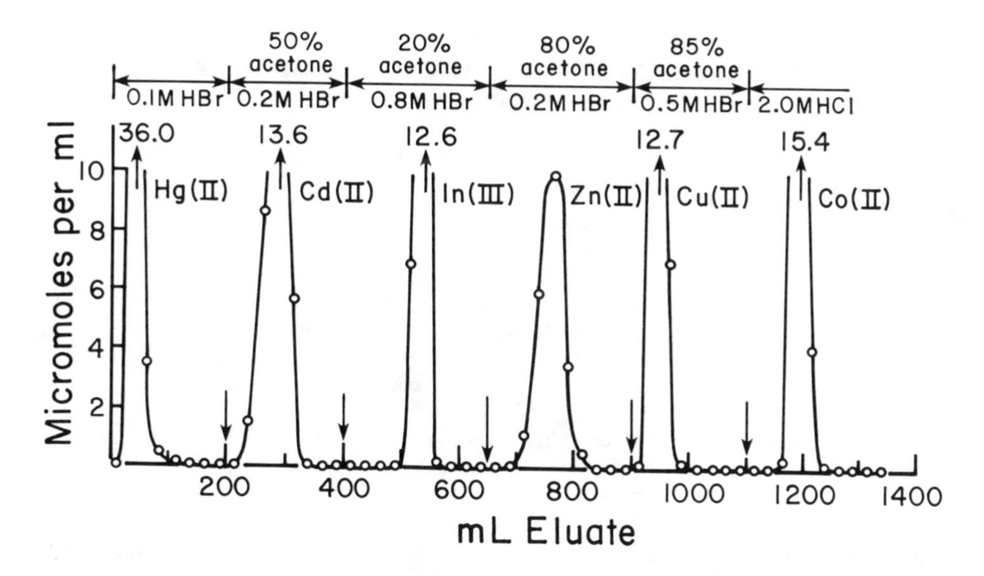

FIGURE 9. Cation-exchange separations using hydrobromic acid in aqueous acetone. Column, 2.0 × 19 cm; resin, AG50W-× 8, 200 to 400 mesh; flow rate, 2 ml/min for 300 min, then 3 ml/min. (From Strelow, et al., *Anal. Chim. Acta,* 76, 377, 1975. With permission.)

TABLE 2
Distribution Coefficients of Uranium(VI) in Organic Solvent-Water Mixtures, 0.6 *M* in Nitric Acid

Solvent	0	40	60	80	90
Ethanol	40	120	220	535	750
2-Propanol	40	160	360	890	large
Acetone	40	130	360	210	140
Tetrahydrofuran	40	130	240	50	40
Methyl glycol	40	110	207	430	310
Acetic acid	40	160	350	480	large

Note: On Dowex®-50 cation-exchange resin, crosslinking unspecified. Large = over 1000.

Abstracted from Korkisch, J., *Sep. Sci. Technol.,* 1, 159, 1966.

organic solvent, then falls slightly as the concentration of organic solvent rises to higher values. Methyl glycol, $CH_3OCH_2CH_2OH$, has an ether oxygen as well as a hydroxyl group. The fall in distribution coefficient at high concentrations of organic solvent is associated with the ether and ketone functions.

Table 3, which like Table 2 is abbreviated with the numbers rounded off, shows the special behavior of uranium(VI) towards THF and a cation-exchange resin. Only bismuth shows a large drop in adsorption at high THF concentrations, indicating the formation of uncharged nitrate-complex ion pairs. Other cations (and the original publication lists 20 cations plus the lanthanides) show continuous increases in distribution coefficient up to very high values as the proportion of THF increases.

The interpretation offered is that an ion pair is formed, $H^+ UO_2(NO_3)_3^-$, with its hydrogen ion coordinated to the ether or ketone oxygen, thus; $:CO\text{–}H^+$. A large cation is thus formed that will pair with the large anion. The ion pair stays in the solution phase, just as we

TABLE 3
Distribution Coefficients of Metal Ions in THF-Water Mixtures, 0.6 *M* in Nitric Acid

	Per cent THF by volume				
Metal Ion	**0**	**40**	**60**	**80**	**90**
U(VI)	40	130	240	50	40
Ca(II)	38	100	145	760	large
Al(III)	130	260	480	880	large
Bi(III)	135	630	590	195	50
Zn(II)	45	160	700	large	large
Fe(III)	170	290	550	large	large

Note: On Dowex®-50 cation-exchange resin. Large = over 1000.

Abstracted from Korkisch, J., *Sep. Sci. Techol.*, 1, 159, 1966.

postulated with the solvated iron-chloride ion pair, $H^+FeCl_4^-$, to account for the fact that in the presence of HCl and acetone or THF, iron(III) was not bound by a cation-exchange resin or anion-exchange resin.

These phenomena have parallels in solvent-extraction chemistry. Uranium(VI) is extracted from aqueous nitric acid and nitrate solutions by solvents that are ethers, esters or ketones; MIBK is very effective. Iron (III), along with Au(III), Ga(III), and Tl(III), is extracted from aqueous hydrochloric acid by the same solvents.

IX. EXTRACTION WITH TRIBUTYL PHOSPHATE

Tributyl phosphate is an electron donor or Lewis base, like the ethers and ketones, and it forms very stable complexes with metal ions and ion pairs with metal salts. For this reason it extracts many metal salts from aqueous solutions. It is used a great deal for extracting uranium, thorium, and the actinides. It also extracts iron(III) and other ions of high charge. To separate complex mixtures of metal ions one might begin by extracting with tributyl phosphate, then use ion exchange to separate the extracted ions from one another. Tributyl phosphate is extremely insoluble in water, however. To use ion exchange one must add another solvent that will bring tributyl phosphate and water together in the same phase. Koch and Korkisch[29] found that methyl glycol was very effective. They constructed triangular phase diagrams to show the miscibility range of the three-component system, tributyl phosphate, methyl glycol, and 12 *M* aqueous hydrochloric acid, and chose the mixture, 30% tributyl phosphate-60% methyl glycol-10% 12 *M* HCl, as their standard mixture to be passed through the anion-exchange column. This composition falls well within the range of homogeneous mixtures.

The basic separation procedure is the following:

1. Make the solution to be analyzed 6 *M* in hydrochloric acid and shake in a separating funnel with two or three portions of tributyl phosphate (TBP), either undiluted or diluted with kerosene. Combine the tributyl phosphate extracts.
2. Mix the TBP solution with methyl glycol and 12 *M* aqueous hydrochloric acid so as to give the 3:6:1 volume ratio just mentioned.
3. Pass the mixed solution through a small column of strong-base anion-exchange resin (Dowex®-50 × 8) that has been washed with the ternary solvent mixture. Flush the column with more of this mixture, and collect the effluent and washings.
4. Wash the column with 90% methyl glycol- 10% 12 *M* HCl to remove excess TBP. (This step is often omitted).

TABLE 4
Anion-Exchange Partition and Tributyl Phosphate Extraction

Element	Partition ratio, TBP, and 6 *M* HCl	Partition ratio, TBP, and 6 *M* HNO_3	Anion-exchange distribution ratio, ml/g
U(VI)	90	200	1600
Th	0.02	80	2.3
Bi	0.6	0.1	large
Au	12,000	150	small
Fe(III)	8,500	small	small
Ge(IV)	80	—	small
Mo(VI)	400	0.15	0.8
Cd	20	small	2.3
Zn	10	small	small
Cu	1	1.5	400
Co	1	small	large

Note: Anion exchanger was Dowex® 1 × 8; solvent for ion exchange was 30% tributyl phosphate, 60% methyl glycol, 10% 12 *M* aqueous HCl by volume.

From Korkisch, W. and Koch, W., *Mikrochim. Acta,* 245, 1973. TBP partition ratios from Marcus, Y. and Kertes, A. S., *Ion Exchange and Solvent Extraction of Metal Complexes,* Wiley, New York, 1969.

5. Pass an aqueous hydrochloric acid solution of the concentration needed to elute the elements desired, e.g., 6 *M* HCl elutes lead, 1 *M* HCl elutes uranium(VI).

To see how this scheme would work, we first look at a table of distribution ratios for metal salts between aqueous hydrochloric acid solutions and TBP. Extensive tables and charts giving the distributions between TBP and solutions of hydrochloric, nitric and sulfuric acids of different concentrations appear in a reference work by Marcus and Kertes.[30] Table 4 presents some of this information. From 6 *M* HCl, uranium(VI) is efficiently extracted (D = 90), and Fe(III), Au(III), Ga(III), and Tl(III) have even larger distribution ratios, 10^3 to 10^4. The alkali and alkaline-earth ions, aluminum, manganese, nickel, and thorium have D values of 0.1 and less, so they are virtually not extracted. Copper, cobalt, and tin(II) have D values close to one, so that if these elements are present in large amounts they will be carried over into the ion exchange step, step 3. Differences in distribution ratios are not enough in themselves to give effective, clear-cut separations between elements, unless the differences are very large, for solvent extraction, as performed here, is a one-step or two-step process, unlike the multi-step process of column chromatography.

Anion-exchange distribution coefficients for the TBP-extractible ions, between the 3:6:1 solvent mixture and Dowex®-1 anion-exchange resin, are given in Table 4. In this operation the only ion having a greater attachment to the resin than uranium is mercury. Other ions that are strongly adsorbed are Co, Cu, Bi, Pb, Sb(III), and some of the platinum metals. Several important ions, including Fe(III) and Zn(II), are very weakly adsorbed by the resin. Germanium and gold are very weakly adsorbed.

Thus, we could separate an imaginary mixture of iron(III), gold(III), germanium(IV), lead, copper(II), and uranium(VI) as follows:

Extract all these elements by tributyl phosphate from 6 *M* HCl. Mix the TBP extract with methyl glycol and 12 *M* HCl as directed. Pass this solution through the anion-exchange resin: Fe(III), Au(III), and Ge(IV) are not retained. Wash the resin to remove TBP, then pass 6 *M* HCl. Lead and copper are eluted. Pass 1 *M* HCl; uranium is eluted.

Korkisch and colleagues[31,32] have used this scheme with minor variations to recover and measure small traces of uranium in sea water, ground water, and geological materials. The final measurement was made colorimetrically with thiocyanate or, for greater sensitivity, by the fluorescence of a pellet made by melting with sodium fluoride or a sodium carbonate-potassium carbonate-sodium fluoride mix. In both these procedures it is necessary to separate uranium carefully from interfering elements, such as molybdenum (for the colorimetric thiocyanate method) and iron, manganese, and copper (for the fluorescence method).

Instead of tributyl phosphate, trioctylphosphine oxide was used as an extractant for uranium in one of Korkisch's procedures. It was used as an 0.1 *M* solution in diethyl ether. The extract was mixed with methyl glycol and 12 *M* HCl in the proportion (by volume) 50:45:5, then passed through the anion-exchange resin. Molybdenum was stripped from the column by a mixture of methyl glycol, 30% hydrogen peroxide, and 12 *M* HCl, in the ratio 90:5:5, then the column was washed with 6 *M* HCl before eluting uranium with 1 *M* HCl. This method was specially adapted to traces of uranium in sea water and manganese nodules. Uranium was recovered free of metals that would interfere in its fluorimetric or colorimetric determination.[33]

To separate iron from other elements that are extracted by TBP, like gold and germanium, one may reduce Fe(III) to Fe(II), which is not extracted.

For uranium in natural waters, where large amounts of iron are not present, the extraction with TBP can be omitted. In one procedure,[34] 1 l of water is mixed with 10 ml concentrated hydrochloric acid, 5 g of ascorbic acid and 20 g of potassium thiocyanate. It is passed through a column containing 4 g of the anion-exchange resin Dowex®-1, which is in the thiocyanate form. Then the column is washed with the solvent mixture (5 parts tetrahydrofuran to 4 parts methyl glycol to 1 part 6 *M* HCl). Iron, molybdenum, copper, zinc, vanadium, and mercury are washed out. Then the elements sought are eluted one after the other, as follows: 6 *M* HCl elutes cobalt; 1 *M* HCl elutes uranium; 0.15 *M* HBr elutes zinc; 2 *M* HNO_3 elutes cadmium.

Uranium in the eluate is measured by fluorescence of a sodium fluoride bead; the other metals are measured colorimetrically. Their concentrations in drinking water are in the range 1 to 5 mg/l. The method also works for sea water.[35]

Uranium is measured in urine by making it 8 *M* hydrochloric acid and passing through Dowex®-1. The column is washed with 6 *M* HCl, followed by the mixture of THF-methyl glycol-aqueous HCl described above. These washings remove all interfering metal ions, while uranium remains on the column. It is eluted with 1 *M* hydrochloric acid and measured fluorimetrically.[36]

To measure thorium in natural water the water is mixed with ascorbic acid and citric acid, then passed through a column of Dowex®-1 anion-exchange resin. Uranium and thorium are adsorbed as anionic citrate complexes. Thorium is eluted with 8 *M* hydrochloric acid, accompanied by other elements. The eluate solution is therefore evaporated to dryness, taken up in 8 *M* nitric acid, and passed through a separate small column of Dowex®-1 resin, which retains the anionic thorium nitrate complex very strongly. After washing the column with dilute nitric acid, thorium is eluted with 6 *M* HCl. Meanwhile, uranium and iron are still on the first column. The column is washed with a mixture of MIBK, acetone, and 1 *M* HCl, ratio 1:8:1 by volume, which removes iron. Uranium is eluted by 1 *M* HCl.[37]

Korkisch and his colleagues have described a great many separations that use ion exchange in mixed solvents to recover and measure trace metals, including Al, As, Au, Be, Bi, Cd, Co, Cu, Ga, Ge, In, Mn, Mo, Ni, Pb, Re, Th, Ti, V, and U. Two review articles[38,39] describe the detection and measurement of toxic metals in natural waters. Methods develped before 1969 are included in a book by Korkisch, *Modern Methods for the Separation of Rarer Metal Ions*.[40]

To show what can be done by combining solvent extraction with ion exchange we cite the measurement of traces of nickel in natural waters.[41] Dimethylglyoxime is added to the water at pH 7.5, and the yellow-red nickel-dimethylglyoxime complex is extracted into chloroform. The chloroform extract is mixed with tetrahydrofuran and methanolic HCl, then passed through a column of cation-exchange resin. Nickel is retained. Chloroform is washed out of the column by methanolic HCl, then nickel is eluted with aqueous 6 *M* HCl.

Table 8 lists another way to separate nickel from other elements using cation exchange. Nickel is eluted by dimethylglyoxime in 95% acetone that is 0.5 *M* in HCl.

Moving away from trace analysis and trace concentration, methods have been developed for the analysis of "yellow cake" (sodium uranate) and nuclear raw materials. A feature of these methods is the choice of atomic absorption spectrometry for the measurement step. Atomic absorption is sometimes thought to be free from interferences, so that separations should be unnecessary. But Korkisch points out that elements do interfere with one another, and to get maximum accuracy they should be presented in pure solutions or at least without large amounts of interfering ions. Vanadium and molybdenum interfere with each other's absorbance in the air-acetylene flame, and the large amounts of iron and aluminum present in many geological samples enhance the absorbance of vanadium. Titanium, too, interferes. To separate these elemets, Korkisch and Gross[42] used a column of anion-exchange resin in 6 *M* HCl, in which vanadium is not absorbed but iron and molybdenum are.

For many purposes it suffices to remove major constituents like iron and (in yellow cake) uranium before atomic absorption, without going to the trouble of separating minor elements from one another.

Another subject for the analytical genius of Korkisch and collaborators has been manganese nodules. These are found on the ocean floor and consist largely of the oxides of iron and manganese, together with smaller amounts of metals of potential economic value, notably nickel. After dissolving samples in hydrochloric and hydrofluoric acid and driving off excess HF, Korkisch[43] proceeded to separate groups of elements by anion exchange thus:

Solvent, 50% MIBK, 40% 2-propanol, 10% 12*M* HCl; iron passes.
Solvent, 6 *M* aqueous HCl; Mn, Cu, Co, Pb pass.
Solvent, 1 *M* aqueous HCl; Zn, Cd pass.
Solvent, 2 *M* nitric acid; U passes.

The second and third fractions were analyzed by atomic absorption, the fourth (uranium) by fluorescence of a sodium fluoride melt.

Another paper in the same series[44] describes the isolation of thallium, vanadium, and molybdenum by anion exchange and their measurement by atomic absorption. Thallium was strongly adsorbed as Tl(III) and then eluted as Tl(I) by passing sulfurous acid. It is noteworthy that absorbances measured after separation were 2 to 16% lower than those made on the original solution, which was rich in iron and manganese. For a really spectacular instance of the need for a chemical separation before using atomic absorption spectrometry we may cite the measurement of gallium in manganese nodules.[45] Absorbances measured in the raw 6 *M* hydrochloric acid solution, containing iron and manganese, were about ten times as great as the true absorbances measured with the gallium solution following anion-exchange separation. (Needless to say, the investigators carefully checked the complete recovery of the small amounts of gallium).

TABLE 5
Tables of Distribution Coefficients

Cation Exchange

In hydrochloric acid, aqueous, Strelow, F. W. E., *Anal. Chem.*, 32, 1185, 1960.
In hydrochloric acid, aqueous, with macroporous resin, Strelow, F. W. E. *Anal. Chem.*, 56, 1053, 1984.
In hydrochloric acid-methanol, macroporous resin, Strelow, F. W. E., *Anal. Chim. Acta*, 160, 31, 1984.
In hydrochloric acid-methanol, Strelow, F. W. E., van Zyl, C. R., and Bothma, C. J. C., *Anal. Chim. Acta,* 45, 81, 1969; also Strelow, F. W. E., in *Ion Exchange and Solvent Extraction,* Vol. 5, Marinsky, J. A. and Marcus, Y., Eds., Dekker, New York, 1973, chap 2.
In hydrochloric acid-acetone, Strelow, F. W. E., Victor, A, H., Van Zyl, C. R., and Eloff, C., *Anal. Chem.*, 43, 870, 1971.
In hydrochloric acid-acetone, macroporous resin, Strelow, F. W. E., *Talanta,* 35, 385, 1988.
In hydrochloric acid-thiourea, Weinert, C. H. S. W., Strelow, F. W. E., and Böhmer, R. G., *Talanta,* 30, 413, 1983.
In hydrobromic acid-acetone, Strelow, F. W. E., Hanekom, M. D., Victor, A, H., and Eloff, C., *Anal. Chim. Acta,* 76, 377, 1975.
In perchloric acid, Strelow, F. W. E. and Sondorp, H., *Talanta,* 19, 1113, 1972.
In nitric and sulfuric acids, Strelow, F. W. E. and Rethemeyer, R., and Bothma, C. J. C., *Anal. Chem.*, 37, 106, 1965.
In tartaric acid and ammonium tartrate, Strelow, F. W. E. and Van der Walt, T. N., *Anal. Chem.*, 54, 457, 1982.

Anion Exchange

In sulfuric acid, Strelow, F. W. E. and Bothma, C. J. C., *Anal. Chem.*, 39, 595, 1967.
In hydrobromic and nitric acids, Strelow, F. W. E., *Anal. Chem.*, 50, 1359, 1978.
In oxalic acid, with and without nitric and hydrochloric acids, Strelow, F. W. E., in *Ion Exchange and Solvent Extraction,* Marinsky, J. A. and Marcus, Y., Eds., Marcel Dekker, New York, 1973, chap 2.

X. APPLICATIONS OF ANION AND CATION EXCHANGE TO METAL SEPARATIONS

A. STUDIES OF STRELOW AND CO-WORKERS

At the National Chemical Research Laboratory in Pretoria, South Africa, F. W. E. Strelow and collaborators have made fundamental studies of the distribution of metal ions between ion-exchange resins and solutions and on the basis of these studies they have devised methods for metal-ion separations, designed especially for the complex mixtures found in rocks and minerals. A review of the earlier part of this work appeared in 1973.[46] The aim was to develop, not routine methods, but rather authoritative reference methods that would separate each metal ion completely and in pure form, unaccompanied by other constituents. As far as possible each element would have its own eluent, and the eluents would be mineral acids, easily removed from the eluates by evaporation. In some cases hydrogen peroxide was added; the metals Ti, V, Mo, and W form peroxide complexes, (so does Cr(VI), but this complex is too unstable to be separated by ion exchange). A number of separations were worked out with oxalic acid, which is easily destroyed by oxidation. Other organic acids are less easily destroyed; however, the hydroxy acids citric, tartaric, lactic, malic, and 2-methyllactic acid may be used to modify cation-exchange selectivities.[47] Thiourea greatly increases the cation-exchange retention of platinum metals and copper while decreasing retention of zinc and other elements;[48] the effects reflect the differences in stabilities of metal-thiourea complexes inside and outside the resin, discussed in Chapter 6. Separations were done in open glass columns with gravity flow, using resins of 200 mesh and finer, and they were slow; often the runs lasted several hours.

Table 5 is a list of papers by Strelow and colleagues that include tables of distribution coefficients for many elements in many solutions over a wide concentration range, using both cation and anion exchange. Excerpts from these tables are given in our Tables 1, 6, and 7. In some of their work the authors used macroporous resins. They had found that ion

TABLE 6
Cation Exchange in HCl-Aqueous Ethanol

	Ethanol %										
	0	0	0	60	60	60	60	80	80	80	80
	HCl, *M*										
	0.2	0.5	1.0	0.2	0.5	1.0	2.0	0.2	0.5	1.0	2.0
D, ml/g											
Na		13.5	7		80	36	25		250		
K		29	14		200	94	54		840		
Mg	350	74	20		234	71	24	2000	600	160	55
Ca			41			310	90			1600	
Cu	380	65	17.5	1060	176	23	5	1040	195	25	0.9
Zn	360	64	16	300	18	3.5	0	48	4	2.4	0
Cd	84	6.5	1.6	35	1.4	0	0	15	0	0	0
Al			61			250	45			500	120
Fe(III)	3400	225	33	6000	360	47	5	1400	150	7	1
U(VI)	250	58	19	850	180	58	16	1270	260	70	15

Note: The exchanger was Dowex® 50W × 8.

From Strelow, F. W. E., van Zyl, C. R., and Bothma, C. J. C., *Anal. Chim. Acta,* 45, 81, 1969; condensed from data in six tables.

TABLE 7
Comparison of Macroporous and Gel-Type Cation-Exchange Resins

	HCl, M			
	0.5	0.5	1.0	1.0
	Resin type			
	8%	MP	8%	MP
D, ml/g				
Na	13.5	26	—	14
K	29	70	14	35
Mg	74	90	20	25
Ca	(80)	850	41	210
Cu	65	117	17	28
Zn	64	99	16	20
Cd	6.5	15	1.6	2
Al	—	790	61	90
Fe(III)	225	800	33.5	90
U(VI)	58	200	19	70

Note: Solvent: water.

From Strelow, F. W. E., *Anal. Chim. Acta,* 160, 31, 1984, and Strelow, F. W. E., van Zyl, C. R., and Bothma, C. J. C., *Anal. Chim. Acta,* 45, 81, 1969; condensed from several tables.

exchange in the customary 8% crosslinked cation-exchange resins was very slow, and they experimented with resins of lower crosslinking, 4% and even 2% (see Table 8). Lower crosslinking gave faster exchange, sometimes more complete recovery, but usually less selectivity. Macroporous resins gave stronger retention and higher selectivity than 8% cross-linked gel-type resins (see Table 7). (Note especially the Ca/Mg selectivity, which is even greater in 60% ethyl alcohol than in water; see Table 6 and the reference there cited.) At the same time they gave faster mass transfer and ion exchange. Macroporous resins, as we have seen (Chapter 2), consist of very small clusters of highly crosslinked polymer connected by zones and filaments of much lower crosslinking, with large pores and vacancies between them. High crosslinking is conducive to high selectivity (Chapter 3).

Kawazu[49] has compared two macroporous cation-exchange resins with each other and a gel-type resin for the separation of 12 metal ions in hydrochloric acid-water-acetone mixtures, and noted that the behavior of the two macroporous resins is not identical. As we have seen, different macroporous resins can differ widely in their structure and properties.

A special objective of Strelow's studies was to devise methods for separating small amounts of minor elements from large amounts of accompanying major elements, generally in silicate rocks. In these separations it was best to choose conditions under which the major elements would not be adsorbed by the resin, or adsorbed only weakly. Then a small column could be used, which need only be big enough to hold back the element sought, plus any accompanying minor elements. Table 8 summarizes Strelow's publications on the separation of minor elements from complex mixtures by cation and anion exchange. For each mixture two eluents are listed. The original mixture of major and minor elements is placed in the column with the first eluent, which (ideally) washes the unwanted major elements through the column. Then, after a rinse with this eluent, the second eluent is passed. This brings out the element or elements sought, while leaving unwanted, accompanying elements on the column, to be stripped or cleaned out of the column later.

TABLE 8
Minor and Trace Metals Separated by Ion Exchange

Metal sought	Separated from	Exchanger	Eluents	Ref.
Ag	Zn, Cd, Cu, Ni	C (MP)	(a) 2 *M* HNO_3 (b) 0.5 *M* HBr in 90% acetone	52
Bi	Tl, Ag	C (× 4)	(a) 0.3 then 0.6 *M* HNO_3: Tl elutes first (b) 0.2 *M* HBr in 20% acetone	53
Bi	Cd, Pb, Zn, In, Ga, Fe, Cu, U	A (× 4)	(a) 2 *M* HNO_3 − 0.03 *M* HBr (b) 0.1 *M* NH_4NO_3 − 0.05 *M* DETPA	54
Ca	Mg, Al, Fe	C (MP)	(a) 3 *M* HCl in 50% methanol (b) 5 *M* HNO_3	55
Co, Ni, Cu, Mn	Te	C	(a) 0.5 *M* HCl in 70% acetone (b) 3 *M* HCl	64
Fe	Cr, Cu	A (× 2)	(a) 9 *M* HCl (b) 6 *M* HCl − 0.05 *M* NaI	56
Ga	Zn, Cu, In, Fe	C (× 4)	(a) 0.5 *M* HBr in 80% acetone (b)3 *M* HCl	57
In	Te, Cd	C	(a) 0.5 *M* HNO_3 (+ HBr) (b) 0.2 *M* HBr in 40% acetone (elutes Cd) (c) 1 *M* HCl (elutes In)	65
Mg	Al	C (× 4)	(a) 0.5 *M* oxalic acid (b) 2 *M* HCl	58
Ni	Zn, Co, Cu, Fe	C (× 4)	(a) 0.5 *M* HCl in 95% acetone (b) 0.5 *M* HCl, 0.5M dimethyl-glyoxime in 95% acetone	59
Pb	Bi, Sn, Cd, In	C (MP)	(a) 0.5 *M* HCl in 50% methanol (b) 3 *M* HCl	60
Pb	Zn, In, Ga, Fe, Cu, Co, Mn, U	A (× 4)	(a) 0.2 *M* HBr (b) 2 *M* HNO_3	61
Th	Silicate rocks, La, Zr, Hf	C (× 4)	(a) 6 *M* HBr (b) 5 *M* HNO_3	66
Ti	Mo	C (× 4)	(a) 0.01 *M* HNO_3 − 0.15% H_2O_2 (Mo passes, Ti elutes later)	62
V	Mo	C	(a) 0.1 *M* citrate − 0.01 *M* HNO_3 (b) 2 *M* HNO_3	63
U	Fe, Zr, etc., silicate rocks	A	(a) 2 *M* HCl in 60% acetone (b) 0.1 *M* HCl	51
U	Cu, Fe, etc.	C	(a) HCl-HBr in aq. acetone (b) 2.5 *M* HNO_3	51

Note: MP = macroporous, × 4, × 2 show resin crosslinking. Eluent (a) removes accompanying elements listed in column 2; eluent (b) removes element sought. DEPTA = diethylenetriamine pentaacetic acid.

A wider objective is of course the complete analysis of a mixture, generally the mixture of salts obtained by dissolution of a silicate rock in hydrofluoric, hydrochloric, and perchloric acids, followed by evaporation to leave only the perchlorate salts plus a small excess of perchloric acid. A procedure was published[50] that separated ten major and minor elements in silicate rocks, each in pure form that permitted their accurate determination, using one cation-exchange resin column, containing 200 to 400 mesh resin and of bed volume 90 ml, which would accommodate the digest from 0.5 g of rock. The elution sequence was as follows:

> 300 ml 0.01 *M* nitric acid-0.15% hydrogen peroxide elutes vanadium(V); 950 ml 0.5 ml 0.5 *M* nitric acid-0.05% hydrogen peroxide elutes first Na, then K; 300 ml 0.5 *M* sulfuric acid-0.05 *M* hydrogen peroxide elutes Ti, then Zr (but not Hf); 350 ml 0.2 *M* HCl in 85% acetone elutes Fe(III); 300 ml 0.75 *M* HCl in 90% acetone elutes Mn(II); 400 ml 1.25 *M* perchloric acid elutes Mg; 450 ml 1.25 *M* nitric acid elutes Ca; 250 ml 3 *M* HCl elutes Al.

Altogether, 3.3 l of eluent are used at a flow rate 3 ml/min, taking 20 h. Recoveries for most elements were 100.0 + 0.1%. The final quantitative measurements were made by EDTA titration, spectrophotometry, or atomic absorption spectrometry.

A very accurate method for determining trace amounts of uranium in rocks, which uses both anion and cation exchange, was described by Strelow and van der Walt.[51] The rock is dissolved by heating with a mixture of hydrofluoric, hydrochloric, and sulfuric acids, and the mixture is evaporated to remove all excess hydrofluoric acid, leaving only a little sulfuric acid. The residue is dissolved in 7 *M* HCl plus hydrogen peroxide and boiled, then passed through AGl × 4 strong-base anion-exchange resin in the chloride form. The 6 *M* HCl elutes zirconium; 2 *M* HCl in 60% acetone elutes iron(III), but with persistent tailing. Therefore, a little iron accompanies uranium, which is eluted next with 0.1 *M* aqueous HCl. The next step is to make the uranium fraction 82% in acetone and 0.2 *M* in hydrochloric acid, and pass the solution through AG50W × 8 cation-exchange resin. Washing the column with 82% acetone-0.2 *M* HCl removes the remaining iron; 86% acetone-0.5 *M* HBr removes copper; then 2.5 *M* aqueous nitric acid brings out the uranium quantitatively and in pure form. The final determination was made spectrophotometrically. This method is claimed to give higher purity and greater accuracy than the method of Korkisch and Hazan[24] described above.

Trace amounts of thorium in rocks were separated by a simple procedure and determined spectrophotometrically by reaction with Arsenazo III.[66] The salts obtained by dissolution of the rock with hydrofluoric, hydrochloric, and perchloric acids were dissolved and passed through a small column of sulfonated polystyrene cation-exchange resin with 4% crosslinking. This crosslinking was important; thorium could not be completely eluted from an 8% crosslinked resin. The column was first washed with 6 *M* aqueous HBr. This removed all metal ions except those of thorium, gold(III), and thallium (III). Washing with 0.5% HBr in 90% acetong removed gold and thallium and any remaining iron(III). Finally, thorium was eluted in high purity by passing 5 *M* nitric acid. Some 10 mg of thorium were recovered from 1 to 2 g of rock with coefficient of variation 1%.

XI. CONCLUSION

Many other methods have been described for the separation of metals and nonmetals from one another on open ion-exchange columns, using special reagents suited to the task at hand. Most of them are tedious and require knowledge and skill of the operator. The analyst of today will seldom need to use them, given the developments in atomic spectroscopy and ion chromatography, but they enshrine much interesting chemistry and should not be forgotten. It is quite possible that the methods we have been describing may be adapted to high-performance ion chromatography as pumps and fittings become available that can handle strong, corrosive acids. New macroporous resins that show minimal volume changes have made it easier to adapt open-column methods to closed columns and small resin particles. Summaries of the older open-column methods are found in the biennial *Fundamental Reviews* of the journal *Analytical Chemistry* before 1982, and in Reference 79, Chapter 2.

REFERENCES

1. **Gjerde, D. T. and Fritz, J. S.,** *Ion Chromatography,* 2nd ed., Huethig, Heidelberg, 1987, chap. 5.
2. **Kraus, K. A. and Nelson, F.,** Anion-exchange studies of the fission products, Vol. 7, Proc. 1st UN Conf. Peaceful Uses Atomic Energy, 1955, 113.
3. **Jentzch, D.,** Anwendung von Ionenaustauschern in der analytischen Chemie, *Fresenius Z. Analyt. Chem.,* 152, 134, 1956.
4. **Wilkins, D. H.,** The determination of nickel, cobalt, iron and zinc in ferrites, *Anal. Chim. Acta,* 20, 271, 1959.
5. **Jones, S. L.,** Analysis of silver solder by anion exchange and EDTA titration, *Anal. Chim. Acta,* 21, 532, 1959.
6. **Schindewolf, U.,** Flüssige Anionenaustauscher, *Z. Elektrochem.,* 162, 335, 1958.
7. **Horne, R. A.,** The adsorption of zinc(II) on anion-exchange resins. I. The secondary cation effect, *J. Phys. Chem.,* 61, 1651, 1957.
8. **Robinson, R. A. and Stokes, R. H.,** *Electrolyte Solutions,* 2nd ed., Butterworths, London, 1959.
9. **Marcus, Y. and Maydan, D.,** Anion exchange of metal complexes. VIII. The effect of the secondary cation; the zinc-chloride system, *J. Phys. Chem.,* 63, 979, 1963.
10. **Korkisch, J. and Gross, H.,** Atomic-absorption determination of lead in geological materials, *Talanta,* 21, 1025, 1974.
11. **Nelson, F., Rush, R. M., and Kraus, K. A.,** Anion-exchange studies. XXVII. Adsorbability of a number of elements in HCl-HF solutions, *J. Am. Chem. Soc.,* 82, 339, 1960.
12. **Faris, J. P.,** Adsorption of the elements from hydrofluoric acid by anion exchange, *Anal. Chem.,* 32, 521, 1960.
13. **Huffman, E. H. and Lilly, R. C.,** The anion-exchange separation of zirconium and hafnium, *J. Am. Chem. Soc.,* 71, 4147, 1949.
14. **Hague, J. L. and Machlan, L. A.,** Determination of titanium, zirconium, niobium and tantalum in steels, *J. Res. Natl. Bur. Standards,* 62, 11, 1959.
15. **Greenland, L. P.,** Simultaneous determination of tantalum and hafnium in silicates by neutron activation analysis, *Anal. Chim. Acta,* 42, 365, 1968.
16. **Faris, J. P.,** Anion-exchange characteristics of elements in nitric acid medium, *Anal. Chem.,* 36, 1157, 1964.
17. **Strelow, F. W. E. and Bothma, C. J. C.,** Anion exchange and a selectivity scale for elements in sulfuric acid media with a strongly basic resin, *Anal. Chem.,* 39, 595, 1967.
18. **Strelow, F. W. E., Rethemeyer, R., and Bothma, C. J. C.,** Ion exchange selectivity scales for cations in nitric acid and sulfuric acid with a silfonated polystyrene resin, *Anal. Chem.,* 37, 106, 1965.
19. **Nelson, F., Murase, T., and Kraus, K. A.,** Ion exchange procedures. I. Cation exchange in concentrated HCl and $HClO_4$ solutions, *J. Chromatogr.,* 13, 503, 1964.
20. **Nelson, F. and Michaelson, D. C.,** Ion exchange procedures. IX. Cation exchange in HBr solutions, *J. Chromatogr.,* 25, 414, 1966.
21. **Fritz, J. S. and Rettig, T. A.,** Separation of metals by cation exchange in acetone-water-HCl, *Anal. Chem.,* 34, 1562, 1962.
22. **Fritz, J. S. and Abbink, J. E.,** Cation-exchange separation of small amounts of metal ions from cadmium, zinc and iron(III), *Anal. Chem.,* 37, 1274, 1965.
23. **Fritz, J. S. and Pietrzyk, D. J.,** Nonaqueous solvents in anion-exchange separations, *Talanta,* 8, 143, 1962.
24. **Hazan, I. and Korkisch, J.,** Anion-exchange separation of iron, cobalt and nickel, *Anal. Chim. Acta,* 32, 46, 1965.
25. **Korkisch, J. and Gross, H.,** Selective cation-exchange separation of cobalt in HCl-acetone solutions, *Sep. Sci. Technol.,* 2, 169, 1967.
26. **Korkisch, J. and Klakl, H.,** Anion-exchange behaviour of the platinum metals and gold in HCl-organic solvent media, *Talanta,* 15, 339, 1968.
27. **Strelow, F. W. E., Hanekom, M. D., Victor, A. H., and Eloff, C.,** Distribution coefficients and cation-exchange behaviour of elements in HBr-acetone media, *Anal. Chim. Acta,* 76, 377, 1975.
28. **Korkisch, J.,** Combined ion exchange-solvent extraction (CIESE): a novel separation technique for inorganic ions, *Sep. Sci. Technol.,* 1, 159, 1966.
29. **Koch, J. and Korkisch, W.,** Anionenaustauschtrennungen der mit Tributylphosphat extrahierbaren Elemente. V. Trennungsmöglichkeiten in TBP-HCl- bzw. TBP-Methylglykol-HCl-haltigen Lösungsmittelsystemen, *Mikorchim. Acta,* 245, 1973.
30. **Marcus, Y. and Kertes, A. S.,** *Ion Exchange and Solvent Extraction of Metal Complexes,* Wiley-Interscience, New York, 1969.
31. **Koch, W. and Korkisch, J.,** Anionenaustauschtrennungen der mit Tributylphosphat extrahierbaren Elemente. IV. Trennung des Urans von einigen Elementen, *Mikrochim. Acta,* 225, 1973.

32. **Korkisch, J. and Koch, W.,** Anionenaustauschtrennungen der mit TBP extrahierbaren Elemente. VII. Anwendung zur Bestimmung des Urans in geologischen Proben, *Mikrochim. Acta,* 865, 1973.
33. **Koch, W. and Korkisch, J.,** Bestimmung geringer Uranmengen nach Konzentrierung durch Extraktion und Anionenaustausch in einem tri-*n*-octyl-phosphinoxidhaltigen Lösungsmittelsystem, *Mikrochim. Acta,* 157, 1973.
34. **Korkisch, J. and Goedl, L.,** Anwendung von Ionenaustauschverfahren zur Bestimmung von Spurenelementen in natürlichen Wässern, *Talanta,* 21, 1035, 1974.
35. **Korkisch, J. and Steffan, I.,** Determination of uranium in sea water after anion-exchange separation, *Anal. Chim. Acta,* 77, 312, 1975.
36. **Korkisch, J. and Steffan, I.,** Bestimmung des Urans in Urinproben nach dessen Abtrennung durch Anionenaustausch, *Mikrochim. Acta,* 273, 1973.
37. **Korkisch, J. and Krivanec, H.,** Application of ion-exchange separations to determine trace elements in natural waters. IX. Uranium, thorium, *Talanta,* 23, 295, 1976.
38. **Korkisch, J.,** Toxic metals in natural waters. Pergamon Series in Environmental Science, 1980, 3 *(Anal. Tech. Environ, Chem.),* 449.
39. **Korkisch, J.,** Analysis of naturally occurring waters for toxic metals using combined ion exchange and solvent extraction, *Pure Appl. Chem.,* 50, 371, 1978.
40. **Korkisch, J.,** *Modern Methods for the Separation of Rarer Metal Ions,* Pergamon Press, Oxford, 1969.
41. **Korkisch, J., Steffan, I., and Staniek, H.,** Separation of nickel by solvent extraction and cation exchange, *J. Indian Chem. Soc.,* 59, 1331, 1982.
42. **Korkisch, J. and Gross, H.,** Determination of vanadium and molybdenum by atomic absorption spectrometry, *Talanta,* 20, 1153, 1973.
43. **Korkisch, J., Hübner, H., Steffan, I., Arrhenius, G., Fisk, M., and Frazer, J.,** Chemical analysis of manganese nodules. Part I. Determination of seven main and trace constituents after anion-exchange separation, *Anal. Chim. Acta,* 83, 83, 1976.
44. **Korkisch, J., Steffan, I., and Arrhenius, G.,** Chemical analysis of manganese nodules. Part III. Determination of thallium, molybdenum and vanadium after anion-exchange separation, *Anal. Chim. Acta,* 94, 237, 1977.
45. **Korkisch, J., Steffan, I., Nonaka, J., and Arrhenius, G.,** Chemical analysis of manganese nodules. Part V. Determination of gallium after anion-exchange separation, *Anal. Chim. Acta,* 109, 181, 1979.
46. **Strelow, F. W. E.,** Application of ion exchange to element separation and analysis, in *Ion Exchange and Solvent Extraction,* Vol. 5, Marinsky, J. A. and Marcus, Y., Eds., Marcel Dekker, New York, 1973, chap. 2.
47. **Strelow, F. W. E. and Weinert, C. H. S. W.,** Comparative distribution coefficients and cation-exchange behaviour of the alkaline-earth elements with various complexing agents, *Talanta,* 17, 1, 1970.
48. **Weinert, C. H. S. W., Strelow, F. W. E., and Böhmer, R. G.,** Cation exchange in thiourea-hydrochloric acid solutions, *Talanta,* 30, 413, 1983.
49. **Kawazu, K.,** Comparison of efficiency of cation-exchange resins in the chromatographic separation of metal ions with aqueous acetone-hydrochloric acid solution, *J. Chromatogr.,* 137, 381, 1977.
50. **Strelow, F. W. E., Liebenberg, C. J., and Victor, A. H.,** Accurate determination of ten major and minor elements in silicate rocks based on separation by cation exchange chromatography on a single column, *Anal. Chem.,* 46, 1409, 1974.
51. **Strelow, F. W. E. and van der Walt, T. N.,** Highly accurate determination of trace amounts of uranium in standard reference materials by spectrophotometry after complete separation by anion and cation exchange, *Fresenius Z. Anal. Chem.,* 306, 110, 1981.
52. **Strelow, F. W. E.,** Separation of silver fron zinc, cadmium, copper, nickel and other elements in nitric acid with a macroporous resin, *Talanta,* 32, 953, 1985.
53. **Meintjies, E., Strelow, F. W. E., and Victor, A. H.,** Separation of bismuth from gram amounts of thallium and silver by cation-exchange chromatography in nitric acid, *Talanta,* 34, 401, 1987.
54. **Strelow, F. W. E. and van der Walt, T. N.,** Quantitative separation of bismuth from lead, cadmium and other elements by anion-exchange chromatography with hydrobromic acid-nitric acid elution, *Anal. Chem.,* 53, 1637, 1981.
55. **Strelow, F. W. E.,** Quantitative separation of calcium from magnesium, aluminum, iron and many other elements by cation-exchange chromatography in methanolic HCl on a macroporous resin, *Anal. Chim. Acta,* 127, 63, 1981.
56. **van der Walt, T. N., Strelow, F. W. E., and Haasbroek, F. J.,** Separation of iron-52 from chromium cyclotron targets on 2% crosslinked anion-exchange resin in hydrochloric acid, *Talanta,* 32, 313, 1985.
57. **Strelow, F. W. E.,** Quantitative separation of gallium from zinc, copper, indium, iron(III) and other elements by cation-exchange chromatography in HBr-acetone medium, *Talanta,* 27, 231, 1980.
58. **van der Walt, T. N. and Strelow, F. W. E.,** Determination of magnesium in alumina ceramics by atomic absorption spectrometry after separation by cation exchange, *Anal. Chem.,* 57, 2889, 1985.

59. **Victor, A. H.,** Separation of nickel from other elements by cation-exchange chromatography in dimethylglyoxime/hydrochloric acid/acetone media, *Anal. Chim. Acta,* 183, 155, 1986.
60. **Strelow, F. W. E.,** Separation of traces and large amounts of lead from gram amounts of bismuth, tin, cadmium and indium by cation exchange in HCl-methanol using a macroporous resin, *Anal. Chem.,* 57, 2268, 1985.
61. **Strelow, F. W. E.,** Separation of traces and minor amounts of lead from large amounts of zinc, indium, gallium and other elements on a low crosslinked anion-exchange resin, *Anal. Chim. Acta,* 183, 307, 1986.
62. **Strelow, F. W. E.,** Determination of traces of titanium down to sub-ppm levels in molybdenum metal and compounds by ion exchange chromatography and spectrophotometry, *Anal. Chem.,* 58, 2408, 1986.
63. **Strelow, F. W. E.,** Determination of traces of vanadium in molybdenum metal and compounds by ion exchange chromatography/spectrophotometry, *Anal. Chem.,* 59, 1907, 1987.
64. **Strelow, F. W. E.,** Separation of traces of Mn, Co, Ni and Cu from gram amounts of tellurium by cation exchange chromatography, *Anal. Chim. Acta,* 212, 191, 1988.
65. **Strelow, F. W. E.,** Separation of traces of indium from gram amounts of tellurium and cadmium by cation exchange chromatography on a single column, *S. Afr. J. Chem.,* 40, 179, 1987.
66. **Victor, A. H. and Strelow, F. W. E.,** Highly accurate determination of thorium in silicate rocks by cation exchange chromatography and spectrophotometry, *Anal. Chim. Acta,* 138, 285, 1982.

Chapter 9

COLLECTION OF TRACES

I. GENERAL CONSIDERATIONS

Ions present in very low concentrations in water or aqueous solutions are collected and concentrated by passing a large volume of solution through a small column of an ion-exchange resin under conditions of high selectivity, so that the ions sought, the analyte ions, are retained in preference to more abundant accompanying species. Of course, the accompanying ionic species are always absorbed to some extent, for they compete with the analyte ions by ion exchange. Selectivity is gained by adding suitable complex-forming or reactant ions to the exchanger or to the solution, or by using exchangers that have specially selective functional groups, generally chelating groups that bind metal ions selectively.

The ability of the exchanger to retain analyte ions depends on the distribution coefficient, D. This can be measured by batch experiments. The quantity D determines the breakthrough volume, V_b, which is the same as the elution volume that we met in Chapter 3. It does not, however, uniquely determine the quantity of the analyte ions that is retained. This quantity depends on the concentration of the substance in solution as well as on D. It equals Dmc_o, where m is the mass of exchanger in the column in grams, D is expressed in liters per gram, and c_o is the concentration in the solution. Figure 1 shows the breakthrough volume and also the loading of the column for two concentrations of the analyte ion. The breakthrough volume is the same for both concentrations, but the quantity retained is proportional to the concentration of the analyte in the solution. These statements, of course, assume that the distribution coefficient D is independent of concentration, which in turn assumes that the loading is small compared to the ion-exchange capacity of the column.

The shape of the breakthrough curve (concentration vs. volume) is described by this equation:[1]

$$\frac{c}{c_0} = \frac{1}{2} - \mathrm{erf}\left[\sqrt{\frac{N}{VV_b}}\,(V_b - V)\right]$$

where the error function, erf(t), is:

$$\mathrm{erf}(t) = \frac{1}{\sqrt{2\pi}}\int_0^t e^{-t^2/2}\,dt$$

Here, N is the number of theoretical plates in the column.

The common strong-acid cation exchangers with functional sulfonic acid groups do not have enough selectivity to be used in the concentration of traces, except with very dilute solutions or in circumstances where incomplete recovery can be tolerated. They may be used to separate analyte ions that have already been concentrated, as a group, by another method. The acrylic resins with functional carboxyl groups have good selectivity for divalent ions against univalent ions, and for heavy-metal ions like those of copper; they have been used to concentrate these ions from water. The resin must be used in the sodium or calcium form, not the hydrogen form, for the –COOH groups are not ionized to any appreciable extent and the resin in its hydrogen-ion form does not swell; it undergoes ion exchange only

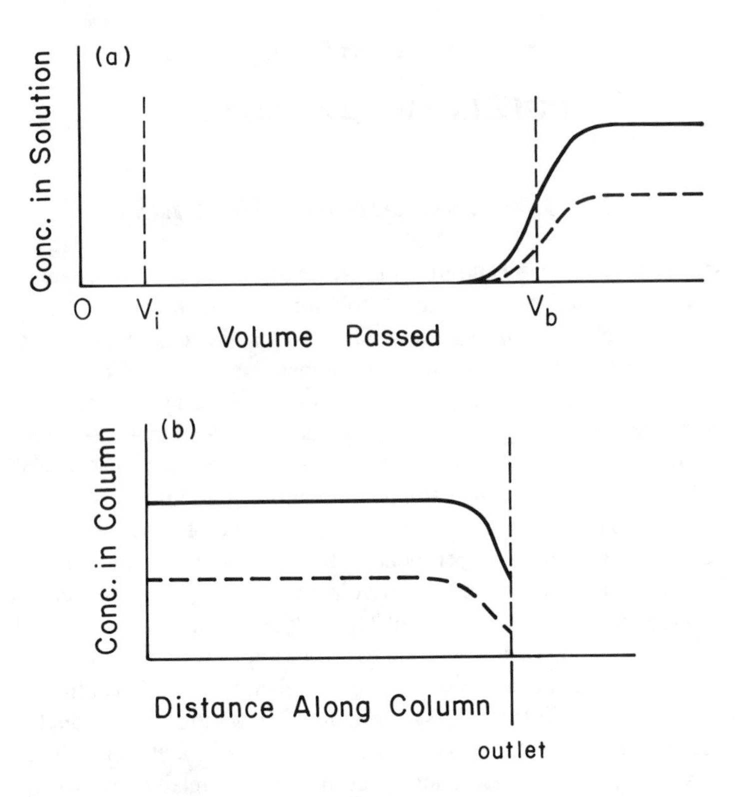

FIGURE 1. Concentration profiles while loading a column to "breakthrough", (a) isostere, or concentration at column exit vs. volume of eluent passed; (b) isochrone, or concentration in the stationary phase versus distance from column inlet at the time of breakthrough. Dashed line and continuous line represent two influent concentrations.

slowly, and as –COOH is converted to ionized –COONa, the resin swells and might burst a packed glass column.

Strong-base anion-exchange resins with functional quaternary ammonium ions are intrinsically more selective than sulfonate cation exchangers, and very high selectivity for metals may result if complexing anions are present. We have seen this effect in Chapter 8. There are many published procedures for the collection of trace metal ions that add a complexing anion before passing the test solution through a column of anion-exchange resin. Thiocyanate causes the adsorption of gallium, uranium and other metals (Chapter 8, Reference 35). In a very dilute solution an anion-exchange resin in its sulfate form adsorbs uranium(VI) with a distribution coefficient of 10^3.[2]

Most procedures for the collection of trace metals by ion exchange use an exchanger with a chelating functional group, usually the iminodiacetate group, though many special tailor-made chelating resins are used as well. Some of these were described in Chapter 2. Here again, the exchangers must be converted from their undissociated acid form into an ionized form like the sodium salt before they are packed into a column.

Most procedures for the collection of trace metals by ion exchange use columns; however, it is possible to use filter papers impregnated with fine particles of resin. Some methods use strips of ion-exchanging membranes, generally in batch techniques. Other batch techniques use single resin beads.

As a rule the trace metals are stripped from the exchangers by suitable eluents like strong acids, and thus concentrated into a small volume; they are then determined by an appropriate analytical method like atomic absorption spectrometry. In some cases, however, an analytical method is used that does not require stripping of the metal; for example, the resin (or other exchanger) is compressed into a disc and then analyzed by X-ray fluorescence, or by activation analysis, or another radiochemical technique. Examples of all these procedures will now be described.

II. USE OF STANDARD CATION- AND ANION-EXCHANGE RESINS

Ordinary sulfonated polystyrene cation exchangers have enough selectivity to retain certain metal ions if the solution from which they are adsorbed is very dilute, so that there is little competition from other ions. As an example we cite the collection of silver ions from water from rain and snow.[3] One or two liters of water was acidified with acetic acid and passed through a small column, 0.7 × 2 cm, of Dowex®-50 × 8. Retained silver was eluted with 10 ml 0.1 *M* thiocyanate (which forms a slightly soluble anionic silver complex) and then measured by neutron activation analysis; as little as 1 ng of silver could be detected. The procedure was used to see the effects of cloud seeding with silver iodide.

Technetium is collected from river waters[4] as TcO_4^- on a small column of anion-exchange resin after first passing through a cation exchanger to remove common cations. After a complicated workup including solvent extraction the technetium is collected in individual anion-exchange resin beads and presented for mass spectometric analysis. Completeness of recovery is ascertained by isotopic dilution.

The use of cation-exchange resins for trace enrichment of metal ions from very dilute solutions, like reactor cooling water, was described in Chapter 4 (Cassidy and Elchuk, Reference 20).

Sometimes the actual trace collection is done by another method, generally solvent extraction, and then the individual species of interest are separated by ion exchange. Examples are found in the work of Korkisch. Thus, beryllium is extracted from water into a solution of acetylacetone in chloroform. Other elements are extracted as well, and they are separated from beryllium by mixing the chloroform extract with tetrahydrofuran and methanolic hydrochloric acid in the proportions 3:6:1, then passing through a cation-exchange resin; beryllium is retained, then eluted selectively with 6 *M* HCl.[5] In this way an evaporation step, which would cause the loss of the beryllium-acetylacetone complex, is avoided. A similar procedure was used to extract traces of beryllium from rocks after opening up with hydrofluoric and perchloric acids.[6] Traces of nickel were extracted from water by a solution of dimethylglyoxime in chloroform, which was then mixed with tetrahydrofuran and methanolic HCl and treated by ion exchanger (see Reference 41, Chapter 8). A general procedure for heavy metals is to extract them as a group with sodium diethyldithiocarbamate into chloroform, then evaporate the chloroform, destroy the excess of organic reagent, dissolve in a tetrahydrofuran:methyl glycol:aqueous HCl mixture, 5:4:1, and pass through an anion exchanger. Cobalt, copper, manganese, and lead are eluted with 6 *M* HCl, uranium by 1 *M* HCl, cadmium and zinc, successively, with 2 *M* nitric acid.[7] These procedures illustrate the method of "combined ion exchange and solvent extraction" described in Chapter 8.

Several examples of anion exchange in the presence of complex-forming inorganic anions were cited in Chapter 8. The use of hydrobromic acid to adsorb lead on an anion exchanger, which was described for analysis of geological materials, also applies to natural waters; the water is made 0.15 *M* in HBr and passed through a column of Dowex®-1 × 8; lead is retained and later stripped with 6 M HCl; concentrations of lead down to a few μg/l are recovered in this way.[8] Silver in fresh water is measured by making the water 1 M in HCl (thus forming the ions $AgCl_2^-$), passing through a column of Dowex®-1 resin, 2.5 × 1.5 cm, eluting the silver with nitric acid in 90% acetone and then using flame atomic absorption, a method that is very sensitive for silver. Sensitivity was enhanced by extracting the silver into methyl isobutyl ketone and ammonium pyrrolidine carbodithioate (APDC). The concentration of silver found in natural fresh water is about 0.2 μg/l.[9]

Sea water is already 0.5 *M* in chloride ions, and when it is passed through a column of strong-base anion-exchange resin some of the trace metal ions are adsorbed, including cadmium, which is present in the sub-parts per billion range; recovery is over 80%.[10] By making the water (sea water or fresh water) 1 *M* in ammonium thiocyanate and then passing through the anion-exchange resin, gallium is adsorbed; it is eluted by 1 *M* NaOH and measured by its color produced with Rhodamine B. Some 0.01 μg/l are found.[11]

In some cases an anion or cation is attached to the exchanger that is strongly held and in its turn holds on to the trace ions of the solutions that flow through the column. A cation-exchange resin loaded with zirconium ions retains traces of fluoride ions.[12] Another way to concentrate traces of fluoride ions[13] is to take an anion-exchange resin and load it with the anions of "alizarin complexone", also called "alizarin fluorine blue". This compound is an anthraquinone that has a sulfonate ion on one side of the molecule and an iminodiacetate chelating group on the other. Lanthanum ions are placed on the chelating group. They have a great affinity for fluoride ions, attaching them as LaF^{2+} and producing a color change in so doing. Fluoride is stripped with 1 *M* NaOH and measured with an ion-selective electrode. By this technique, as low as 0.002 μg of fluoride ions per liter of water can be detected.

An anion-exchange resin loaded with the ions of 8-hydroxyquinoline-5-sulfonate has the sulfonate ions oriented towards the fixed positive ions of the resin, leaving the 8-hydroxyquinoline unit free to form chelate compounds with metal ions. Many ions can be captured by the loaded resin, including Ca, Sr, Ba, Cu, Zn, Cd, Mn, Co, and Ni. They have been measured *in situ* by neutron activation.[14,15]

III. REACTIVE ION EXCHANGE

A general way in which to use standard strong-acid or strong-base ion-exchange resins selectively and for the collection of trace ions has been developed by Janauer and colleagues, who call it reactive ion exchange. A counter-ion is placed on the resin that will react chemically with the ion to be adsorbed; or, the adsorbed ions of interest are stripped selectively with an ionic species that reacts with them in a special way. Thus, chromate or dichromate ions are retained by an anion-exchange resin and stripped by back-flushing with a solution of ferrous ions; these reduce the anions CrO_4^{2-} to the cations Cr^{3+}, which readily leave the column.[16] A cation-exchange resin loaded with Cu^{2+} is used to collect traces of ferro- and ferricyanide ions from water; these ions combine with cupric ions to form insoluble salts that are retained in the resin.[17] Alternatively, the ferrocyanide ions may be held on an anion-exchange resin, which is then used as a selective filter to collect traces of Cu, Zn, and Cd from water, each of which forms an insoluble ferrocyanide.[18] In both of these operations macroporous resins are more effective than microporous. A resin loaded with reducing ions, Fe(II) and Sn(II) as cations or SO_3^{2-} as an anion, collects reducible ions like MnO_4^-, CrO_4^{2-}, and VO_3^-.[19] Quantities of ions in the sub-parts per billion range can be recovered.

As reactive ion exchange we can include the concentration of traces of amino acids from sea water on a column of chelating ion-exchange resin carrying copper(II) ions by Siegal and Degens in 1966.[20] Amino acids were recovered from waste water on a nickel-loaded chelating resin.[21] In both cases the acids were stripped from the resin by passing a concentrated ammonia solution. The mechanism was metal-ligand coordination and ligand exchange (see Chapter 6).

IV. IMINODIACETATE CHELATING RESINS

The standard chelating resin which is sold under the names Chelex®-100 and Dowex®-

1 is crosslinked polystyrene with functional iminodiacetate groups, $-CH_2N(CH_2COOH)_2$. It has roughly the same affinities for metal ions that EDTA has, i.e., singly charged alkali-metal cations are held most weakly, then the divalent alkaline-earth cations, then divalent and trivalent transition-metal and heavy-metal ions. The ions of mercury(II), uranium(VI), and copper(II) are held most strongly; the equilibrium constant for the exchange reaction, $Cu^{2+} + CaRes = Ca^{2+} + CuRes$ is about 500.[22] Thus, heavy metal ions are adsorbed strongly when natural waters, including sea water, are passed through columns of this resin. The iminodiacetate group is weakly acidic, and therefore the hydrogen-form resin swells in water only weakly and it exchanges ions slowly. To use the resin in a column for removing trace heavy-metal ions from water the resin should first be converted to its sodium, ammonium, or calcium form. The resin swells considerably when it is converted from the hydrogen form or a heavy-metal form to the fully ionized sodium form, a fact that must be kept in mind when using and regenerating the resin in columns. The large volume changes that occur when one ionic form replaces another prevent this chelating resin from being used in high-performance chromatography, but it is used a great deal for the collection of trace metals.

The standard procedure is to pack a small glass or plastic column with the resin and pass a large volume of water, preferably adjusted to pH 5 or higher to keep the resin ionized and swollen. (At higher pH values the hydrolysis of doubly charged heavy-metal ions becomes significant; pH 5.5 is considered optimum). The adsorbed metal ions are stripped with a strong acid, say 2 *M* nitric acid. Volume enrichment factors of 20 to 100 are easily achieved. The recovered metal ions may be determined by any method of choice, such as EDTA titration or atomic absorption spectrometry (flame or electrothermal) or inductively coupled plasma emission spectrometry. Instead of stripping the metals from the column, one may remove the resin, dry it, mix it with cellulose as a binder, and compress it to a pellet that is the target for X-ray fluorescence. The pellet is coherent and easily handled.[23] The resin may also be analyzed by neutron activation.[24]

V. TRACE METALS IN SEA WATER

A stringent test of trace recovery is its success with sea water. Sea water is 0.46 *M* in sodium ions, 0.05 M in magnesium ions, and 0.01 *M* in calcium, and it contains organic matter, mainly humic and fulvic materials, that can bind metals in chelated forms; some of the organic matter is colloidal and some is particulate. Zinc, copper, lead, and cadmium are present in the low parts-per-billion (μg/l) range and below. A typical analysis is shown in Table 1. Many investigators have used ion exchange to recover and preconcentrate heavy metals in sea water, and most of them have used Chelex®-100. An early study was made by Riley and Taylor in 1968.[25] Their column was 2 × 10 cm; through it they passed ten l of sea water. They eluted the adsorbed heavy metals by passing 30 ml 2 *M* nitric acid, and measured the concentrations by flame atomic absorption.

In any method of trace analysis it is important to find the degree of recovery. One way to do this is to add to the sample a known, small amount of an isotopic tracer, a radioactive isotope if one is available, otherwise a stable isotope that can be measured by mass spectrometry. Another way is to add a measured "spike" or quantity of the element sought and to see if this quantity can be completely recovered, by measuring the concentration of the element before and after adding the spike. Yet another way is to use an analytical method that is so sensitive that it can be used on the original material without preconcentration. Such a method is differential-pulse anodic stripping voltammetry. Another very sensitive method is graphite-furnace atomic absorption spectrometry. Yet another is inductively coupled plasma-mass spectrometry. All these methods have been used to check recoveries of metals from sea water and to verify the reliability of various preconcentration methods.

TABLE 1
Trace Metals in Sea Water

Element	In costal water Sample 1	In costal water Sample 2	In open ocean
Cd	0.024 ± 0.004	0.023 ± 0.001	0.033 ± 0.002
Pb	0.22 ± 0.06	0.018 ± 0.001	0.09 ± 0.01
Zn	0.41 ± 0.05	0.28 ± 0.01	0.30 ± 0.03
Cu	0.96 ± 0.04	0.22 ± 0.02	0.11 ± 0.02
Fe	1.03 ± 0.04	6.9 ± 0.02	0.18 ± 0.04
Mn	0.68 ± 0.05	1.13 ± 0.02	0.023 ± 0.003
Ni	0.31 ± 0.04	0.34 ± 0.01	0.27 ± 0.02
Co	0.015 ± 0.007	not det.	not det.

Note: Micrograms per liter with standard deviations shown.

Abstracted from Reference 30, Tables 3 and 4.

Using anodic stripping voltammetry, Florence and Batley[26] found that a significant fraction of Cu, Pb, Zn, and Cd was not removed by the chelating resin, nor was it extracted with ammonium pyrrolidine dithiocarbamate (APDC). To find the total metal they heated the original sea water for some time at pH below 2. This treatment liberated the metal ions from their complexes with organic matter. Other workers[27,28] showed that there were labile complexes, from which the metal could be removed by chelating ion exchange and by electrolysis at a mercury drop cathode (the first step in anodic stripping voltammetry), and nonlabile or inert complexes from which the metals could be removed, but slowly. Slow passage of the sample through the resin bed is emphasized.[29] It is customary to preserve sea water samples for later analysis by making them 0.025 *M* in nitric acid before storage;[30] this treatment would decompose most metal-organic complexes. The nitric acid is neutralized with ammonia before analysis.

The role of organic material in natural waters should always be kept in mind. Most of the organic material is humic and fulvic acids, that are well known to form complexes with metal ions.[31] One way to study these complexes and find their stability is to use ion exchange. In the early days of ion exchange, Schubert and Fronaeus measured formation constants of strontium-citrate and copper(II)-acetate complexes by bringing solutions into contact with cation-exchange resins and measuring the distribution of the metal ions, assuming for a start that only the free (or hydrated) metal ions exchanged with the resin while the uncharged or anionic complexes did not. Recent studies along these lines have been made by Cantwell and associates[32] with a view to clarifying metal-organic ligand association in natural waters and waste waters. They take a very small column of strong-acid cation-exchange resin (Dowex®-50W × 8) and pass enough of the water in question to give complete breakthrough and bring the solution and resin into equilibrium, first adding to the water a large, measured excess of an inert electrolyte like sodium nitrate. After equilibrium is reached the cations are stripped from the column and the amounts of the metal ions measured by atomic absorption. The studies include fulvic and humic ligands as well as citrate, glycinate, phthalate, and others. The authors are able to measure the concentrations of "free" copper and nickel ions.

Graphite-furnace (or "electrothermal") atomic absorption spectrometry, which is extremely sensitive, has been used to compare concentrations in the original sea water with those found by preconcentrating by ion exchange,[33,34] and these studies, too, have shown that recovery by ion exchange is essentially complete, more nearly complete than by extraction with APDC. Chromium was poorly recovered, perhaps because it was present in its anionic forms, perhaps because Cr(III) complexes are well known to be kinetically inert.

A standard procedure for concentrating trace heavy-metal ions in sea water is due to Kingston et al. and is as follows:

> Chelex®-100, 200/400 mesh, is used in the ammonium form in a bed of volume 3 to 4 ml. It is washed before use by passing two 5-ml portions of 5 *M* nitric acid, then one 5-ml portion of 4 *M* HCl, then water, 2 *M* ammonia, and again water. The sea water is brought to pH 5.4 with ammonia if it is strongly acid, otherwise with ammonium acetate. The desired volume of sea water (100 ml or more) is passed through the resin at 0.8 ml/min. After the sea water has passed the column is rinsed with water, then four 10-ml portions of 1 *M* ammonium acetate at pH 5.2, then water again. This treatment removes most of the retained sodium, calcium, and magnesium ions from the sea water. Now the trace metals are eluted with two 5-ml portions of 2.5 *M* nitric acid. The resin bed is washed and regenerated with ammonia, after which blank experiments are run.

Later experimenters,[35] who used graphite-furnace atomic absorption spectrometry, found that it took much more nitric acid to remove the heavy metals than Kingston had reported, and developed a combination of batch shaking and column elution to adsorb and desorb the heavy metals. Another minor improvement over the technique of Kingston is to start with the chelating resin in its calcium form. At the start of passing sea water the ammonium-form resin shrinks because divalent ions replace ammonium, and the principal divalent ion that enters the resin is the calcium ion. It is true that magnesium is five times as abundant as calcium in sea water, but the iminodiacetate group binds calcium ions a great deal more strongly than it does magnesium ions.

Inductively coupled plasma emission spectrometry, which is not as sensitive as graphite-furnace atomic absorption but is much faster and better adapted to multi-element analysis, has been used for trace elements in sea water following preconcentration on Chelex®-100. Sensitivity was improved by using ultrasonic nebulization; however, this extra complication may not have been worth while. Results by this technique agreed well with those obtained by graphite-furnace atomic absorption.[36]

Smits et al.[37] compared the efficacy of eight procedures for the recovery of six trace metals, Ba, Mn, Co, Zn, Eu, and Cs from natural waters (including sea water and estuarine water) as well as solutions made from distilled water. One procedure used Chelex®-100 (rather, Dowex®A-1, which is chemically similar); another used a chelating cellulose-based exchanger; another used bonded silica with dithiocarbamate groups. Recoveries of Co and Zn from sea water by ion exchange (chelating exchangers) were 85%. Again the importance of enough contact time was emphasized.[37]

In Chapter 2 we described the synthesis of chelating ion exchangers made from polystyrene and from porous glass and silica. A silica-bonded 8-hydroxyquinoline was prepared expressly for the recovery of trace metals by the following series of reactions:[38]

$$\text{Silica} + (\text{EtO})_3\text{SiC}_3\text{H}_6\text{NH}_2 = \text{Si–O–SiC}_3\text{H}_6\text{NH}_2$$

$$\text{Plus ClC}_6\text{H}_4\text{NO}_2\text{=Si–O–SiC}_3\text{H}_6\text{NHC}_6\text{H}_4\text{NO}_2$$

$$\text{Reduced} = \text{Si–O–SiC}_3\text{H}_6\text{NHC}_6\text{H}_4\text{NH}_2$$

$$\text{Diazotized} = \text{Si–O–SiC}_3\text{H}_6\text{NHC}_6\text{H}_4\text{N}_2^+$$

$$\text{Plus oxine in ethanol} = \text{Si–O–SiC}_3\text{H}_6\text{NHC}_6\text{H}_4\text{N:N-oxine}$$

where "oxine" is 8-hydroxyquinoline, bound thus:

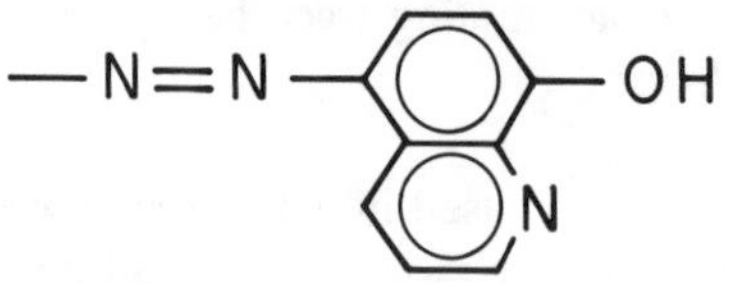

—N=N—(8-hydroxyquinoline structure: OH, N)

The same series of reactions is used to attach 8-hydroxyquinoline to macroporous cross-linked polystyrene, except that one starts by nitrating the polystyrene to attach the $-NO_2$ group.[39] The ion-exchange capacities of these products are rather low, 0.06 mmol/g for the porous silica product, but the oxine group gives a much higher selectivity for heavy metals than does the iminodiacetate group, and the columns can be correspondingly smaller. Typical columns contain 0.6 g of polymer or bonded silica.[40]

Advantages claimed for silica-bonded oxine over iminodiacetate chelating resin are that the columns permit faster flow and that the greater selectivity makes it unnecessary to wash out the ions of sodium, calcium, and magnesium before eluting the heavy metals. The recoveries obtained with silica-bonded and polymer-bonded oxine exchangers agree well with those obtained by other methods. A problem that has arisen with silica-based exchangers is the presence of trace heavy-metal impurities in the silica.

A way to attach 8-hydroxyquinoline or any other functional group to an adsorbent is by way of a long hydrocarbon chain that carries the functional group. Isshiki et al.[41] impregnated the macroporous polystyrene resin Amberlite XAD-4 with the compound 7-dodecenyl-8-hydroxyquinoline and used a 500-mg column of this impregnated or permanently-coated resin (see Chapter 5) to recover several trace metals from sea water. Earlier, Parrish[42] had used the same compound to impregnate Amberlite XAD-7, an acrylic polymer. Obviously this long-chain substituted hydroxyquinoline could have been loaded on to a column of C-18 bonded silica, and another functional group could have been used instead of 8-hydroxyquinoline.

A novel use of 8-hydroxyquinoline, covalently bound to silica through the $-CH_2CH_2CH_2-$ group, was described in Chapter 2, Reference 55. Adsorption of Al(III) on this material causes fluorescence, as does the combination of Al(III) with 8-hydroxyquinoline in solution. Silica is transparent to UV light, and there is no organic group in the bonded material to absorb UV except for 8-hydroxyquinoline itself. Excitation is at 365 nm, emission at 520 nm; quantities of aluminum in the low nanogram range can be detected. Presumably other metals can be detected and measured also.

In summary, preconcentration on a chelating resin or a chelating bonded-silica exchanger is a reliable way to extract heavy metals from sea water. Not only does the extraction step increase sensitivity by reducing the volume of water that contains the trace metals, but it separates the trace metals from the large amounts of sodium chloride and other salts that would interfere with measurement by inductively coupled plasma emission spectrometry, ICP-mass spectrometry, or graphite-furnace atomic absorption.

VI. OTHER CHELATING EXCHANGERS

A host of chelating functional groups have been attached to bonded silica and to polymers by chemical bonds and the products used to collect traces of metal ions from water. Among the functional groups are arsonic acid,[43] dithiocarbamate,[44] imineisocyanate,[45] and arsenazo.[46] An important chelating polymer that has been made commercially is "Srafion NMRR", with the isothiouronium group $= S(NH)NH_2$; this group is selective for gold and the platinum metals. The polymer need not be polystyrene; hydrophilic vinyl polymers (Fractogel® TSK)[47] and acrylic polymers (Spherons®)[48] have been used. Spheron® Thiol, Spheron® Oxine, and Spheron® ACDA have been made, where ACDA, 2-amino-1-cyclopentene-1-dithiocarboxylic acid, is the group

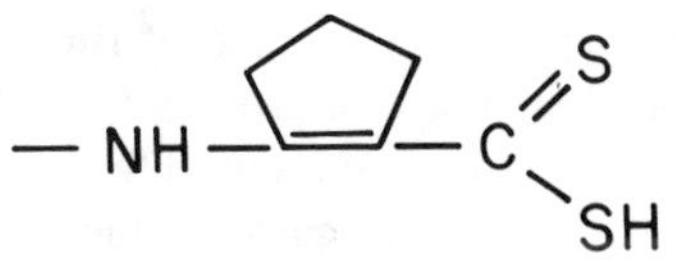

S
— NH — — C
SH

Preparation of ACDA-bonded silica is described by Seshardri, and Kettrup,[49] who also gives it selectivity order for metal ions, which is: Ag strongest, then Hg, Pd, Pt, Cu, Cd, Fe, and Zn. An interesting point is that one can start with silica gel-impregnated filter paper and attach ACDA to it, thus making a chelating paper that will filter out heavy metals in the way described below.

The metal-loaded Spheron® polymers were used to adsorb herbicides from water by ligand exchange. The polyethyleneimime-polyphenylene isocyanate resin mentioned above recovered Cu, Cd, Pb, and Zn from Dead Sea brine, surely a test of selectivity. Another advantage to some of these materials is the very small volume changes when one ion is substituted for another; as we have noted, Chelex®-100 shows very large volume changes, which lead to trouble even in small open columns.

A simple way to use selective organic ligands, which is getting away from ion exchange, is to add a solution of the organic reagent to the sea water or water being tested, with due regard for pH, and then pour the water through a small column of non-ionic macroporous resin, such as XAD-4 or XAD-7. The metal-ligand complex, which is uncharged, is held on the column. The column is rinsed with water and then eluted with a solvent such as a 1:1 chloroform-methanol mixture. This extracts the chelated metal-ligand complex. The solvent is evaporated and the complex decomposed by heating with nitric and perchloric acids, the remaining metal salt is dissolved in dilute nitric acid, and the metal is determined by a method of choice, like graphite-furnace atomic absorption spectrometry. Ishiki and Nakayama[50] used 14 ligands, including APDC, PAR, dithizone, 2-nitroso-1-naphthol, and dibenzoylmethane, and two XAD resins, to recover cobalt from sea water. With the same general technique Fritz et al.[51] used the water-soluble reagent ammonium *bis*-(carboxymethyl)dithiocarbamate and the resin XAD-4 to concentrate trace metals for measurement by inductively coupled plasma-mass spectrometry.

General reviews of trace metal collection, that include chelating ion exchange and immobilized chelating ligands as well as other preconcentration methods, are given by Leyden and Wegscheider[52] and Nickless.[53]

VII. BATCH METHODS

If the distribution coefficients are high enough, trace metals can be collected by adding the adsorbent to the test solution and shaking for an hour or so, then separating the resin. Several early papers describe trace concentration by this means. An advantage of the batch method over the column method is that it is faster, and another advantage is that it uses less resin. Chelex®-100 is generally used.[54] The order of affinities of different cations for Chelex®-100 resin is described in the reference just cited. Over the pH range 1 to 3 the strongest adsorption is shown by bismuth (90% extracted by pH 1), followed by Cu, Fe(III), Pb, Ni, Cd, Zn, Co, Ba, Sr, and Fe(II). All these ions except Fe(II) are 90% adsorbed, or better, below pH 3.

A more recent paper[55] described the extraction of several elements from sea water by shaking 0.5 g of Chelex®-100 with 1 l of sea water adjusted to pH 6. After 3 h of shaking the resin is removed and extracted with acid. Concentration factors of 100 were obtained, and detection limits ranged from 6 to 180 ng/l. The mode of measurement was inductively coupled plasma emission spectrometry.

The batch method was used to collect manganese from sea water.[56] What could be classed as a batch method, perhaps, is the method of filtering the solution through circles of resin-impregnated paper, or filters made of ground-up chelating resin in a porous matrix. This method is attractive if X-ray fluorescence is used for measuring the metals.[57] The silica-bonded ACDA, described above, was fabricated in the form of filter discs and used by Kettrup to filter out trace metal ions preparatory to measurement by X-ray fluorescence. Of course, recovery is less than 100%. Allowance can be made for incomplete recovery by using isotope dilution. For example, a strong-base anion-exchange resin carrying the anion of 8-hydroxyquinoline-5-sulfonic acid was added in moderate excess to a solution containing zinc. At the same time, the radioactive tracer Zn-65 was added in a known amount. The solution and resin were shaken together for 2 h, then the resin was separated and its activity measured. The process was repeated after adding a known amount of nonradioactive zinc to another portion of the solution to be analyzed. The more zinc there is in the test solution, the more the radioactive zinc is diluted. Zinc in 100 ml of river water was found to be 0.12 ±0.02 ppm.[58]

Taking advantage of the great sensitivity of graphite-furnace atomic absorption, Koide et al. have devised batch techniques that use single resin beads. A bead of strong-base anion-exchange resin, 0.5 mm diameter and 4% crosslinked, is shaken for a few hours with 0.5 to 1 ml of sea water plus a complexing agent, cyanide or thiocyanate, without preconcentration. Tests with radioactive tracers showed that recovery of most trace metals is 95% or better. The individual resin-loaded beads are placed directly in the graphite furnace for atomic absorption measurements.[59] Copper was determined in water by these authors[60] using several beads of cation-exchange resin, 0.8 mm diameter, with 50 ml water. The resin and water sample were shaken for a few minutes, then the resin was removed and dried, and individual resin beads were placed in the atomic-absorption furnace. Two nanograms per liter could be detected. Technetium was concentrated into single beads for mass spectrometric determination. The bead was placed on the filament of the ionization chamber.[4]

To monitor trace metals in polluted river water Lochmüller et al.[61] suspended strips of cation-exchange membrane in the river, and when the membranes had reached a steady state, he took them out and measured the heavy metals by proton-induced X-ray emission. Tests showed that the X-ray intensity was related reproducibly to the concentrations of lead, cadmium, and other metals in the river water and was, of course, affected inversely by the calcium-ion concentration. (Magnesium and sodium ions had an effect too, by simple ion exchange, but these effects were smaller.)

Other investigators[62] mounted a disc of polyacrylate ion-exchanger membrane on the end of a rod which was rotated in contact with the solution to be analyzed. The heavy metals entered the membrane and formed a layer only a nanometer thick. This surface layer was examined by X-ray photoelectron spectroscopy. Metal ions in the parts-per-million range could be detected and measured.

Another use of ion-exchange membranes in trace analysis is by Donnan dialysis.[63] A disc of cation-exchange resin membrane is mounted at the lower end of a vertical glass tube some 2 cm in diameter. Inside the tube is placed 0.1 *M* lithium chloride. The disc and the end of the tube are immersed in the solution under test, which is much more dilute that 0.1 *M*. Trace cations migrate through the membrane into the inner solution and are replaced by an equivalent amount of lithium ions that go out into the outer (test) solution. Suppose the trace cation is singly charged, M^+; then, when equilibrium is reached, ratio (M:Li) inside = ratio (M:Li) outside.

The result is that the cation M becomes more concentrated inside the membrane than outside. The practical utility of this process is diminished by fact that the analytes sought are mixed with a large concentration of an unwanted salt.

VIII. ION-EXCHANGE RESIN SPOT TESTS

Spot tests are qualitative tests for elements and compounds that are made by mixing a drop or two of the test solution with a selective reagent that produces a colored product. They are made more sensitive by making the reaction occur within a bead of ion-exhange resin. The resin must, of course, be transparent and have little or no color of its own.

A number of these tests have been devised, largely by Fujimoto and Nakatsukasa[64] and followers. A typical test is to place one or two beads of cation-exchange resin in a drop of solution on a white porcelain spot plate (glazed porcelain having round depressions about 1 cm diameter and a few mm deep) and add a small amount of 1,10-phenanthroline, along with hydroxylamine as a reducing agent. If iron is present the red iron(II)-phenanthroline complex is formed, and being a cation, and what is more having an aromatic structure, it is absorbed and concentrated into the resin bead, which may be observed with a low-power microscope. The resin should have low crosslinking to allow entry of the reagent. Recently macroporous resins have served well. The test is much more sensitive than simply mixing the solutions with no resin bead. Zinc is detected by placing a bead of Dowex®-1 × 1 (1% crosslinked strong-base anion exchanger) in a drop of test solution at pH 9, then adding the reagent Zincon. In a few minutes a blue color develops in the resin. The detection limit is 4 ng.[65] A test for copper uses a bead of anion-exchange resin that is loaded with the reagent bathocuproine disulfonate (sulfonate of 4,7-diphenyl-1,10-phenanthroline); at pH 4, copper(I) forms an orange-red complex. A general test for several transition metals uses PAR, 4-(2-pyridylazo)resorcinol and a cation-exchange resin.[66] Detection limits are around 1 ng. Tests have been devised for silver, gold, and uranium.[67-69] It is possible to make these tests semi-quantitative by using, not a single bead, but a small bed of resin beads in a glass column and measuring the light absorbance.

REFERENCES

1. **Glueckauf, E.,** Theory of chromatography, IX. Theoretical plate concept in column separations, *Trans. Faraday Soc.*, 51, 34, 1955.
2. **Danielsson, A., Rönnholm, B., Kjellström, L. E., and Ingman, F.,** Fluorimetric method for the determination of uranium in natural waters, *Talanta*, 20, 185, 1973.
3. **Warburton, J. A. and Young, L. G.,** Determination of silver in precipitation down to 10^{-11} M concentrations by ion exchange and neutron activation analysis, *Anal. Chem.*, 44, 2043, 1972.
4. **Anderson, T. J. and Walker, R. L.,** Determination of picogram quantities of technetium-99 by resin bead mass spectrometric isotope dilution, *Anal. Chem.*, 52, 709, 1980.
5. **Korkisch, J., Sorio, A., and Steffan, I.,** Atomic-absorption determination of beryllium in liquid environmental samples, *Talanta*, 23, 289, 1986.
6. **Korkisch, J. and Sorio, A.,** The determination of beryllium in geological and industrial materials by atomic absorption spectrometry after cation-exchange separation, *Anal. Chim. Acta*, 82, 311, 1976.
7. **Korkisch, J. and Sorio, A.,** Determination of 7 trace elements in natural waters after separation by solvent extraction and anion-exchange chromatography, *Anal. Chim. Acta*, 79, 207, 1975.
8. **Korkisch, J. and Sorio, A.,** Anwendung von Ionenaustauschverfahren zur Bestimmung von Spurenelementen in natürlichen Wässern. V. Blei, *Talanta*, 22, 273, 1975.
9. **Chao, T. T., Fishman, M. J., and Ball, J. W.,** Determination of traces of silver in waters by anion exhange and atomic absorption spectrophotometry, *Anal. Chim. Acta*, 47, 189, 1969.
10. **Hiiro, K., Kawahara, A., and Tanaka, T.,** Ion-exchange enrichment method for the parts-per-billion level of cadmium in sea water, *Bunseki Kagaku*, 22, 1210, 1973.
11. **Kiriyama, T. and Kuroda, R.,** Anion-exchange enrichment and spectrophotometric determination of traces of gallium in natural waters, *Fresenius Z. Anal. Chem.*, 332, 338, 1988.
12. **Kokubu, N., Hayasida, Y., Kobayasi, T., and Yamasaki, A.,** Determination of fluoride content in various sodium chloride samples by zirconium-loaded cation exchange resin, *Chem. Abstr.*, 94, 149586u, 1981.

13. **Okabayashi, Y., Oh, R., Nakagawa, T., Tanaka, H., and Chikuma, M.,** Selective collection of fluoride ions on an anion-exchange resin loaded with alizarin fluorine blue sulfonate, *Analyst,* 113, 829, 1988.
14. **Akaiwa, H., Kawamoto, H., Ogura, K., and Kogure, S.,** 8-Quinolinol-5-sulfonic acid-loaded resin as a preconcentrating agent in the neutron activation analysis of the chalcophile elements, *Radioisotopes,* 28, 681, 1979.
15. **Akaiwa, H., Kawamoto, H., and Nakata, N.,** Neutron activation analysis of chalcophile elements using a chelating agent-loaded resin, *J. Radioanal. Chem.,* 36, 59, 1977.
16. **Pankow, J. F. and Janauer, G. E.,** Analysis of chromium traces in natural water, *Anal. Chim. Acta,* 69, 97, 1974.
17. **Ramseyer, G. O. and Janauer, G. E.,** Selective separations by reactive ion exchange. II. Preconcentration and determination of complex iron cyanides in water, *Anal. Chim. Acta,* 77, 133, 1975.
18. **De Layette-Mills, J., Karm, L., Janauer, G. E., Chan, P. K., and Bernier, W. E.,** Selective separations by reactive ion exchange. IV. Preconcentration of cadmium and zinc, *Anal. Chim. Acta,* 124, 365, 1981.
19. **Lin, J. W. and Janauer, G. E.,** Selective separations by reactive ion exchange. III. Preconcentration and separation of oxo anions, *Anal. Chim. Acta,* 79, 219, 1975.
20. **Siegel, A. and Degens, E. T.,** Concentration of dissolved amino acids from saline waters by ligand exchange chromatography, *Science,* 151, 1098, 1966.
21. **Hemmasi, B.,** Ligand exchange chromatography of amino acids on nickel-Chelex 100, *J. Chromatogr.,* 87, 513, 1973.
22. **Rosset, R.,** in *Ion Exchange in Anaytical Chemistry,* Rieman, W. and Walton, H. F., Eds., Pergamon Press, Oxford, 1970.
23. **Leyden, D. E., Patterson, T. A., and Alberts, J. J.,** Preconcentration and X-ray fluorescence determination of copper, nickel and zinc in sea water, *Anal. Chem.,* 47, 733, 1975.
24. **Nadkarni, R. A. and Morrison, G. H.,** Determination of the noble metals in geological meterials by neutron activation analysis, *Anal. Chem.,* 46, 232, 1974.
25. **Riley, J. P. and Taylor, D.,** Chelating resins for the concentration of trace elements from sea water, *Anal. Chim. Acta,* 40, 479, 1968.
26. **Florence, T. M. and Batley, G. E.,** Trace metal species in sea water. I. Removal of trace metals from sea water by a chelating resin, *Talanta,* 23, 179, 1976.
27. **Abdullah, M. I., El-Rayis, O. A., and Riley, J. P.,** Reassessment of chelating ion-exchange resins for trace metal analysis of sea water, *Anal. Chim. Acta,* 84, 363, 1976.
28. **Figura, P. and McDuffie, B.,** Characterization of the calcium form of Chelex-100 for trace metal studies, *Anal. Chem.,* 49, 1950, 1977.
29. **Figura, P. and McDuffie, B.,** Use of Chelex resin for determination of labile trace metals in aqueous ligand media and comparison of the method with anodic stripping voltammetry, *Anal. Chem.,* 51, 120, 1979.
30. **Sturgeon, R. E., Berman, S. S., Willie, S. N., and Desaulniers, J. A. H.,** Preconcentration of trace elements from sea water with silica-immobilized 8-hydroxyquinoline, *Anal. Chem.,* 53, 2337, 1981.
31. **Aike, G. R., McKnoght, D. W., Wershaw, R. L., and MacCarthy, P., Eds.,** *Humic Substances in Soil, Sediment and Water,* Wiley-Interscience, New York, 1985.
32. **Sweileh, J. A., Lucyk, D., Kratochvil, B., and Cantwell, F. F.,** Specificity of the ion-exchange atomic-absorption method for free Cu(II) specues determination in natural water, *Anal. Chem.,* 59, 586, 1987.
33. **Sturgeon, R. E., Berman, S. S., Desaulniers, J. S. H., Mykytiuk, A. P., McLaren, J. W., and Russell, D. S.,** Comparison of methods for the determination of trace elements in seawater, *Anal. Chem.,* 52, 1585, 1980.
34. **Kingston, H. M., Barnes, I. L., Brady, T. J., Rains, T. C., and Champ, A. M.,** Separation of eight transition elements from alkali and alkaline earth elements in estuarine and sea water with chelating resin and their determination by atomic absorption spectrometry, *Anal. Chem.,* 50, 2064, 1978.
35. **Sturgeon, R. E., Berman, S. S., Desaulniers, A., and Russell, D. S.,** Preconcentration of trace metals from sea water for determination by graphite furnace atomic absorption spectrometry, *Talanta,* 27, 85, 1980.
36. **Berman, S. S., McLaren, J. W., and Willie, S. N.,** Simultaneous determination of 5 trace metals in seawater by inductively coupled plasma-atomic emission spectrometry with ultrasonic nebulization, *Anal. Chem.,* 52, 488, 1980.
37. **Smits, J., Nelissen, J., and Van Grieken, R.,** Comparison of preconcentration procedures for trace metals in natural waters, *Anal. Chim. Acta,* 111, 215, 1979.
38. **Hill, J. M.,** Silica gel as an insoluble carrier for the preparation of selective chromatographic adsorbents, *J. Chromatogr.,* 76, 455, 1973.
39. **Willie, S. N., Sturgeon, R. E., and Berman, S. S.,** Comparison of 8-quinolinol-bonded polymer supports for the preconcentration of trace metals from sea water, *Anal. Chim. Acta,* 149, 59, 1983.

40. **McLaren, J. W., Mykytiuk, A. P., Willie, S. N., and Berman, S. S.,** Determination of trace metals in seawater by inductively coupled plasma-mass spectrometry with preconcentration on silica-immobilized 8-hydroxy-quinoline, *Anal. Chem.*, 57, 2907, 1985.
41. **Isshiki, K., Tsuji, F., Kuwamoto, T., and Nakayama, E.,** Preconcentraion of trace metals from seawater with 7-dodecenyl-8-quinolinol impregnated macroporous resin, *Anal. Chem.*, 59, 2491, 1987.
42. **Parrish, J. R.,** Macroporous resins as supports for a chelation liquid ion exchanger in extraction chromatography, *Anal. Chem.*, 49, 1189, 1977.
43. **Fritz, J. S. and Moyers, E. M.,** Concentration and separation of trace metals with an arsonic acid resin, *Talanta,* 23, 590, 1976.
44. **Barnes, R. M. and Genna, J. S.,** Concentration and spectrochemical determination of trace metals in urine with a poly(dithiocarbamate) resin and inductively coupled plasma-atomic emission spectrometry, *Anal. Chem.*, 51, 1065, 1979.
45. **Horvath, Zs. and Barnes, R. M.,** Carboxymethylated polyethyleneimine-polyphenylene isocyanate chelating ion-exchange resin preconcentration for inductively coupled plasma spectrometry, *Anal. Chem.*, 58, 1352, 1986.
46. **Chwastowska, J.,** Synthesis and analytical properties of an arsenazo I resin, *Chem. Anal. (Warsaw),* 32, 643, 1987.
47. **Landing, W. M., Haraldsson, C., and Paxeus, N.,** Vinyl polymer agglomerate based transition metal cation-chelating ion-exchange resin containing the 8-hydroxyquinoline functional group, *Anal. Chem.*, 58, 3031, 1986.
48. **Nielen, M. W. F., van Ingen, H. E., Valk, A. J., Frei, R. W., and Brinkmann, U. A. Th.,** Metal-loaded sorbents for selective on-line sample handling and trace enrichment in liquid chromatography, *J. Liquid Chromatogr.*, 10, 617, 1987.
49. **Seshadri, T. and Kettrup, A.,** Synthesis and characterization of a silica-gel ion-exchanger bearing 2-amino-1-cyclopentene-1-dithiocarboxylic acid (ACDA) as chelating agent, *Fresenius Z. Anal. Chem.*, 310, 1, 1982.
50. **Isshiki, K. and Nakayama, E.,** Selective concentration of cobalt in sea water by complexation and sorption on macroporous resins, *Anal. Chem.*, 59, 291, 1987.
51. **Plantz, M. R., Fritz, J. S., Smith, F. G., and Houk, R. S.,** Separation of trace metal complexes for analysis of samples of high salt content by inductively coupled plasma-mass spectrometry, *Anal. Chem.*, 61, 149, 1989.
52. **Leyden, D. E. and Wegscheider, W.,** Preconcentration for trace element determination in aqueous samples, *Anal. Chem.*, 53, 1059A, 1981.
53. **Nickless, G.,** Trace metal determination by chromatography, *J. Chromatogr. Chromatogr. Rev.*, 313, 129, 1985.
54. **Blount, C.W., Leyden, D. E., Thomas, T. L., and Guill, S. M.,** Application of chelating ion-exchange resins for trace element analysis of geological samples using X-ray fluorescence, *Anal. Chem.*, 45, 1045, 1973.
55. **Cheng, C. J., Akgi, T., and Haraguchi, H.,** Simultaneous multi-element analysis for trace metals in sea water by inductively coupled plasma-atomic emission spectrometry after batch preconcentration on a chelating resin, *Anal. Chim. Acta,* 198, 173, 1987.
56. **Smith, R. G.,** Improved ion-exchange technique for the concentration of manganese from sea water, *Anal. Chem.*, 46, 607, 1974.
57. **Van Grieken, R. E., Bresseleers, C. M., and Vanderborght, B. M.,** Chelex-100 ion-exchange filter membranes for preconcentration in X-ray fluorescence analysis of water, *Anal. Chem.*, 49, 1326, 1977.
58. **Akaiwa, H., Kawamoto, H., and Ogura, K.,** Substoichiometric determination of zinc by the isotope-dilution method using a chelating agent-loaded resin, *Talanta,* 24, 394, 1977.
59. **Koide, M., Lee, D. S., and Stallard, M. O.,** Concentration and separation of trace metals from seawater using a single anion exchange bead, *Anal. Chem.*, 56, 1956, 1984.
60. **Takada, T. and Koide, T.,** Atomic absorption spectrometric determination of trace copper in water by sorption on an ion-exchange resin and direct atomization of the resin, *Anal. Chim. Acta,* 198, 303, 1987.
61. **Lochmüller, C. H., Galbraith, J. W., and Walter, R. L.,** Trace metal analysis in water by proton-induced X-ray emission analysis of ion-exchange membranes, *Anal. Chem.*, 46, 440, 1974.
62. **Czuha, M. and Riggs, W. M.,** X-ray photoelectron spectroscopy for trace metals determination by ion-exchange absorption from solution, *Anal. Chem.*, 47, 1836, 1975.
63. **Cox, J. A. and Di Nunzio, J. E.,** Donnan dialysis of cations, *Anal. Chem.*, 49, 1272, 1977.
64. **Fujimoto, M. and Nakatsukasa, Y.,** Microanalysis with the aid of ion-exchange resins. Determination of nanogram amounts of iron with ''Ferron'', *Anal. Chim. Acta,* 26, 427, 1962.
65. **Fujimoto, M. and Nakatsukasa, Y.,** Mikroanayse mit Hilfe von Ionenaustauschern. Nachweis von Nanogrammen Zn(II), *Mikrochim. Acta,* 551, 1968.
66. **Fujimoto, M. and Iwamoto, T.,** Mikroanalyse mit Hilfe von Ionenaustauschern. Die Harztüpfelmethode zur Nachweis von Nanogrammen der Schwermetalle mit PAN oder PAR, *Mikrochim. Acta,* 655, 1963.

67. **Grdinic, V. and Luterotti, S.,** Microdetection of silver with *P*-dimethylaminobenzilidine-rhodanine by the "Resin Spot Test" technique, *Fresenius Z. Anal. Chem.*, 308, 461, 1981.
68. **Christova, R. and Ivanova, M.,** A highly selective and sensitive method for identification of gold, *Mikrochim. Acta,* 349, 1976 II.
69. **Grdinic, V.,** Specific microdetection of uranium with 4-(2-pyridylazo)-resorcinol in ion-exchange resin beads, *Fresenius Z. Anal. Chem.*, 307, 205, 1981.

Chapter 10

MISCELLANEOUS ANALYTICAL USES OF ION EXCHANGE

In this final chapter we shall review some simple and common uses of ion exchange, and also some unusual applications. We shall start with methods that can be performed in small glass columns open to the air.

I. MEASUREMENT OF TOTAL ELECTROLYTE CONCENTRATION

In analyzing natural waters from lakes, rivers, and wells, and also domestic water supplies, anions and cations are generally determined separately, but it is sometimes helpful to know the total ionic concentration. The cations usually present are Na^+, Mg^{2+}, and Ca^{2+}. The anions are HCO_3^-,Cl^-, and SO_4^{2-}, plus silicate. Two separate portions of water are taken for analysis. To the first a drop or two of methyl red indicator is added, and the water is titrated with standard hydrochloric acid until the indicator starts to turn from yellow to organe. At this stage, most of the bicarbonate ions have been converted to carbonic acid, but not all. The solution is boiled for a minute to drive off the carbon dioxide. It is then cooled, and the titration with acid is resumed carefully until the methyl red indicator turns sharply from yellow to red. The total consumption of standard acid gives the bicarbonate concentration.

Interrupting the titration at the right point, when a little bicarbonate still remains, takes some practice, and the analyst may prefer to add an excess of acid, then boil out the carbon dioxide, cool, and back-titrate with standard base.

The second portion of the water to be analyzed is passed through a small glass column holding a bed of sulfonated polystyrene cation exchanger in the hydrogen form. Convenient dimensions are 1 cm wide and 15 cm long; a convenient particle size is 100 to 200 mesh. Smaller particles cause unduly slow flow. Needless to say, the resin should be well washed with distilled or deionized water before passing the sample to be analyzed. The sample is passed at about 10 ml/min and the column is rinsed with 2 to 3 bed volumes (bulk volumes, not void volumes) of pure water. The combined effluent, sample plus rinsings, is boiled to remove carbon dioxide, then cooled and titrated with standard sodium hydroxide, using phenolphthalein indicator. This second titration gives the total concentration of strong-acid anions, i.e., chloride and sulfate, plus nitrate if any is present. The first titration gives the bicarbonate-ion concentration, plus carbonate if any is present. The sum of the two titrations gives the total electrolyte concentration of the water, under the assumption that no weak-acid ions other than carbonate and bicarbonate are present. If the color change of phenolphthalein in the second titration is not sharp, it is a hint that weak acids are present. Waters from oil wells in certain geological formations have been known to contain borate.

Silicate ions may complicate this analysis. Dissolved silica is very common in natural waters. It is present in part as colloidal silica, but more usually as silicate anions, which are not simply $HSiO_3^-$, but chains of silicon and oxygen atoms such as $HO–Si(OH)_2–Si(OH)_2–O^-$ and longer chains.

A. REMOVAL OF INTERFERING IONS

This procedure was one of the very earliest applications of ion exchange to chemical analysis. It was used by Samuelson and was described in Chapter 1. An example of its use is the replacement of cations by hydrogen ions before titrating sulfate with barium perchlorate and an adsorption indicator. The method was introduced by Fritz and Freeland.[1] The point is that all metallic cations are adsorbed on precipitated barium sulfate to some extent, and cause errors in titrating sulfate ions with barium ions using an adsorption indicator. Passing the sample through a small bed of strong-acid cation-exchange resin in the hydrogen form replaces other cations by H^+ and avoids this error.

In this application and the last, it is important to use a small column of resin, one that is large enough to convert, say, two or three samples before it needs regeneration, but not excessively large. A mistake that is often made is to use a column that is too large. A large column requires long flow times and large volumes of rinse water, which dilute the analyte unduly.

B. DETERMINATION OF PHOSPHATE IN PHOSPHATE ROCK

This procedure, too, was mentioned in Chapter 1. Metal ions, particularly Fe^{3+}, interfere with the determination of phosphate. Cation exchange may be used to remove the metal ions and replace them with hydrogen ions. A sample of phosphate rock is treated with a small excess of nitric acid; the solution is kept hot for a while to coagulate silica, then it is diluted and filtered, and passed through a column of cation-exchange resin in the hydrogen form. The bed is rinsed and the rinsings added to the column effluent. Now the solution, which contains phosphoric and nitric acids and no cations other than hydrogen, is titrated potentiometrically with standard sodium hydroxide, using a glass electrode. The pH is plotted against the volume of base. The curve has two inflection points, one near pH 4 where H_3PO_4 is converted to $H_2PO_4^-$, the other near pH 9 where $H_2PO_4^-$ is converted to HPO_4^{2-}. The volume of base used between these two inflection points gives the quantity of phosphoric acid. Excess nitric acid is neutalized before the first inflection is reached. Obviously the amount of excess nitric acid should be no more than is necessary. Too much nitric acid tends to displace the metal ions from the ion-exchange resin, or in other words, prevents complete retention of metal ions by the resin. If done properly, this method for phosphate is highly accurate.

C. PREPARATION OF SOLUTIONS OF ACIDS AND BASES

Often an acid or a base is required in the laboratory and only a salt is available. The commonest example is a quaternary ammonium hydroxide. Solutions of quaternary ammonium hydroxides are used in ion-pair chromatography, described in Chapter 5; they are used as regenerants in membrane suppressors (Chapter 4); they are also used to titrate weak acids in nonaqueous solvents. They can easily be made by passing a solution of a salt (for example, tetramethylammonium bromide) through a column of a strong-base anion-exchange resin in its hydroxide form.

When the hydroxide ions in this column are exhausted the column must be regenerated by passing a solution of sodium or potassium hydroxide. At this point one is made painfully aware of the weak affinity of a strong-base anion exchanger for hydroxide ion, and its much stronger affinity for chloride and bromide. A large excess of the regenerant must be passed before these ions are displaced and the column is ready to be used again.

Methyl mercuric hydroxide, CH_3HgOH, is made from methyl mercuric halides by anion exchange. Long-chain acids like dodecyl sulfonic acid are made from their sodium salts by passing the solutions through a strong-acid cation exchanger in its hydrogen form.

D. DEIONIZED WATER

The water used in chemical laboratories and in manufacturing is generally deionized rather than distilled. To make it, tap water is passed through a bed of hydrogen-form cation-exchange resin, and then through a bed of hydroxide-form anion-exchange resin. In the first column the dissolved salts are converted to their acids; in the second column the acids are neutralized by the exchangeable hydroxide ions and their anions are adsorbed by the exchanger, leaving pure water. When salts start to break through the beds must be regenerated, the first with dilute sulfuric acid, the second with sodium hydroxide solutin. More efficient removal of dissolved salts (i.e., to a lower residual concentration) is obtained in a mixed bed, where the cation-exchange and anion-exchange resin beads are intimately mixed together. The problem arises when the mixed bed must be regenerated. For small-scale use in the laboratory it may be more economical to throw the exhausted mixed bed away. To regenerate the mixed resins they are first unmixed by backflushing; the quaternary base anion-exchange resins are less dense than the sulfonic acid cation-exchange resins (the difference depends on the crosslinking) and so float on top; the top and bottom halves of the bed are regenerated separately. After rinsing free of excess regenerants the resins are mixed by blowing air, and the mixed bed is ready to deionize more water.

More popular than the mixed bed is reverse osmosis. The raw water (water as it comes from the municipal supply) is forced against its osmotic pressure through a membrane that is impermeable to salts. The exact mechanism of the salt exclusion is not clear, but one factor is the Donnan equilibrium (see Chapter 3). It is only necessary to have fixed ionic groups in the pores of the membrane where water passes. The Donnan equilibrium excludes salt ions that have the same charge as the fixed ions, generally negative ions, but if ions of one charge are prevented from passing, ions of the other charge are prevented too, for electrical neutrality must be maintained.

E. SINGLE-BEAD MICROSTANDARDS

At the opposite extreme from water purification, in terms of the amount of exchanger used, is the use of single beads of ion-exchange resins as standards of quantity for sensitive analytical measurements. Beads of styrene-divinylbenzene copolymer, properly made, are homogeneous and almost perfect spheres. Recently such beads have been prepared in zero gravity (in the space shuttle) and distributed by the National Bureau of Standards as standards of size. When fully sulfonated the beads keep their uniformity. Their diameter can be measured microscopically with good accuracy. If their density and exchange capacity are known—and these too can be measured with high accuracy—then a resin bead can be used as a standard of mass of the counter-ion species, for calibration in activation analysis or graphite-furnace atomic absorption spectrometry.[2,3] A dry bead of sulfonated polystyrene, 10 μm in diameter, holds 8×10^{-11} g of sodium ions.

F. ISOTOPE SEPARATION

Stable isotopes are used in chemical analysis as internal standards in gas chromatography-mass spectrometry and inductively coupled plasma-mass spectrometry, and to check recoveries in analytical separations. They have other important uses as well. They are separated from natural mixtures in several ways, one of which is ion exchange.

Consider the isotopic exchange of lithium ions between a solution and a cation-exchange resin:

$$Li^6(\text{solution}) + Li^7(\text{resin}) = Li^7(\text{solution}) + Li^6(\text{resin})$$

The lighter lithium isotope, Li^6, is held more strongly by the resin. The equilibrium constant or isotope separation factor is about 1.003 for an 8% crosslinked sulfonated polystyrene resin. The value depends on temperature and crosslinking, being less at higher temperature and more at higher crosslinking.[4] A similar relation is found with the calcium isotopes; here, Ca^{40} is more strongly held by the resin than Ca^{44}, and the separation factor is 1.0002 to 1.0005.[5,6] A reason for the isotopic discrimination, which is very plausible, is that the ions are less strongly hydrated in the resin; hence the effect of crosslinking. The zero point energy of vibration of the metal ion-oxygen bond is lower for the heavier isotope, and therefore the heavier isotope is preferred in the water phase, where there is more hydration.

The distribution of boron isotopes between an anion-exchange resin and an aqueous boric acid solution shows a similar effect. Here the separation factor is much larger, about 1.01 to 1.02, depending on the type of anion exchanger. The selectivity of the ion-exchange process is compounded by the isotope effect in the ionization equilibria. The ionization of boric acid is slightly different for the two boron isotopes, B^{10} and B^{11}, and it is the anion that is held by the resin, not the undissociated boric acid. There are several different ionic species, which complicates the interpretation. A similar situation exists with the ammonium ion-ammonia equilibrium. Here it is the heavier isotope, N^{15}, that is held more strongly, and the separation factor between N^{15} and N^{14} is 1.025.[7] The method of doing the separation is as follows. A long column of cation-exchange resin is converted to the hydrogen form by passing acid and rinsing with water, then a suitable volume of dilute ammonia solution (ammonium hydroxide) is introduced. A band of ammonium ions is formed. Now a solution of sodium hydroxide is passed. The band moves forward and as it moves, ammonium ions are in equilibrium with ammonia. The heavy isotope, N^{15}, concentrates in the rear of the band and is eluted last.

To use these small differences in affinity to separate isotopes the method of displacement chromatography is used. One takes a long column of resin and introduces a band of the ions whose isotopes are to be separated, then one passes a displacing ion that drives the band along the column. The more strongly bound isotope lags behind and concentrates in the rear of the band, or the trailing edge; the less strongly bound isotope moves ahead and concentrates at the front of the band, or the leading edge. For example, a lithium acetate solution is introduced into a column of cation-exchange resin that has been equilibrated with hydrochloric acid. Then a solution of sodium acetate is passed. The sodium ions drive the lithium ions down the column. The further the band of lithium ions travels, the more is the isotope separation, until eventually diffusional mixing overcomes the ion-exchange separation. Distances of travel of hundreds of meters are common. One does not use a column of that length; rather, the band is recycled through the same column (or a few columns in series) many times.

Boron-10, whose concentration in natural boron is 19.8%, was enriched to 98.3% in boric acid after travelling 754 m along a column of weak-base anion-exchange resin, with pure water as the eluent.[8] Boron-10 has an important use in nuclear power production. It is a very strong neutron absorber, while boron-11 hardly absorbs neutrons at all.

Isotopes of uranium and many other elements have been separated by displacement ion-exchange chromatography. Oxidation-reduction equilibria and complex-forming equilibria are used to maximize the isotope effect.

G. SPECIATION

Reference was made in Chapter 4 to the ability of ion-exchange chromatography to separate and measure individual ionic species of an element, while spectroscopic methods, like atomic absorption, give only the total amount of the element of interest. Often quoted is the separation of arsenic acid (eluted first), arsenious acid, methyl arsonic acid, and dimethyl arsonic acid on a strong-acid cation exchanger, with water eluent.[9] Another very interesting example is the analysis of anionic species of technetium by chromatography on a strong-base anion exchanger with sodium acetate eluent.[10] Reduction of pertechnetate ion, TcO_4^-, by sodium borohydride produces several species in different oxidation states. This mixture is used in nuclear medicine as a diagnostic tool; technetium enters specific organs of the body and because it is radioactive, it reveals anomalies. Individual technetium species, separated by ion exchange, behave in specific ways.

H. THE DISCOVERY OF MENDELEVIUM, ELEMENT 101

It is proper to recall the application of ion exchange to the discovery of a new element. Ion exchange entered into the discovery because elution orders in both the lanthanide and the actinide series follow a regular pattern. The lanthanide ions are adsorbed on a strong-acid cation-exchange resin, then eluted with a solution of a complexing agent, citrate or α-hydroxyisobutyrate, which lifts the most stable complexes off the resin first, and the least stable last. Lutecium, with the largest atomic number, has the smallest cation of the lanthanide ions and therefore forms the most stable complex. The remaining lanthanide ions come off the column in reverse order of atomic number.

The same sequence is found in the actinides. Elements 100, 99, and 98 are eluted in that order from a column of cation-exchange resin. Thus, Ghiorso et al.[11] set up this experiment: they bombarded element 99, einsteinium, with fast helium ions on a gold target, dissolved the target in nitric-hydrochloric acid, removed the gold by anion exchange, added the remaining solution to a small column of Dowex®-50 cation-exchange resin, and eluted with ammonium α-hydroxyisobutyrate, counting neutrons emitted from the drops that came out of the column. Elements 100, 99, and 98 appeared in the expected sequence, with element 100 first. After the void volume, but before the peak for element 100, came the new element, atomic number 101. In the first experiment, five events of spontaneous fission, corresponding to *five atoms* of this element were detected. The ion-exchange elution order helped to establish its identity. The new element was named mendelevium, for Dmitri Ivanovitch Mendeleef, the founder of the periodic system.

II. CONCLUSION

It has been said that the first half of the 20th century was the era of nuclear physics, and the second half is the era of molecular biology. Ion exchange had a part in the discoveries of nuclear physics and chemistry, and it has had a part in revealing the makeup of nucleic acids and the double helix. Today the separation of proteins and peptides, both analytical and preparative, is an area of rapid growth. Ion exchange is not the only way to make these separations, but it is certainly a powerful way. Analytical ion exchange has come far since the pioneer work of Samuelson.

REFERENCES

1. **Fritz, J. S. and Freeland, M. Q.,** Direct titrimetric determination of sulfate, *Anal. Chem.*, 26, 1593, 1954.
2. **Freeman, D. H. and Paulson, R. A.,** Chemical microstandards from ion-exchange resins, *Nature*, 218, 563, 1968.
3. **Freeman, D. H.,** Precise studies of ion-exchange systems using microscopy, in *Ion Exchange: A Series of Advances*, Vol. 1, Marinsky, J. A., Ed., Marcel Dekker, New York, 1966, chap. 5.
4. **Lee, D. A. and Begun, G. M.,** The enrichment of lithium isotopes by ion-exchange chromatography. I. The influence of degree of crosslinking, *J. Am. Chem. Soc.*, 81, 2332, 1959.

4a. **Fujine, S., Saito, K., and Shiba, K.,** The effects of temperature and the use of macroreticular resins in lithium isotope separation by displacement chromatography, *Sep. Sci. Technol.*, 18, 15, 1983.

5. **Heumann, K. G. and Klöppl, H.,** Calciumisotopenseparation und Bestimmung der Reaktions enthalpie beim Isotopenaustausch an einen stark sauren Kationenaustauscher, *Z. Anorg. Allg. Chem.*, 472, 83, 1981.
6. **Russell, W. A. and Papanastassiou, D. A.,** Calcium isotope fractionation in ion-exchange chromatography, *Anal. Chem.*, 50, 1151, 1978.
7. **Spedding, F. H., Powell, J. E., and Svec, H. J.,** A laboratory method for separating nitrogen isotopes by ion exchange, *J. Am. Chem. Soc.*, 77, 6125, 1955.
8. **Aida, M., Fujii, Y., and Okamoto, M.,** Chromatographic enrichment of ^{10}B by using weak-base anion-exchange resin, *Sep. Sci. Technol.*, 21, 643, 1986.

9. **Dietz, E. A. and Perez, M. E.,** Purification and analysis methods for methylarsonic acid and hydroxydimethylarsine oxide, *Anal. Chem.*, 48, 1088, 1976.
10. **Pinkerton, T. C., Heineman, W. R., and Deutsch, E.,** Separation of technetium hydroxyethylidene diphosphonate complexes by anion-exchange high-performance liquid chromatography, *Anal. Chem.*, 52, 1106, 1980.
11. **Ghiorso, A., Harvey, B. G., Choppin, G. R., Thompson, S. G., and Seaborg, G. T.,** New element mendelevium, atomic number 101, *Phys. Rev.*, 98, 1518, 1955.

INDEX

A

B

D

E

K

L

M

N

O

P

Q

R

S

T

U

V

W

X

Y

Z